T0271037

Gas Turbines

This long-awaited, physics-first, design-oriented text describes and explains the underlying flow and heat transfer theory of secondary air systems. An applications-oriented focus throughout the book provides the reader with robust solution techniques, state-of-the-art three-dimensional computational fluid dynamics (CFD) methodologies, and examples of compressible flow network modeling. It clearly explains elusive concepts of windage, nonisentropic generalized vortex, Ekman boundary layer, rotor disk pumping, and centrifugally driven buoyant convection associated with gas turbine secondary flow systems featuring rotation. The book employs physics-based, design-oriented methodology to compute windage and swirl distributions in a complex rotor cavity formed by surfaces with arbitrary rotation, counterrotation, and no rotation. This text will be a valuable tool for aircraft engine and industrial gas turbine design engineers as well as graduate students enrolled in advanced special topics courses.

Bijay K. Sultanian is founder and managing member of Takaniki Communications, LLC, a provider of web-based and live technical training programs for corporate engineering teams, and an adjunct professor at the University of Central Florida, where he has taught graduate-level courses in turbomachinery and fluid mechanics since 2006. Prior to founding his own company, he worked in and led technical teams at a number of organizations, including Rolls-Royce, GE Aviation, and Siemens Power and Gas. He is the author of *Fluid Mechanics: An Intermediate Approach* (2015) and is a Life Fellow of the American Society of Mechanical Engineers.

Cambridge Aerospace Series

Editors: Wei Shyy and Vigor Yang

Gas Turbines

Internal Flow Systems Modeling

BIJAY K. SULTANIAN

CAMBRIDGE
UNIVERSITY PRESS

University Printing House, Cambridge CB2 8BS, United Kingdom

One Liberty Plaza, 20th Floor, New York, NY 10006, USA

477 Williamstown Road, Port Melbourne, VIC 3207, Australia

314-321, 3rd Floor, Plot 3, Splendor Forum, Jasola District Centre, New Delhi - 110025, India

103 Penang Road, #05-06/07, Visioncrest Commercial, Singapore 238467

Cambridge University Press is part of the University of Cambridge.

It furthers the University's mission by disseminating knowledge in the pursuit of
education, learning and research at the highest international levels of excellence.

www.cambridge.org
Information on this title: www.cambridge.org/9781107170094
DOI: 10.1017/9781316755686

First published 2018

A catalogue record for this publication is available from the British Library

Library of Congress Cataloging in Publication data
Names: Sultanian, Bijay K.
Title: Gas turbines : internal flow systems modeling / Bijay K. Sultanian.
Description: Cambridge, United Kingdon ; New York, NY, USA : Cambridge University Press, 2018. |
Series: Cambridge aerospace series | Includes bibliographical references and index.
Identifiers: LCCN 2018010102 | ISBN 9781107170094 (hardback)
Subjects: LCSH: Gas-turbines–Fluid dynamics–Mathematics. | Gas flow–Mathematical models. |
BISAC: TECHNOLOGY & ENGINEERING / Engineering (General).
Classification: LCC TJ778 .S795 2018 | DDC 621.43/3–dc23
LC record available at https://lccn.loc.gov/2018010102

ISBN 978-1-107-17009-4 Hardback

To my dearest friend Kailash Tibrewal, whose mantra of "joy in giving" continues to inspire me; my wife, Bimla Sultanian; our daughter, Rachna Sultanian, MD; our son-in-law, Shahin Gharib, MD; our son, Dheeraj (Raj) Sultanian, JD, MBA; our daughter-in-law, Heather Benzmiller Sultanian, JD; and our grandchildren, Aarti Sultanian, Soraya Zara Gharib, and Shayan Ali Gharib, for the privilege of their unconditional love and support!

Contents

Preface

Albert Einstein famously said, "Education is not the learning of facts, but the training of the mind to think." I have concluded over a 30-plus-year career spent bridging the gap between academic research and practical gas turbine design that current and future gas turbine, heat transfer, and secondary air systems (SAS) or internal air systems (IAS) design engineers (both academic and practitioners) must not only learn how and what to do but, more importantly, question why things are done the way they are so that we can find ways to do them better.

I have found the best approach to training the mind is summarized by another famous Einstein quote, "Everything should be made as simple as possible, but not simpler." The beauty of simplicity is that it makes learning contagious. For a topic as complex as fluid dynamics, a traditional, mathematics-first approach has failed many, as it tends to make the study of engineering far more complex, and less intuitive, than the physics-first approach that I use here. It is a technique that I have developed over a career spent learning from giants, practicing with the best and brightest, and teaching the future leaders of our industry.

Few people may know that I spent the first ten years of my professional career without ever solving for a rotating flow. It wasn't until 1981, when I began my PhD at Arizona State University, that I first became fascinated with a new and emerging technology: computational fluid dynamics, or CFD. Although CFD has since become a ubiquitous tool used by hundreds of industries, back then, in order to incorporate CFD into my research, I had to pick an obscure topic that I had to teach myself: the numerical prediction of swirling flow in an abrupt pipe expansion.

In the fall of 1985, I started my first postdoc job at Allison Gas Turbines (now Rolls-Royce). At Allison, I continued to develop prediction methods for turbulent swirling flows in gas turbine combustors using advanced turbulence models. For example, we successfully developed a low Reynolds number turbulence model to predict heat transfer across a mixed axial-radial flow in a rotor cavity. Even more exciting, we were able to predict nonentraining Ekman boundary layers on the rotor disks between the inner source region and the outer sink region. These were my first practical applications of the CFD-based modeling in gas turbine secondary air systems I first researched during my PhD.

Three years later, when I joined the heat transfer and secondary flow group at GE Aircraft Engines (GEAE), now called GE Aviation, I came across a new operational term – windage. Windage was a significant factor in calculating the thermal boundary conditions for gas turbine parts in contact with secondary air flows. Even though we had design tools to compute windage temperature rise (it always increased coolant air temperatures used in convective boundary conditions), there was no precise definition of what "windage" actually was.

Some of my colleagues thought of it as viscous dissipation, whereas others saw it resulting from friction on both stator and rotor surfaces. Just as I had done during my PhD, I went back to first principles (angular momentum and steady flow energy equations) to define windage in as simple terms as I could (not simpler). Not only did the simplicity of my definition of windage made it incredibly contagious among my colleagues at GEAE, but by using this new framework for windage, I was also able to develop one of my most successful design tools – BJCAVT. BJCAVT was the first program that could automatically compute windage and swirl distributions in a complex rotor cavity formed by surfaces under three conditions: arbitrary rotation, counterrotation, and no rotation. This program was extensively validated by numerous engine measurements; and in addition to being very user-friendly, it was solution-robust, always unconditionally converging in a few iterations with no user intervention. BJCAVT became widely popular and an integral part of GEAE design practice, initially at GEAE for all aero engines and later at GE Power Systems for all power generation gas turbines. Four years later, as a result of BJCAVT and other unique developments at GEAE, I was given my most prestigious managerial award in 1992 with the following citation:

On behalf of Advanced Engineering Technologies Department, it gives me great pleasure to present to you this Managerial Award in recognition of your significant contributions to the development of improved physics-based heat transfer and fluid systems analysis methodologies of rotating engine components. These contributions have resulted in more accurate temperature and pressure predictions of critical engine parts permitting more reliable designs with more predictable life characteristics.

A few years later, Professor Tom Shih, a world authority on gas turbine internal cooling and CFD, invited me to coauthor a book chapter on Computations of Internal and Film Cooling. It is important to note that internal cooling design of high-temperature turbine airfoils derive its inlet boundary conditions from a SAS model of the gas turbine engine. These cooled airfoils are also simulated in the SAS model as resistive elements through pressure ratio versus effective area curves. In terms of the flow and heat transfer physics, a lot is common between airfoil internal cooling and SAS modeling; both are simulated in design through complex, locally one-dimensional flow networks. Internal cooling, however, entails one simplification. When the coolant air enters the rotating serpentine passages of a blade, it always assumes the state of solid-body rotation with the blade. In a rotor-rotor or rotor-stator cavity, however, the coolant air rotation in the bulk may in general be different from those of the rotor surfaces forming the cavity.

The interaction of windage and vortex temperature change in a rotor cavity is found to be a significant source of confusion in gas turbine design. Since most design codes have a built-in calculation of temperature change in an isentropic forced vortex, this change is inadvertently added to the windage temperature rise in the cavity. In 2004, to unravel the mystery of these and other related concepts, I was invited to Siemens Energy, Orlando, to give a lecture to a team of heat transfer and SAS engineers.

In 2006, I joined Siemens Energy full time to develop advanced tools for internal cooling design of turbine airfoils. At that time, the University of Central Florida (UCF) invited me to join the faculty as an adjunct professor for teaching a graduate course, "EML5402 – Turbomachinery," in the fall semester. I merrily accepted the invitation. This opportunity allowed me to bring my years of industry experience into the classroom to train the next generation of engineers to handle the challenges of designing future gas turbines. The following year, UCF asked me to teach the core graduate course, "EML5713 – Intermediate Fluid Mechanics," in the spring semester. To my surprise, I found that many students pursuing their graduate studies in the thermofluids stream didn't have a grasp on the first-principal fundamentals of fluid mechanics, particularly in the control volume analysis of various conservation laws and one-dimensional compressible flow in a duct featuring arbitrary area change, friction, heat transfer, and rotation. Hardly anyone in the class could physically explain (without using the Mach number equations of Fanno flows) why the Mach number of a subsonic compressible flow in a constant-area duct increases downstream due to wall friction, which is known to slow things down! Similarly, in this duct, if one eliminates friction (a practically difficult task!) and heats the flow, why does the total pressure decrease and the Mach number increase in the flow direction? Unlike incompressible flows, which are formally taught in most courses on fluid mechanics, compressible flows feature other nonintuitive behavior like choking when the flow velocity equals the speed of sound and the formation of a normal shock in the supersonic regime. All bets are off when such flows also involve duct rotation.

During the course of my teaching graduate courses at UCF, I realized that many students needed help in understanding the key foundational concepts of fluid mechanics. At the same time, the course on turbomachinery dealing with the design and analysis of primary flowpath aerothermodynamics inspired in me to develop a follow-up course dealing with secondary air systems modeling, which is the subject of this book. Since fluid mechanics is a prerequisite core course for advanced courses in the thermofluids stream, I decided to write my first textbook using a physics-first approach. That 600-page book, *Fluid Mechanics: An Intermediate Approach*, was published in July 2015 by Taylor & Francis.

While at Siemens Energy, I also realized that the engineers working on SAS modeling and internal cooling design needed some help on understanding the flow and heat transfer physics of various components of their models and not just follow their operational design practices. In 2007, I began a twenty-hour lecture series within Siemens titled "Physics-Based Secondary Air Systems Modeling." The response to this series was overwhelming, as more than 60 engineers globally joined these online lectures. Encouraged by this experience at Siemens, I taught a two-day preconference workshop on "Physics-Based Internal Air Systems Modeling" in conjunction with the ASME Turbo Expo 2009 in Orlando. I later taught this workshop in an eight-hour format at ASME Turbo Expo 2016 in Seoul, South Korea, and ASME Turbo Expo 2018 in Oslo, Norway. Teaching these workshops and publishing a graduate-level textbook on fluid mechanics gave me the confidence needed to finally write this textbook.

The book is the culmination of three decades of continuous learning in gas turbine industry and a decade of teaching graduate-level courses in turbomachinery and fluid mechanics at UCF. It has taken this long for me to study the fascinating, and sometimes counter-intuitive, world of gas turbine secondary flow systems to the point that I can present the most complex topics in a simplified way that will make learning these topics contagious.

I suggest the following syllabus for a three-credit graduate course (Turbomachinery II) in a sixteen-week semester:

Week 1: Chapter 1 (Overview of Gas Turbines for Propulsion and Power Generation)

Weeks 2–5: Chapter 2 (Review of Thermodynamics, Fluid Mechanics, and Heat Transfer)

Weeks 6–8: Chapter 3 (1-D Flow and Network Modeling)

Weeks 9–11: Chapter 4 (Internal Flow around Rotors and Stators)

Week 12: Chapter 5 (Labyrinth Seals)

Weeks 13–16: Chapter 6 (Whole Engine Modeling)

However, the course instructors are free to fine-tune this syllabus and reinforce it with their notes and/or additional reference material to meet their specific instructional needs. The book features a number of worked-out examples, chapter-end problems, and projects, which may be assigned as a team-project for students to work on during the entire semester.

Acknowledgments

This is my long-awaited second dream book! A contribution of this magnitude would not have been possible without the perpetual love and support of my entire family to which I shall forever remain indebted.

My dream to study such a book originated during my twelve-year career at GE where I was so fortunate to have participated in the design and development of world's two largest and most efficient gas turbines: GE 90 to propel planes and steam-cooled 9H/7H to generate electricity. The challenges of heat transfer and cooling/sealing flow designs in these machines were beyond anything I had experienced before. Among all my distinguished colleagues at GE, three individuals stand out: Mr. Ernest Elovic and Mr. Larry Plemmons at GE Aircraft Engines (GEAE) and Mr. Alan Walker at GE Power Generation. They are my true professional heroes. I owe my most sincere gratitude to Mr. Elovic and Mr. Plemmons (deceased) who introduced me to the concept of "physics-based" design predictions. Because it has become an integral part of my conviction, I have used the term "physics-based" very often in this book. I cannot wait to send Mr. Walker and Mr. Elovic each a printed copy of this book with my best compliments and highest regards!

A gift of knowledge is the greatest gift one can give and receive. Mr. Alan Walker gave me such a gift by sponsoring me to complete the two-year Executive MBA program at the Lally School of Management and Technology. While I remain greatly indebted to Mr. Walker for this unprecedented recognition, I also thank him for keeping my technical skills vibrant through my direct involvements in the redesign of gas turbine enclosure ventilation system for the first full-speed no-load (FSNL) testing of the 9H machine, robust design of a high-pressure inlet bleed heat system, CFD-based high-performance exhaust diffuser designs in conjunction with a joint technology development program with Toshiba, Japan, and development of other innovative methods and tools for concurrent design engineering of steam-cooled gas turbines.

I wish to thank Professor Ranganathan Kumar who invited me to teach graduate courses at UCF in 2006 as an adjunct faculty. Without this teaching opportunity my dream books would not have become textbooks. I continue to cherish a highly referenced book-chapter on Computations of Internal and Film Cooling that Professor Tom Shih and I coauthored at the turn of the twenty-first century.

I owe many thanks to my longtime friends Dr. Ray Chupp and Dr. John Blanton for reviewing Chapter 5 and Dr. Kok-Mun Tham and Dr. Larry Wagner for reviewing Chapter 6 and suggesting several improvements in these chapters.

I offer my sincere gratitude to Steve Elliot, my editor at Cambridge University Press, who believed in my book proposal and, more important, in my passion to complete this book. I thoroughly enjoy all my interactions with him. I wish to thank my content manager Mark Fox and all the staff at the Press for their exemplary support and professional communications during the entire book production process.

Last but not least, I will remain eternally grateful to all the readers, and more so to those who will be inspired to write someday a better textbook on this topic, making this one obsolete.

About the Author

Dr. Bijay (BJ) K. Sultanian, PhD, PE, MBA, ASME Life Fellow is a recognized international authority in gas turbine heat transfer, secondary air systems, and Computational Fluid Dynamics (CFD). Dr. Sultanian is the founder and managing member of Takaniki Communications, LLC (www.takaniki.com), a provider of high impact, web-based, and live technical training programs for corporate engineering teams. Dr. Sultanian is also an adjunct professor at the University of Central Florida, where he has been teaching graduate-level courses in turbomachinery and fluid mechanics since 2006. As an active member of IGTI's Heat Transfer Committee since 1994, he has instructed a number of workshops at ASME Turbo Expos. His graduate-level textbook, *Fluid Mechanics: An Intermediate Approach*, was published in July 2015.

During his three decades in the gas turbine industry, Dr. Sultanian has worked in and led technical teams at a number of organizations, including Allison Gas Turbines (now Rolls-Royce), GE Aircraft Engines (now GE Aviation), GE Power Generation (now GE Water & Power), and Siemens Energy (now Siemens Power & Gas). He has developed several physics-based improvements to legacy heat transfer and fluid systems design methods, including new tools to analyze critical high-temperature gas turbine components with and without rotation. He particularly enjoys training large engineering teams at prominent firms around the globe on cutting-edge technical concepts and engineering and project management best practices.

During his initial ten-year professional career, Dr. Sultanian made several landmark contributions toward the design and development of India's first liquid rocket engine for a surface-to-air missile (Prithvi). He also developed the first numerical heat transfer model of steel ingots for optimal operations of soaking pits in India's steel plants.

Dr. Sultanian is a Life Fellow of the American Society of Mechanical Engineers (1986), a registered Professional Engineer (PE) in the State of Ohio (1995), a GE-certified Six Sigma Green Belt (1998), and an emeritus member of Sigma Xi, The Scientific Research Society (1984).

Dr. Sultanian received his BTech and MS in Mechanical Engineering from the Indian Institute of Technology, Kanpur (1971) and the Indian Institute of Technology, Madras (1978), respectively. He received his PhD in Mechanical Engineering from Arizona State University (1984) and his MBA from the Lally School of Management and Technology at Rensselaer Polytechnic Institute (1999).

.

1   Overview of Gas Turbines for Propulsion and Power Generation

1.0   Introduction

This is the age of gas turbines with their ever-growing contributions to people's living standard and well-being. As a great technological marvel, perhaps next only to the inventions of electricity and light bulb, gas turbines have become indispensible in commercial aviation, shrinking the travel time around the globe in hours rather than days and weeks as was the case in the early 1990s by sea. Almost all modern military fighter jets with high maneuverability deploy gas turbine engines. Even in liquid rocket propulsion, gas turbines are used to pump liquid fuel and oxidizer to the combustion chamber at high pressure. Nonflying gas turbines, where weight considerations are important only to reduce material cost, have revolutionized the means of power generation both on land and sea. Their impressive applications portfolio includes utility and industrial power generation, combined heat and power (CHP), oil and gas, and mechanical drive. Gas turbines are a strong candidate of choice where fast power is needed in the distributed power generation for commercial buildings and facilities. Their fuel flexibility is leveraged in applications involving biogas, biomass, waste gas, and waste to energy to produce utility steam. In view of the growing demand for energy around the world, it is highly unlikely that wind turbines and other forms of turbomachinery using renewable will make gas turbines obsolete as they (gas turbines) did to piston-powered reciprocating machines in the early part of the last century. In fact, in the foreseeable future, the world demand for gas turbines for both propulsion and power generation is expected to grow monotonically.

Bathie (1996), Soares (2014), and Saravanamuttoo et al. (2017) present the history of gas turbines for aircraft propulsion and for various power-generation applications. Historically, the gas turbine technology has cascaded from military engines to commercial engines to large and small engines used for power generation. Aeroderivative gas turbines used for land and marine applications are directly derived from aircraft engines. For reasons of high reliability and safety, the development of the most of aviation gas turbines has been evolutionary rather than revolutionary, each engine being an upgrade of a previous successful engine or a conglomeration various technologies from other engines.

Olson (2017) touts the advanced technologies associated with the GE9X™ engine, shown in Figure 1.1. Designed specifically for the Boeing 777X airplane, the GE9X is the most fuel-efficient and quietest jet engine GE has ever produced. This engine is designed to deliver 10 percent improved aircraft fuel burn versus the GE90–115B-

Figure 1.1 Cutaway view of the GE9X engine (with permission from GE Aviation).

powered 777–300ER and 5 percent improved specific fuel consumption versus any twin-aisle engine available. Additional design features include an approximate 10:1 bypass ratio, 60:1 overall pressure ratio, and 8 db margins to Stage 5 noise limits. As to the cooling technology, the GE9X engine features ceramic matrix composite (CMC) materials in the combustor and high-pressure turbine for twice the strength, a third of the weight, and greater thermal management capabilities than their metal counterparts. The low-pressure turbines of the GE9X use enhanced titanium aluminide (TiAl) airfoils, which are stronger, lighter, and more durable than their nickel-based counterparts. Achieving this milestone for gas turbines in the history of aviation would have been beyond any forecast fifty years ago!

GE is not the sole manufacturer of such large aircraft engines. Other original equipment manufacturers (OEMs) include Pratt & Whitney (P&W) and Rolls-Royce (R-R), which in their portfolio have similar class of engines, being marketed in a close global competition with GE. As a result of the enormous cost, which runs into hundreds of millions of dollars, associated with the development of a new large gas turbine for commercial aviation and a decade-long breakeven point, the entire market for such engines has remained divided among these three companies (GE, P&W, and R-R) with no new major OEM seen on the horizon.

Unlike aircraft engines, the gas turbines used to generate electricity can operate under a simple cycle (i.e., the Brayton cycle, discussed in Chapter 2) or jointly with a steam turbine using a heat recovery steam generator (HRSG), which yields significantly improved combined-cycle thermal efficiency ($\eta_{\text{th}_{\text{CC}}} \approx \eta_{\text{th}_{\text{GT}}} + \eta_{\text{th}_{\text{ST}}} - \eta_{\text{th}_{\text{GT}}}\eta_{\text{th}_{\text{ST}}}$). Until the early 1990s, the simple-cycle thermal efficiency, which depends on the overall pressure ratio and turbine inlet temperature (TIT), was limited to 33–38 percent, and the combined-cycle efficiency was in the range of 51–58 percent. Around the turn of this century, OEMs including GE and Siemens embarked on the new line of high-efficiency gas turbines called H-class with a TIT of around 1427°C, which is around 100°C higher than the previous class of gas turbines. These new gas turbines for electricity generation were developed to break the perceived barrier of the combined-cycle efficiency of 60 percent. GE designated their H-class machines as 9H for 50 Hz (3000 rpm) and 7H for 60 Hz (3600 rpm) applications. To maintain the same turbine tip speed, 9H gas turbines are hence larger in size than their 7H counterpart. Siemens, by contrast, designated their machines as SGT5–8000H and SGT6–8000H for 50 Hz and 60 Hz electricity generation, respectively.

For the initial development of the H-class gas turbines, GE used a somewhat revolutionary design philosophy of introducing closed-loop steam cooling in the first stage turbine stator and rotor system, including internal cooling of vanes (nozzles) and blades (buckets). At this time, GE remains the only OEM that has successfully introduced steam cooling for a rotating gas turbine component. After an extensive validation process, GE installed their first 9H combined-cycle gas turbine at Baglan Bay in 2003. Since then, the plant has been reliably providing up to 530 MW to the UK national grid, operating at over 60 percent combined-cycle efficiency.

Siemens, by contrast, used an evolutionary approach to the design and development of their H-class gas turbines and tested their first SGT5–8000H at full load in Ingolstadt, Germany, in 2008. The gas turbine unit performed at 40 percent efficiency and as a part of a combined-cycle system reached a world efficiency-record of 60.75 percent. This plant has been providing power to the German grid since the end of the testing period.

While maturing their steam-cooled gas turbine technology, GE simultaneously launched the development of the traditional air-cooled H-class machines under the designation 9HA and 7HA. According to Vandervort, Wetzel, and Leach (2017), in April 2016, under the auspices of the Guinness Book of World Records, a 9HA.01 GTCC set a world record for the combined cycle efficiency of 62.22 percent while producing more than 605 MW of electricity. In June 2016, GE and Électricité de France (EDF, Electricity of France) officially inaugurated the first 9HA combined-cycle power plant in Bouchain, France, and achieved a combined cycle efficiency of over 62 percent. A cutaway view of GE's 9HA gas turbine is shown in Figure 1.2.

Key gas turbine technologies, their mutual interactions, and their influence on the core components (compressor, combustor, and turbine) are depicted in Figure 1.3. Aerodynamics influences the design and performance of gas turbine primary flow path, which participates directly in the energy conversion process. Modern gas turbine compressors and turbines feature 3-D airfoils, whose details are designed using computational fluid dynamics (CFD) for a nearly isentropic performance.

A device is as strong as its weakest link. All components of a gas turbine must perform in concert for its successful operation. To realize the desired aerodynamic performance, the structural integrity of both compressor and turbine are critically

Figure 1.2 Cutaway view of GE's 9HA gas turbine (with permission from GE).

important, as they involve rotating components at very high temperature. A failure of either of them could be catastrophic. The key drivers of gas turbine technology are: (1) the fuel cost, which in turn drives the technology development for higher efficiency; (2) engine reliability, durability, and availability, which require active life management of each engine from cradle to grave, determining its maintenance intervals and the overall product cost; and (3) environmental regulations against pollution, which drives the combustor technology development.

As the compressor pressure ratio and TIT keeps rising for more efficient gas turbines, heat transfer (cooling), secondary air system (SAS), and materials and coatings constitute today's pacing technologies. SAS delivers gas turbine cooling and sealing flows, which could be around 20 percent of the compressor flow. Note in Figure 1.3 that SAS strongly influences gas turbine heat transfer, which in turn has a weak influence on SAS.

Any reduction in cooling and sealing flows directly translate into higher thermal efficiency for a gas turbine. Advances in materials and coatings technology, such as CMC, has led to increased cooling effectiveness with reduced cooling flow requirements. In addition, many aspects of gas turbine design are already benefitting from the fast-emerging additive manufacturing (also called 3D printing) technology. Earlier designs were almost always constrained by manufacturability. With the widespread use of additive manufacturing, the new paradigm is "if you can design it, we can manufacture it."

1.1   Primary Flow: Energy Conversion

The primary flow of the core engine consists of the flow through low-pressure and high-pressure compressors, combustor, and high-pressure and low-pressure turbines. As the air flows against an adverse gradient, the high-pressure ratio over the compressor is achieved in multiple stages to prevent boundary layer separation over the airfoils. For an

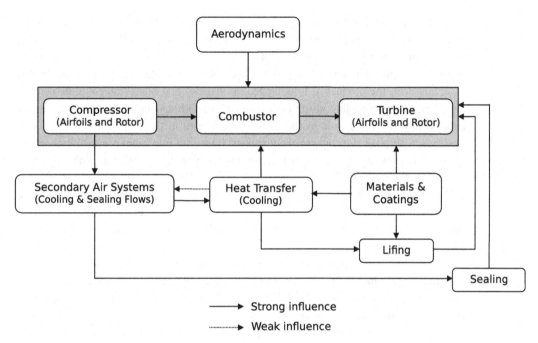

Figure 1.3 Interactions among gas turbine key technologies.

axial-flow compressor, which is found in most modern large gas turbines, the flowpath area continuously decreases downstream with the increase in air density, pressure, and temperature as the work transfer from the compressor blades (rotating airfoils) to airflow occurs continuously in each stage. This transfer of energy into the airflow from compressor is governed by the Euler's turbomachinery equation presented in Chapter 2. In essence, this equation states that for unit air mass flow rate through a blade passage, we obtain the amount of work transfer by subtracting the product of the air tangential velocity (absolute) and blade tangential velocity at the inlet from their product at the outlet. It is interesting to note that Euler's turbomachinery equation deals with velocities in the absolute (inertial) reference frame. Although turbines operate under the most adverse thermal environment, the compressor operating under stall-free and surge-free conditions is the heart of a gas turbine, playing a critical role in its overall operation and performance. The compressor flow is the source of all cooling and sealing flows in a gas turbine with the exception of a steam-cooled gas turbine, where some of the cooling needs are met by steam in a combined cycle operation, as in GE's 9H/7H machines.

The combustor is the place where the primary flow path air receives chemical energy from the fuel through an efficient combustion, significantly raising its temperature (TIT). But for a slight loss of pressure in the combustor, the turbine handles nearly the same overall pressure ratio as the compressor but in fewer stages. The turbine flow path predominantly operates under a favorable pressure gradient with negligible propensity for boundary-layer separation. For a gas turbine engine used in aircraft propulsion, the high-pressure turbine drives the compressor, and the exhaust gases from the turbine are expanded in a nozzle for generating the propulsive thrust. In a power generation application, by contrast, we may have separate turbines; one, which is called a gas generator turbine, drives the compressor; and the other, called a power turbine, rotates a generator to produce electricity. Euler's turbomachinery equation also holds good for computing energy transfer from hot gases to turbine blades and rotor. In this case, however, the product of flow tangential velocity and blade tangential velocity at the outlet is less than that at the inlet of a row of turbine blades. Vanes (nonrotating airfoils) have no direct role in the work transfer both in the compressor and turbine. Their main purpose is to receive the upstream flow with minimum pressure loss and to prepare the flow to enter the downstream blades, which are rotating, with minimum entrance loss.

Heat transfer (cooling) considerations take the center stage in the design of turbines, whose flow path contains hot gases at temperatures close to the melting point of structural material in contact. For an acceptable life and durability of the turbine components during their entire operational envelope, designers must ensure that these components are adequately cooled using compressor air at the required high pressure. A serious uncertainty, however, remains for the temperature distribution in the hot gases exiting the combustor, critically impacting the thermal design of the first stage vanes and possibly the downstream blades.

While the gas turbines for aircraft propulsion are fitted with a nozzle to expand the flow exiting the last stage turbine to ambient pressure with a high exit velocity to produce thrust, shaft-power gas turbines use an exhaust diffuser at the turbine exit. Using additional duct work with minimum pressure loss, the gases from the exhaust diffuser are either ducted to an HRSG in a combined-cycle operation or by-passed to ambient in a simple-cycle operation. The primary role of a diffuser is to render the turbine exit static pressure subambient through the static pressure recovery to the ambient pressure, while minimizing loss in pressure in the downstream duct. The exhaust diffuser thus helps create higher pressure ratio across the turbine, making it more efficient. For a detailed experimental and 3-D CFD investigation of scaled GE's 9E gas turbine exhaust diffuser, which exhausts sideways, see Sultanian, Nagao, and Sakamoto (1999).

Based on the Brayton cycle analysis presented in Chapter 2, we can easily deduce Equation 1.1 for the net specific work output (nondimensional), which when multiplied by the compressor mass flow rate (neglecting fuel mass flow rate into the combustor) yields the total cycle power output, and Equation 1.2 for the thermal efficiency:

$$\frac{w_{\text{net}}}{c_p T_{t_1}} = \left(\frac{T_{t_3}}{T_{t_1}}\right) \eta_{T_i} \left(1 - \frac{1}{\pi^{\frac{\kappa-1}{\kappa}}}\right) - \frac{1}{\eta_{C_i}} \left(\pi^{\frac{\kappa-1}{\kappa}} - 1\right) \tag{1.1}$$

$$\eta_{\text{th}} = \frac{\left(\frac{T_{t_3}}{T_{t_1}}\right) \eta_{T_i} \left(1 - \frac{1}{\pi^{\frac{\kappa-1}{\kappa}}}\right) - \frac{1}{\eta_{C_i}} \left(\pi^{\frac{\kappa-1}{\kappa}} - 1\right)}{\left(\frac{T_{t_3}}{T_{t_1}}\right) - \frac{1}{\eta_{C_i}} \left(\pi^{\frac{\kappa-1}{\kappa}} - 1\right) - 1} \tag{1.2}$$

Under the assumptions of equal pressure ratio across the compressor and turbine with an isentropic efficiency of 0.9 and $\kappa = 1.4$ for the fluid (assumed to be air) in their

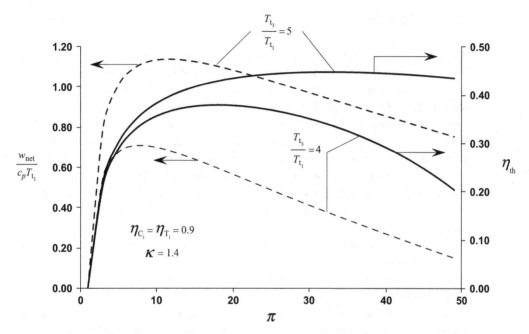

Figure 1.4 Variations of net specific work output and thermal efficiency with pressure ratio and turbine inlet temperature.

primary flow path, Equations 1.1 and 1.2 are plotted in Figure 1.4. We can make the following key observations from this figure:

(1) The net specific work output ($w_{\text{net}}/c_p T_{t_1}$) and the cycle thermal efficiency (η_{th}) depend upon the compressor pressure ratio (π) and the ratio of turbine inlet temperature to compressor inlet temperature (T_{t_3}/T_{t_1}).

(2) For a given compressor pressure ratio, both $w_{\text{net}}/c_p T_{t_1}$ and η_{th} increase with T_{t_3}/T_{t_1}.

(3) For a given value of T_{t_3}/T_{t_1}, the compressor pressure ratio needed to maximize $w_{\text{net}}/c_p T_{t_1}$ is lower than that needed to maximize η_{th}. This explains why the gas turbines used for military planes requiring higher $w_{\text{net}}/c_p T_{t_1}$ tend to operate at a lower pressure ratio, while the gas turbines for civil aviation, where higher thermal efficiency is preferred, operate at a higher pressure ratio.

(4) The curves for $w_{\text{net}}/c_p T_{t_1}$ feature sharper maxima than those for η_{th}. This means that a small variation of compressor pressure ratio around the optimal value will not significantly impact the engine thermal efficiency.

From Equation 1.1, we can easily show that, for a given value of T_{t_3}/T_{t_1}, the optimum pressure ratio that yields the maximum value of $w_{\text{net}}/c_p T_{t_1}$ is given by

$$\pi_{optimum} = \left(\eta_{\text{T}_i} \eta_{\text{C}_i} \frac{T_{t_3}}{T_{t_1}} \right)^{\frac{\kappa}{2(\kappa-1)}} \tag{1.3}$$

which tends to vary as $0.692(T_{t_3}/T_{t_1})^{1.75}$ for a gas turbine with compressor and turbine isentropic efficiency of 0.9.

1.2   Internal Flow System (IFS)

The air flows extracted from the compressor primary flow path for the purpose of cooling various hot components and providing effective sealing between rotor and stator parts is known among gas turbine engineers as secondary flows. The system of such air flows is called secondary air system (SAS). It is also alternatively known as internal air system (IAS). In this textbook, we have further generalized this system where coolant could be different from air, for example, steam in a steam-cooled gas turbine in a combined-cycle operation, calling it internal flow system (IFS). The use of secondary flows instead of internal flows may be confused with the secondary flows of the first and second kind found in duct flows, see Schlichting (1979). For all practical purposes, SAS, IAS, and IFS are used synonymously here. Further note that the primary flow paths of gas turbine compressor and turbine are essentially annuli through which internal flows of air and hot gases occur with interruptions from vanes and blades, the latter being responsible for work transfer into compressor air flow or out of the hot gases flowing through the turbine. In this textbook, we will avoid referring to the primary air flow as an internal flow.

A typical cooling and sealing arrangement of a hypothetical turbine (to avoid disclosing the proprietary design of any OEM) as a part of its internal flow systems is shown in Figure 1.5. The internal air system is also used to balance the pressure distribution on the rotating disk and drum structure in the engine to maintain acceptable bearing loads. Another function of the internal air flow is to ventilate the bearing compartments so as to prevent the buildup of combustible gas mixtures and to carry lubrication oil droplets to the oil separators. A distinguishing feature of gas turbine internal flows is the presence of rotation with its generally nonintuitive behavior. The energy transfer in such flows occurs both from heat transfer and work transfer, which requires interactions with rotating components.

Based on years of research in the flow and heat transfer of gas turbine rotating disk systems at the University of Sussex, largely funded by Rolls-Royce, Owen and Rogers (1989, 1995) were the first to publish a comprehensive monogram in two volumes. Unfortunately, both these volumes have been rendered out of print. More recently, Childs (2011) published a book on rotating flow covering the ones found in gas turbine internal air systems as well as earth's atmosphere. Among others, Kutz and Speer (1994) and Johnson (2010) describe industry-oriented approach to the simulation of secondary air systems involving elements such as restrictors, tappings, seals, vortices, and coverplates. They also briefly discuss the two-phase (oil and air) flow that occurs in bearing chamber vent systems.

1.2.1   Key Components

Gas turbine internal flows are generally driven between the compressor flow path as the source and the turbine flow path as the sink. In each system, the flow has to pass through a number of components such as stationary and rotating orifices, stator–rotor and rotor–rotor cavities, preswirler nozzles, labyrinth and other types of seals, turbine rim seals, and so on. Some of these components are depicted in Figure 1.6, and their brief

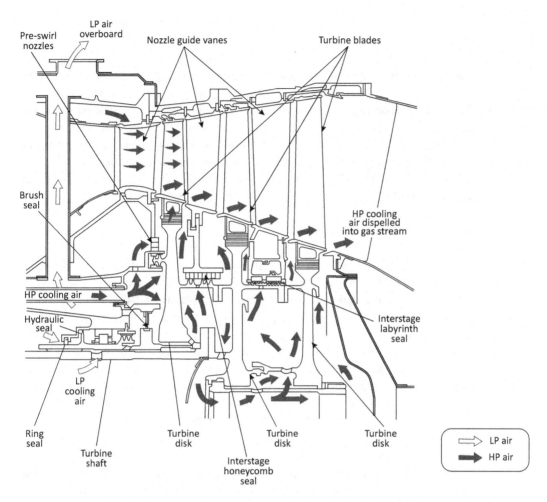

Figure 1.5 Cooling and sealing arrangement of a hypothetical turbine (with permission from Roll-Royce).

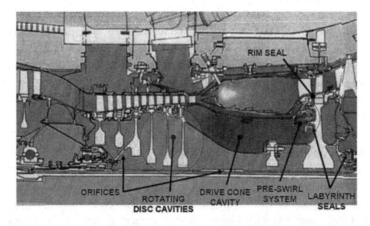

Figure 1.6 Key components of gas turbine internal flow systems (Source: Alexiou and Mathioudakis (2009) with permission from ASME).

description is provided here. We discuss modeling of these components in later chapters.

Ori ce. Orifice, like a resistor in an electrical network, is a basic component of the gas turbine internal flow system. It may belong to stationary part or a rotating part, in which case it is called a rotating orifice. Typically, an orifice is modeled as an adiabatic flow element. For given inlet total pressure and temperature, exit static pressure, and reference flow area, orifice mass flow rate depends upon the discharge coefficient, or loss coefficient, which mainly depends upon its length-to-diameter ratio, ratio of throughflow velocity to total velocity at orifice inlet, pressure ratio across the orifice, and its rotational velocity. Note that both the discharge coefficient and loss coefficient are determined using empirical correlations. In the flow network modeling using an orifice element, its exit dynamic pressure associated with flow velocity is considered lost for the downstream element.

Channel. A channel, also known as pipe, duct, or tube element, offers the most flexibility in internal flow modeling. Unlike an orifice element, the channel allows the simulation of heat transfer effects and conservation of flow linear momentum governed by the momentum equation. The most general formulation of 1-D compressible flow through a channel includes area change, friction, heat transfer, and rotation. Thus, we can possibly simulate an orifice using a channel but not the other way around. While the closed-from analytical solutions are available for individual effects in a channel flow (compressible), namely, isentropic flow with area change, Fanno flow (constant area with no heat transfer and rotation), and Rayleigh flow (constant area with no friction and rotation), when two and more effects are present, their linear superposition is ruled out. In that case, we resort to numerical solution of the resulting nonlinear system of governing equations. For the simulation of general channel flow, we need empirical equations to determine the wall friction factor and heat transfer coefficient.

Vortex. Vortex is a uniquely important component of gas turbine internal flow system and arises due to the presence of rotation. Like solid-body rotation, the forced vortex is characterized by a constant angular velocity. By contrast, a free vortex, which is free from any external torque, keeps its angular momentum constant. In internal flows, because of the presence of walls with friction, a pure free vortex is seldom found. When the flow enters a rotating channel, it immediately assumes the state of solid-body rotation with the channel and thus becomes a forced vortex. In gas turbine cavities, the internal flow features the most complex vortex structure. Such a vortex structure does not fit the definition of either the forced vortex or the free vortex. We call this a general vortex, which can be modeled using a stacked combination of forced vortices. For a radially inward flow in a rotor cavity, the flow shows the tendency of a free vortex with its swirl velocity increasing downstream. For a large radial span, the flow may rotate faster than the adjacent rotor surfaces. The radially outward flow features the opposite behavior, rotating slower than the rotor surface.

The primary effect in a free, or forced, vortex is to increase the static pressure in a radially outward flow and decrease it in a radially inward flow. It also influences the amount of work transfer with the rotor surface in contact. For example, if the fluid is rotating at the same angular velocity as the surface in contact, no rotational work transfer occurs. The pumping flow induced by a rotor disk also depends on the vortex strength of the adjacent fluid. Unlike other IFS components, which can be modeled as a loss element with empirically defined relationship between their mass flow rate and the associated pressure drop, the change in static pressure in a vortex is nearly independent of its mass flow rate. The modeling of a vortex in the flow network therefore requires special consideration.

Cavity. In gas turbine design, cavities are widely encountered by internal flows. But for some three-dimensional features like bolts, these cavities are modeled as axisymmetric. In general, the surfaces forming the cavity may be rotating, counterrotating or stationary. In the absence of a radially inward or outward flow, as is often the case for compressor rotor disk cavities, the internal flow behaves like a forced vortex in solid-body rotation with the disks. In other cases, a complex vortex structure prevails in the cavity flow. From design considerations, for a given coolant mass flow rate, the cavity flow model provides the distributions of swirl velocity, windage, and static pressure. These quantities form essential inputs to accurate heat transfer simulation. The accuracy of rotor axial thrust computation largely depends upon the accuracy of static pressure distribution in rotor cavities.

In the foregoing, we have presented brief descriptions of the four basic components (orifice, channel, vortex, and cavity) of a gas turbine internal flow system. Other components such as labyrinth seal, rim seal, and preswirler may be modeled as a combination of the four basic components. These components may therefore be considered as superelements in a flow network model.

Internal flow systems are generally represented by a complex flow network of various components (elements) interconnected at junctions (chambers or nodes). In each element, the flow is assumed to be locally one-dimensional. A typical flow network of gas turbine internal flow systems is shown in Figure 1.7. Chapter 3 presents a comprehensive discussion on 1-D flow network modeling and the related robust solution method.

1.2.2   Ef ciency Impact on Gas Turbine Components

The cooling flows peeled off from the compressor primary flow path works somewhat similar to regeneratively cooling, robbing energy in the form of heat transfer and work transfer and dumping some of it into the turbine flow path for potential work extraction by the rows of turbine blades. Whereas the equivalence of work transfer and heat transfer holds in terms of energy (the first law of thermodynamics), and whereas the conversion of work into thermal energy is 100 percent, the complete conversion of thermal energy back to work is not possible (the second law of thermodynamics). As a result of heating, wall friction, and mixing with high-momentum primary flow, the internal flows feature monotonic increase in entropy, which results in irreversible loss in total pressure. This entropy generation in the primary and internal flow systems is the main reason behind the efficiency loss during energy conversion in gas turbines.

In compressors, the primary air flow faces an adverse pressure gradient in all its stages. This makes the compressor prone to boundary layer separation, leading to design considerations of improving stall and surge margins. As discussed by Cumpsty (2004), the matching of various stages in a multistage axial-flow compressor remains a serious design challenge. To protect these compressors from stalling and surging conditions,

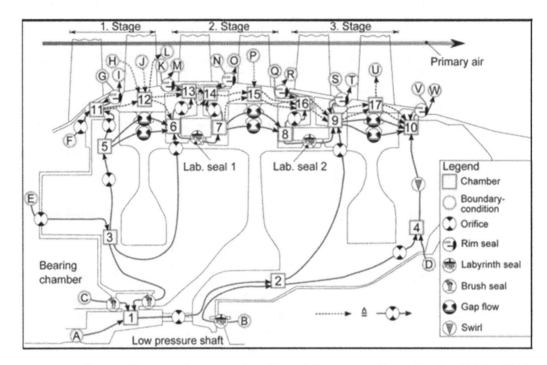

Figure 1.7 Flow network representation of internal flow systems (Source: Brack and Muller (2014) with permission from ASME).

bleed valves are generally used during start-up and shut-down. These valves, which discharge into the gas turbine exhaust duct, are only open during acceleration to the rated speed and deceleration from the rated speed. When we bleed compressor air from a stage to become an internal flow, the air flow through the downstream stages is reduced. Unless it is already factored into the original design, the internal flow extraction will render the latter stages to perform at off-design conditions, reducing their polytropic efficiencies. Thus, for the compressor to perform with minimum loss in its aerodynamic efficiency with acceptable stall and surge margins, the realistic schedule of bleed flow rates and their stage-wise locations must be factored into the original design.

Figure 1.8 shows two categories of internal flows. The first kind, which fully participates in the work output from the turbine, is called nonchargeable, and the second kind, which is ignored for work extraction through the turbine, is called chargeable, directly impacting the overall cycle performance. The only nonchargeable flow shown in the figure corresponds to the cooling flow used for the first-stage vanes. The net effect of this internal flow, which bypasses the combustor, is to lower the turbine inlet total temperature and somewhat lower the corresponding total pressure as a result of entropy increase from mixing and heat transfer. Horlock and Torbidoni (2006) present the calculations of isentropic efficiencies defined by Timko (1980) and Hartsel (1972) for a cooled turbine stage with a polytropic efficiency of 0.9. From their calculations, the linear variation of the cooled-turbine isentropic efficiency with the coolant air flow fraction can be approximated by the following equation:

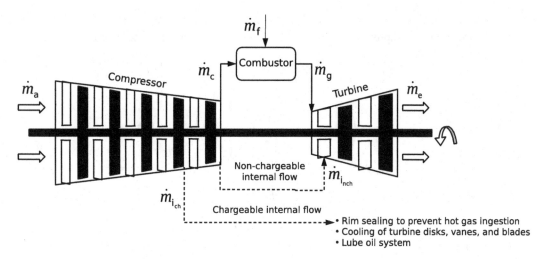

Figure 1.8 Chargeable and nonchargeable internal flows in gas turbine design and cycle performance.

$$\tilde{\eta}_{T_i} = \frac{1 - \left(1/\pi_T\right)^{\frac{\eta_{P_T}(\kappa-1)}{\kappa}}}{1 - \left(1/\pi_T\right)^{\frac{\kappa-1}{\kappa}}} - \frac{1}{3}\left(\frac{\dot{m}_{i_T}}{\dot{m}_g + \dot{m}_{i_T}}\right) \tag{1.4}$$

where $\dot{m}_{i_T}$ is the internal air mass flow rate used for turbine stage cooling and $\dot{m}_g$ is the mass flow rate of hot gases from the combustor.

1.2.3   Penalty on Engine Cycle Performance

Both materials and cooling technologies must advance simultaneously to realize the trend of continuous increase in gas turbine cycle efficiency by increasing the combustor outlet temperature (COT) or turbine inlet temperature (TIT). MacArthur (1999) and Horlock, Watson, and Jones (2001) suggested that with the contemporary materials technology further increases in COT might result in a decrease of cycle efficiency. The penalty of internal flows, which are associated with cooling and sealing, on engine cycle performance can be determined only by a judicious application of both the first and second laws of thermodynamics. While it is straightforward to apply the first law of thermodynamics (steady flow energy equation), the second analysis requires user-specified models for cooling losses. Young and Wilcock (2002a, 2002b) provide such models, which are expressed in terms of irreversible entropy creation rates rather than the loss of total pressure or modified stage efficiency. Wilcock, Young, and Horlock (2005) used these models to compute the effect of turbine blade cooling on the cycle efficiency of shaft-power gas turbines.

To understand the effect of chargeable internal flow on specific work output and cycle efficiency, let us consider a simple gas turbine with the compressor and turbine operating at constant isentropic efficiencies η_{C_i} and η_{T_i}, respectively. In the cycle analysis of this gas turbine, we neglect the fuel mass flow rate and the combustor pressure drop, which implies equal pressure ratio across both compressor and turbine. We obtain the following equations (see Problem 1.4) for the specific work output and cycle thermal efficiency with their dependence on the chargeable internal flow as a fraction of compressor air flow, engine pressure ratio, and turbine inlet temperature:

$$\frac{\tilde{w}_{\text{net}}}{c_P T_{t_1}} = \left(1 - \frac{\dot{m}_{i_{\text{ch}}}}{\dot{m}_a}\right)\left(\frac{T_{t_3}}{T_{t_1}}\right)\eta_{T_i}\left(1 - \frac{1}{\pi^{\frac{\kappa-1}{\kappa}}}\right) - \frac{1}{\eta_{C_i}}\left(\pi^{\frac{\kappa-1}{\kappa}} - 1\right) \tag{1.5}$$

$$\tilde{\eta}_{\text{th}} = \frac{\left(\dfrac{T_{t_3}}{T_{t_1}}\right)\eta_{T_i}\left(1 - \dfrac{1}{\pi^{\frac{\kappa-1}{\kappa}}}\right) - \dfrac{1}{\left(1 - \dfrac{\dot{m}_{i_{\text{ch}}}}{\dot{m}_a}\right)\eta_{C_i}}\left(\pi^{\frac{\kappa-1}{\kappa}} - 1\right)}{\left(\dfrac{T_{t_3}}{T_{t_1}}\right) - \dfrac{1}{\eta_{C_i}}\left(\pi^{\frac{\kappa-1}{\kappa}} - 1\right) - 1} \tag{1.6}$$

where $\dot{m}_{i_{\text{ch}}}$ is the chargeable internal air mass flow rate and $\dot{m}_a$ is the compressor inlet air mass flow rate.

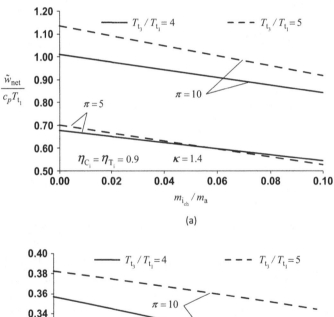

(a)

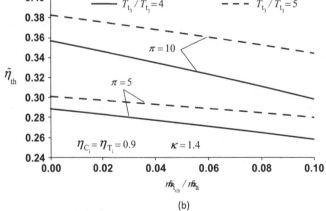

(b)

Figure 1.9 (a) Net specific work output versus chargeable internal flow fraction and (b) cycle thermal efficiency versus chargeable internal flow fraction.

Equation 1.5 is plotted in Figure 1.9a for two values (4 and 5) of turbine inlet to compressor inlet temperature ratio T_{t_3}/T_{t_1} and two values (5 and 10) of compressor, or turbine, pressure ratio π. The figure shows that for all combinations of pressure ratios and temperature ratios the net specific work output of the cycle decreases with increasing chargeable flow rate. The rate of decrease, however, increases with increase in pressure ratio and temperature ratio. For a given pressure ratio, the specific work input into the compressor remains constant, independent of the temperature ratio and the amount of the chargeable flow bled from the compressor. In this case, as we increase the temperature ratio the specific work output of the turbine increases. This in turn increases the deficit in turbine total work output with increase in the chargeable flow. The net effect is that, for a given pressure ratio, as we increase the chargeable flow, the specific net work output of the cycle decreases faster with an increase in temperature ratio, as shown in the figure.

Now for a given value of the temperature ratio T_{t_3}/T_{t_1}, if we increase the pressure ratio, both the compressor work input and turbine work output increase. With increasing chargeable flow, while the compressor total work input remains constant, the turbine work output decreases more than the corresponding value at a lower pressure ratio. The net effect is that, for a given temperature ratio, the net specific work output decreases with increasing chargeable flow faster with an increase in pressure ratio, as shown in Figure 1.9a.

Equation 1.6 is plotted in Figure 1.9b, which shows that the cycle efficiency decreases with chargeable flow for all pressure ratios and temperature ratios. For a given pressure ratio, however, the rate of decreases gets lower with increased temperature ratio. This trend for the cycle efficiency is opposite to the trend we discussed in the foregoing for the net specific work output. For the same temperature ratio, however, the constant pressure ratio lines are converging in the direction of increasing chargeable flow, similar to the trend for the specific net work output. As we increase the pressure ratio, the compressor will demand higher work input, but less thermal energy input in the combustor through added fuel will be needed to achieve the specified turbine inlet temperature. This interplay of various effects is more complex and does not lend itself to simple explanations behind the trends observed in Figure 1.9b for the thermal efficiency. We are left with no choice but to rely on Equation 1.6 to compute various effects.

1.3 Physics-Based Modeling

Physics is the foundation of engineering; mathematics is the language of physics. All engineering solutions must not only be mathematically sound; they must also satisfy all the applicable laws of physics to be realized in the physical world. Fortuitously, all the flow and energy transfer processes in a gas turbine are governed by the conservation laws of mass, momentum, energy (the first law of thermodynamics), while satisfying the entropy constraint as dictated by the second law of thermodynamics. In terms of mathematical equations in tensor notation, we summarize these conservation laws as follows:

Continuity equation:

$$\frac{\partial \rho}{\partial t} + \frac{\partial (\rho u_i)}{\partial x_i} = 0 \qquad (1.7)$$

Momentum equations:

$$\frac{\partial(\rho u_i)}{\partial t} + \frac{\partial(\rho u_j u_i)}{\partial x_j} = -\frac{\partial P_s}{\partial x_i} + \frac{\partial \sigma_{ij}}{\partial x_j} + S_i \tag{1.8}$$

Energy equation:

$$\frac{\partial(\rho e)}{\partial t} + \frac{\partial(\rho u_j e)}{\partial x_j} = \frac{\partial \dot{q}_j}{\partial x_j} - \frac{\partial(P_s u_j)}{\partial x_j} + \frac{\partial(u_i \sigma_{ij})}{\partial x_j} + Q_S \tag{1.9}$$

where e is the specific total energy, σ_{ij} the stress tensor, S_i the momentum source term, $\dot{q}_j$ the diffusive energy flux vector, and Q_S the energy source term. We discuss these equations in Chapter 2 in greater detail, in Chapter 3 for 1-D flow network modeling, and in Chapter 6 in the context of whole engine modeling with emphasis on the turbulence models widely used in gas turbine design applications. Solutions of Equations 1.7–1.9 in their most general form (time-dependent evolution of pressure, temperature, and velocities throughout the flow domain) in gas turbine design is handled by a number of leading CFD commercial codes.

The thermofluids design of a gas turbine is a complex undertaking and far from being handled by a pushbutton technology by simply integrating a number of commercial and in-house codes to execute the design process. The entire design process is multidisciplinary in nature. Generally, a new gas turbine design is handled in three phases: conceptual design, preliminary design, and detailed design. In each of these phases, different degrees of geometric details and approximations are used for modeling and related numerical solutions. In the conceptual design phase, analyses are limited to "back-of-the-envelope" and 1-D modeling. In the preliminary design phase, when the product geometry gets a preliminary definition, the engineers undertake 1-D and 2-D (axisymmetric) modeling. In the detailed design phase, when the product definition needs to be finalized to release engineering drawings for manufacturing, some 3-D analysis is undertaken to fine-tune the design for reduced losses, higher performance, and higher reliability and durability.

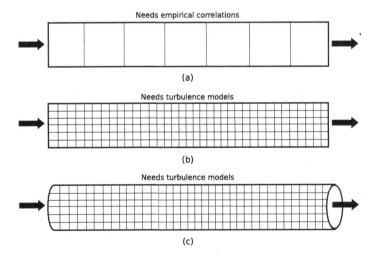

Figure 1.10 Physics-based modeling: (1) 1-D CFD, (b) 2-D CFD, and (c) 3-D CFD.

In thermofluids design engineering, 1-D, 2-D, and 3-D modeling are also called 1-D CFD, 2-D CFD, and 3-D CFD, respectively, and are schematically shown in Figure 1.10. The only requirement for a prediction method, be it 1-D, 2-D, or 3-D, to be physics-based is that it must not violate any of the aforementioned conservation laws. Although each method must validate with the product performance data obtained from in-house testing and field operation, a method that is not physics-based and based entirely on previous empirical data and arbitrary correction factors is undesirable and short-lived as a reliable predictive tool. Physics-based methods tend to make more consistent predictions and, at times, may need only some minor corrections for an acceptable validation with the actual design. These methods are continuously validated using data from gas turbine thermal surveys, and the discrepancies between pretest predictions and test data are meticulously resolved by improved physical modeling and adjustment of correction factors.

In 1-D CFD, shown schematically in Figure 1.10a, the flow domain is divided into large control volumes. Each control volume contains a part of the bounding walls. The prediction yields one-dimensional variation of various flow properties, usually along the flow direction. The modeling is based on the empirical correlations to compute the discharge coefficient, friction factor, loss coefficient, and heat transfer coefficient. Like all CFD analyses, boundary conditions are specified at the inlet, outlet, and walls. The 1-D CFD offers designers maximum flexibility to adjust correction factors to the empirical correlations to improve validation with the component test data.

A 2-D CFD analysis, shown schematically in Figure 1.10b with a two-dimensional computational grid, becomes necessary when the flow properties vary both in the flow direction and in one more direction normal to the flow direction. A three-dimensional axisymmetric flow is commonly handled by 2-D CFD in cylindrical polar coordinates. In this case, for predicting turbulent flows, which most gas turbine internal flows are, one of the turbulence models is used instead of the empirical correlations used in 1-D CFD. As shown in Figure 1.10b, to resolve local variations in flow properties, very small control volumes are used, making the final results grid-independent, that is to say that the further refinement in grid will have little effect on the solution. Except those near a wall, all other control volumes used in a 2-D CFD are without a wall. Knowing the flow properties at many locations (grid points) in a 2-D CFD has the distinct advantage of providing two-dimensional flow visualization, delineating regions of internal flow recirculation and wall boundary layer separation. This information is certainly missing in a 1-D CFD. For computing integral quantities such as the loss in total pressure from inlet to outlet, we need to compute section-average values from the 2-D CFD results. These integral quantities may then be compared with the corresponding 1-D CFD results or test data. A physics-based method to postprocessing CFD results is presented in Chapter 6. It is important to note that when using 2-D CFD, changing the turbulence model is the only option a designer has to improve the validation of the CFD results with the test data. The flow visualization aspect of a 2-D CFD analysis often proves useful in gas turbine design for reinforcing 1-D CFD modeling. It may come as a surprise to some that the integral results obtained from a 2-D CFD analysis may not often be more accurate than those from a 1-D CFD, which directly uses the applicable empirical correlations. In this respect, 1-D CFD tends to be more postdictive; that is, using the empirical correlations obtained from the test data to predict these data, than predictive.

The general methodology to carry-out a 3-D CFD analysis, which becomes necessary to understand the three-dimensional behavior of a flow field, is similar to that used for a 2-D CFD. In this case, we use a three-dimensional grid system, shown schematically in Figure 1.10c, with a number of interconnected three-dimensional control volumes. In each of these control volumes, all the governing conservation equations must be satisfied, albeit approximately due to the numerical nature of the solution obtained. For example, to simulate the hot gas ingestion across a turbine rim seal with ingress and egress driven by a circumferential variation in static pressure at the exit of preceding vanes, one needs to carry out a 3-D CFD analysis, preferably unsteady at that.

Often in gas turbine design, circumferentially periodic geometry and boundary conditions allow one to use a 3-D CFD model for a sector, which yields results faster than a full 360-degree model. Like 2-D CFD, the accuracy of the results from 3-D CFD using a model of high geometric fidelity with a fine mesh depends on the turbulence model used. Designers have little control to further improve these results for a better match with the test data. Note that the current limitations of various statistical turbulence models can only be overcome by using large eddy simulation (LES) and direct numerical simulation (DNS), both of which are a few years away from being a part of routine gas turbine design applications.

Modern gas turbines for aircraft propulsion and shaft-power used for electric power generation and mechanical drives demand the very best of the thermofluids (thermodynamics, heat transfer, and fluid dynamics) sciences and structure and fracture mechanics for component life assessment to meet the upward moving targets of thermal efficiency and specific work output (or propulsive thrust), which primarily depend on compressor pressure ratio and turbine inlet temperature. Among various technologies used in the design, development, and customer-friendly field operation of gas turbines, advances in cooling and materials technologies stand out. The designers must ensure the specified durability, expected reliability, and manufacturability of various gas turbine components. Working as a system, each gas turbine must deliver its overall operational reliability and flexibility while meeting its performance guarantees and regulatory constraints on environmental pollution, noise level, and safety from potential explosion, including containment of critical component failure.

From the foregoing discussion it is clear that the gas turbine design is a multi-disciplinary undertaking, requiring a multiphysics approach to the design and analysis of all its components. Chapter 6 discusses in detail the methodology of an integrated flow, heat transfer, and mechanical modeling of gas turbine components and the engine as a whole.

1.4   Robust Design Methodology

The typical engineering education is built on deterministic evaluations of equations and formulas to yield numerical values of dependent (response) variables for a given set of independent (design) variables. This mindset among engineers persists in the real-world engineering environment. When it comes to manufacturing, we are all expected to include a tolerance on each dimension of a part to be built. Thus, uncertainty and variability is ubiquitous in the engineering world, and the probability of failure of any engineered product is seldom zero, regardless of how well it is designed and built.

Suppose, for example, that we need to calculate the diameter of a sharp-edge orifice to meter and measure the required mass flow rate through a gas pipe line. Such a calculation for a general compressible flow requires the knowledge of pressure and temperature conditions upstream of the orifice, the downstream pressure, and the discharge coefficient, all of which have some uncertainty associated with them. Even when we carry out the calculation at the mean expected values of these input variables, the computed orifice diameter can only be realized within some tolerance, depending upon the capability of the machining process used. In the actual operation, we should expect the final orifice to perform at the mean target value with a finite variance.

Robustness is insensitivity to uncertainty. A component or system design is considered robust if its intended performance is not affected by uncertainties associated with inputs or uncontrollable environmental conditions (noises). While the governing equations are deterministic, the independent variables and boundary conditions have noise associated with them. We must therefore ensure that within the probabilistic variations in various inputs, the output (response) is functionally acceptable. Such a design is called a robust design, as depicted in Figure 1.11 in which the design variables are set such as to yield an increased average performance (response) with minimum variance. As shown in the figure, both designs 3 and 4 have the equal average performance; but, because of its higher robustness, design 4 is preferred to design 3.

Robust design methodology (RDM) means systematic efforts to achieve insensitivity to various noise factors. For a final robust design, a designer must have complete awareness of variation and apply RDM at all stages of the product and process design. Invariably, RDM involves a probabilistic assessment of the design requiring thousands of deterministic runs using random samples of all input variables. In this sense, a robust

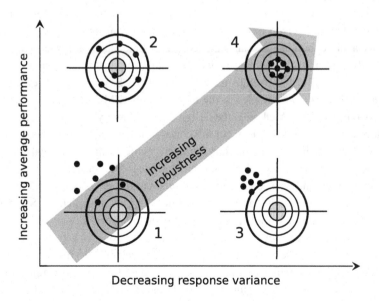

Figure 1.11 Pictorial representation of a robust design.

design is alternatively called a probabilistic design. Because each run in a probabilistic assessment is deterministic, it behooves a designer to develop a physics-based model of the design under consideration with acceptable product validation. Multiples runs are not a substitute for the lack of understanding of the underlying physics of design. It is important to note that a robust design need not necessarily be an optimum design, which could probably fall off the cliff under certain variation of a design variable or environmental noise. For situations in which the product reliability is of utmost concern, robustness in design becomes the most desirable quality, at times, at the expense of some loss in performance.

The main task of a gas turbine internal flow system is to provide the required cooling or sealing flow to the target location at the required pressure and temperature conditions, knowing the source conditions at the compressor bleed location. Brack and Muller (2014) present a probabilistic analysis of the secondary air system (SAS), shown in Figure 1.7, of a three-stage low pressure turbine of a jet engine at takeoff. The study focuses on the robustness of three key functions of SAS: (1) turbine rotor cooling flow, (2) axial bearing loads, and (3) effectiveness toward preventing hot gas ingestion in wheel-space. To determine the uncertainty associated with these SAS deliverables and to identify major drivers of variation, they have used a Latin hypercube sampling method coupled with the correlation coefficient analysis on the 1-D flow network model. In the following sections, we present a more general and widely used approach of Monte Carlo simulation (MCS), often complemented by response surface modeling (RSM), to carry out a probabilistic (robust) design.

1.4.1   Monte Carlo Simulation

Monte Carlo simulation is a powerful statistical analysis method and widely used for solving complex engineering problems involving a number of random variables, which feature various types of probability distribution. The accuracy of MCS does not depend on the problem size or on how nonlinear the engineering models are. Because MCS involves a large number of runs, the cost and time taken for each run, which is necessarily deterministic, determines its application in gas turbine design.

Figure 1.12 shows the overall methodology of MCS. Once we have determined a large number of random samples of the input variables, MCS can be carried out using one of two methods. In Method 1, the design code is directly used for each combination of random input variables x_i to yield the corresponding output (response) variables y_j. We can then obtain the distribution for each y_j for further statistical analysis and assessment of design robustness and reliability. If the design code is too time-consuming to run and we need a quick assessment of design, especially in the preliminary design phase, gas turbine designers resort to Method 2 in which MCS is carried out using a simple surrogate model, often developed by the response surface modeling (RSM) methodology, which we discuss in the next section.

For example, suppose we wish to carry out the probabilistic rim seal design for an acceptable sealing effectiveness in the stage 1 turbine to prevent or minimize hot gas ingestion. Conducting MCS using a 3-D CFD model in this region will be a daunting task, almost impractical in today's design environment. However, one can leverage the

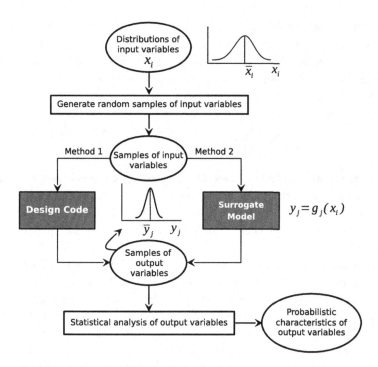

Figure 1.12 Schematic of Monte Carlo simulation.

3-D CFD method in the development of a multispoke orifice model, discussed in Chapter 4, and use it to conduct the required MCS. To further reduce the MCS time, one may use, for the multisurface orifice model, a surrogate model developed using the RSM techniques.

1.4.2 Response Surface Modeling

Two main building blocks of RSM are the design of experiments (DOE) and regression analysis, and the end result is the response surface equation (RSE), which acts like a transfer function as a surrogate for the full-fledged design code. Besides supporting a fast MCS, RSM finds standalone applications in design optimization and sensitivity analysis. Myers, Montgomery, and Anderson-Cook (2016) provide comprehensive details on all aspects of RSM. We provide here some introductory discussion on this topic.

DOE constitutes the first step in RSM. Here one chooses a particular design such as Box-Behnken design (BBD), full central composite design (CCD), fractional CCD, or three-level full factorial design (FF3). These designs provide the points, having coordinates in terms of coded variables, on which to perform actual evaluations of the response. For the physical variable x_i, the coded variable ξ_i, which ranges from −1 to +1, is defined by the equation

$$\xi_i = \frac{x_i - 0.5\left[\max\left(x_i\right) + \min\left(x_i\right)\right]}{0.5\left[\max\left(x_i\right) - \min\left(x_i\right)\right]} \tag{1.10}$$

Craney (2003) recommends the fractional CCD as the best generic choice, which may be replaced by a full CCD for the cases with less than five input variables.

The second step in RSM is to utilize carry out a regression analysis on the evaluated responses. For example, for the response y, a second-order response surface equation (RSE) for k input variables can be written as

$$y = \beta_0 + \sum_{i=1}^{k} \beta_i x_i + \sum_{i=1}^{k-1} \sum_{j=i+1}^{k} \beta_{ij} x_i x_j + \sum_{i=1}^{k} \beta_{ii} x_i^2 \qquad (1.11)$$

where x_i are the input (design) variables. The constant β_0, the coefficients β_i for linear terms, β_{ii} for pure quadratic terms, and β_{ij} for cross-product terms are estimated using linear least squares regression, where an overdetermined system of linear equations is solved, see Sultanian (1980). Equation 1.11 is also known as transfer function, and it must be adequately validated before using it as surrogate model replacing the actual design code. For the Monte Carlo simulation (MCS), each x_i is simulated according to its statistical distribution (probability distribution function) and then the response y is calculated using Equation 1.11. This is repeated a large number of times, resulting in a distribution function for y.

By using the linearization method around the mean of the input variables x_i, it may not often be necessary to carry out MCS for estimating the mean and variance of the response (output) variable y. According to Bergman et al. (2009), the Gauss approximation formula is used in the following way:

$$\text{Var}(y) \approx \sum_{i=1}^{k} c_i^2 \text{Var}(x_i) + \sum_{i \neq j}^{k} c_i c_j \text{Cov}(x_i, x_j) \qquad (1.12)$$

where c_i is the sensitivity coefficient belonging to x_i. Under the assumption that the input variables are independent, Equation 1.12 reduces to

$$\text{Var}(y) \approx \sum_{i=1}^{k} c_i^2 \text{Var}(x_i) \qquad (1.13)$$

In Equations 1.12 and 1.13, the sensitivity coefficient c_i for each x_i is simply obtained as $c_i = \partial y / \partial x_i$ from Equation 1.11

$$c_i = \beta_i + \sum_{i \neq j}^{k} \beta_{ij} \mu_j + 2 \beta_{ii} \mu_i \qquad (1.14)$$

where $\mu_i = E(x_i)$, the mean of x_i. We can also approximate the mean of the output variable as $E(y) \approx y(\mu_i)$ by evaluating Equation 1.11 at the mean value of each input variable.

In Chapter 3, we present the modeling of a general compressible flow orifice whose discharge coefficient depends upon a number of design and operational parameters whose direct and interaction effects are not sufficiently known through empirical correlations. One way to mitigate this situation is to develop a response surface model for the orifice using high-fidelity CFD analyses following the RSM methodology. The resulting RSE will be a valuable method to implement in a general purpose flow network for modeling various internal flow systems of modern gas turbines.

1.5   Concluding Remarks

In this chapter, we have presented a brief overview of the state-of-the-art gas turbines and their pacing technologies behind their widespread applications in aircraft propulsion (aviation) and electric power generation. The bibliography at the chapter will be helpful to the readers who are interested in the history of gas turbines and their primary flow path design fundamentals and performance calculations. Advances in materials and cooling technologies will essentially determine the future growth of gas turbines with higher thermal efficiency and net specific work output. Additive manufacturing (3-D printing) has currently emerged as a disruptive technology to remove major constraints of manufacturing of many critical gas turbine components, thus allowing a more robust design of these components to achieve higher durability and performance at lower cost. The trend of achieving higher efficiency by increasing turbine inlet temperature and compressor pressure ratio is not sustainable unless the penalty associated with the required cooling and sealing flows, which is currently around 20 percent of the compressor air flow, is minimized.

From modeling considerations, we have identified four basic components (orifice, channel, vortex, and cavity) of the internal flow system and suggested that other superelements such as labyrinth seals, turbine rim seals to prevent hot gas ingestion, preswirler system, and others can be modeled combining these basic elements.

As introduced in this chapter, we emphasize throughout this book that the internal flow systems be simulated using the physics-based modeling, be it 1-D, 2-D, or 3-D. The basic idea behind this thermofluids modeling approach is to ensure that the conservation equations of mass, momentum, energy, and entropy are satisfied over each control volume regardless of its size. For large control volumes, used in 1-D CFD, we make direct use of the applicable empirical correlations. For small control volumes used in 3-D CFD, the empirical information is generally buried in the turbulence models used. This view of physics-based modeling is more encompassing than the view of 3-D CFD being the only physics-based analysis method, and it will hopefully inculcate in young design engineers the dying culture of "back of the envelope" calculations in their design activities.

The chapter ends with a brief discussion on the probabilistic or robust design methodology based on Monte Carlo simulation (MCS) either directly using the design code or with a surrogate model developed using the response surface modeling (RSM) methodology.

Worked Examples

Example 1.1 Two compressors are compared for bidding. Compressor A is submitted with a total pressure ratio of 4.5 and isentropic efficiency of 85 percent. Compressor B is submitted with a discharge static pressure of 620 kPa, static temperature of 250°C, and velocity of 150 m/s as tested at the ambient condition of 101 kPa and 18°C. Which compressor is actually more efficient? Assume: $\kappa = 1.4$ and $c_P = 1004\,\mathrm{J/(kg\,K)}$.

Solution

Compressor A
Pressure ratio $(\pi_C) = 4.5$
Isentropic efficiency $(\eta_{CA_i}) = 85\%$
Polytropic efficiency $\left(\eta_{P_{CA}}\right)$ of Compressor A:

$$\kappa = 1.4;\ \frac{\kappa-1}{\kappa} = \frac{1.4-1}{1.4} = 0.2857;\ \frac{\kappa}{\kappa-1} = \frac{1.4}{1.4-1} = 3.5$$

$$\eta_{P_{CA}} = \frac{\ln\left(\pi_C^{\frac{\kappa-1}{\kappa}}\right)}{\ln\left(1+\frac{\pi_C^{\frac{\kappa-1}{\kappa}}-1}{\eta_{CA_i}}\right)} = \frac{\ln\left(4.5^{0.2857}\right)}{\ln\left(1+\frac{4.5^{0.2857}-1}{0.85}\right)} = 87.8\%$$

Compressor B

Inlet total pressure $(P_{t_1}) = 101$ kPa
Inlet total temperature $(T_{t_1}) = 18 + 273 = 291$ K
Exit static pressure $(P_{s_2}) = 620$ kPa
Exit static temperature $(T_{s_2}) = 250 + 273 = 523$ K
Exit velocity $(V_2) = 150$ m/s
Exit total temperature : $T_{t_2} = T_{s_2} + \frac{V_2^2}{2c_P} = 523 + \frac{150\times150}{2\times1004} = 534.2$ K
Exit total pressure : $\frac{P_{t_2}}{P_{s_2}} = \left(\frac{T_{t_2}}{T_{s_2}}\right)^{\frac{\kappa}{\kappa-1}} = \left(\frac{534.2}{523}\right)^{3.5} = 1.077$

$$P_{t_2} = 1.077 \times 620 = 667.75\ \mathrm{kPa}$$

Polytropic efficiency $\left(\eta_{P_{CB}}\right)$ of Compressor B:

$$\frac{T_{t_2}}{T_{t_1}} = \left(\frac{P_{t_2}}{P_{t_1}}\right)^{\frac{n-1}{n}}\ \text{where}\ \frac{n-1}{n} = \frac{\kappa-1}{\kappa\eta_{P_{CB}}}$$

Now

$$\frac{n-1}{n} = \frac{\ln\left(\frac{T_{t_2}}{T_{t_1}}\right)}{\ln\left(\frac{P_{t_2}}{P_{t_1}}\right)} = \frac{\ln\left(\frac{534.2}{291}\right)}{\ln\left(\frac{667.75}{101}\right)} = \frac{0.6075}{1.8888} = 0.3216,$$

which yields

$$\eta_{P_{CB}} = \frac{\left(\dfrac{\kappa - 1}{\kappa}\right)}{\left(\dfrac{n - 1}{n}\right)} = \frac{0.2857}{0.3216} = 88.8\%$$

Thus, Compressor B is more efficient than Compressor A.

Example 1.2 The thermodynamic performance of a gas turbine working on a Brayton cycle is given below:

	Ideal Cycle	Real Cycle
Compressor specific work input (kJ/kg)	350	407
Specific energy input to cycle (kJ/kg)	758	700
Turbine specific work output (kJ/kg)	758	705
Net specific work (kJ/kg)	408	298
Cycle thermal efficiency	54%	42%

If the polytropic efficiency of the compressor is 90 percent, compute the polytropic efficiency of the turbine. Assume $\kappa = 1.4$ and $c_P = 1004 \, \text{J}/(\text{kg K})$ throughout the cycle.

Solution

Compressor

Ideal compressor work $= 350 \, \text{kJ/kg}$

Actual compressor work $= 407 \, \text{kJ/kg}$

Isentropic efficiency $\left(\eta_{C_i}\right) = \dfrac{350}{407} = 0.86$

With the given compressor polytropic efficiency of 0.90, we can compute the compressor pressure ratio π from the following equation from Chapter 2 (using an iterative solution method, e.g., "Goal Seek" in Excel):

$$\eta_{C_i} = \frac{\pi^{\frac{\kappa-1}{\kappa}} - 1}{\pi^{\frac{\kappa-1}{\eta_{Pc} \kappa}} - 1}$$

giving $\pi = 12.75$.

Turbine

Ideal turbine work $= 758 \, \text{kJ/kg}$

Actual turbine work $= 705 \, \text{kJ/kg}$

Isentropic efficiency $\left(\eta_{T_i}\right) = \dfrac{705}{758} = 0.93$

In the given Brayton cycle, we have turbine pressure ratio equal to compressor pressure ratio. The polytropic efficiency of the turbine is then computed from the following equation from Chapter 2:

$$\eta_{P_T} = \frac{\ln\left(\eta_{T_i}\left\{\left(\frac{1}{\pi}\right)^{\frac{\kappa-1}{\kappa}} - 1\right\} + 1\right)}{\ln\left(\left(\frac{1}{\pi}\right)^{\frac{\kappa-1}{\kappa}}\right)}$$

giving $\eta_{P_T} = 0.901$. Note that $\eta_{P_C} > \eta_{C_i}$ and $\eta_{P_T} < \eta_{T_i}$.

Problems

1.1 Using the analysis of Brayton cycle presented in Chapter 2, derive Equation 1.1.

1.2 Using the analysis of Brayton cycle presented in Chapter 2, derive Equation 1.2.

1.3 Derive Equation 1.3.

1.4 Derive Equations 1.5 and 1.6.

1.5 Using the relationships between the polytropic efficiency and isentropic efficiency presented in Chapter 2 and assuming a polytropic efficiency of 0.9 for both the compressor and turbine, re-plot Figure 1.9.

1.6 For a gas turbine with fixed T_{t_3}/T_{t_1}, give a physics-based argument to explain why the compressor pressure ratio needed to maximize $W_{net}/c_p T_{t_1}$ is lower than that needed to maximize η_{th}.

1.7 Show that two compressors, one with 4/1 pressure ratio and overall isentropic efficiency of 84.2 percent, and the other with 8/1 pressure ratio and overall isentropic efficiency of 82.7 percent, have an equal polytropic efficiency of ~87 percent. If these two compressors are operated together in series, calculate the overall isentropic efficiency of the combined unit. Assume $\kappa = 1.4$.

1.8 A gas turbine, schematically shown in Figure 1.13, draws in air from atmosphere at 1 bar and 10°C and compresses it to 5 bar with an isentropic efficiency of 80 percent. The air is heated to 1200 K at constant pressure and expanded through two stages in series back to 1 bar. The high-pressure turbine (gas generator turbine) produces just enough power to drive the compressor. The low-pressure

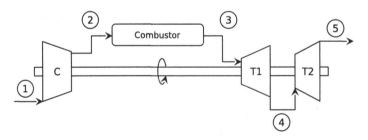

Figure 1.13 A gas turbine engine using a gas generator turbine and a power turbine (Problem 1.8).

turbine (power turbine) is connected to an external load and produces 80 kW of power. The isentropic efficiency is 85 percent for each turbine stage. Assume $\kappa = 1.4$ for the compressor and $\kappa = 1.333$ for both turbines. The gas constant R is 287 J/(kg K) for the entire cycle. Neglect the increase in mass due to the addition of fuel in the combustion chamber. Calculate: (a) the mass flow rate of air, (b) the inter-stage pressure of the turbines, and (c) the thermal efficiency of the cycle.

1.9 Consider the frictionless flow of a calorically perfect gas in a channel with thermal interaction. For the case in which the wall is shaped to keep the Mach number constant, find: (a) an expression for both static and total pressure ratios P_{s_2}/P_{s_1} and P_{t_2}/P_{t_1} in terms of κ, M, and T_{t_2}/T_{t_1} and (b) an expression for the area ratio A_2/A_1 in terms of the same variables. Hint: As we know from the Rayleigh flow (frictionless, constant-area channel flow) that the total pressure changes due to heat transfer. This important relationship between the changes in total pressure due to the changes in total temperature is governed by the following equation:

$$\frac{dP_t}{dT_t} = -\kappa M^2 \left(\frac{P_t}{T_t} \right)$$

1.10 A gas turbine for aircraft propulsion is mounted on a test bed. Air at 1 bar and 293 K enters the compressor at low velocity and is compressed through a pressure ratio of 4 with isentropic efficiency of 85 percent. The air then passes to a combustion chamber where it is heated to 1175 K. The hot gas then expands through a turbine which drives the compressor and has an isentropic efficiency of 87 percent. The gas is then further expanded isentropically through a nozzle leaving at the speed of sound. The exit area of the nozzle is 0.1 m². Assume $\kappa = 1.4$ for the compressor and turbine exhaust nozzle, $\kappa = 1.333$ for the turbine, and $R = 287$ J/(kg K) for the entire cycle. Neglect the increase in mass due to the addition of fuel in the combustion chamber. Sketch the open thermo-dynamic cycle on a $T - s$ diagram and determine: (a) the pressure at the turbine and nozzle outlets, (b) the mass flow rate, and (c) the thrust on the engine mountings.

1.11 In this problem, schematically shown in Figure 1.14, you are asked to compute the loss in total pressure due to the mixing of a low momentum flow (B) with a high momentum flow (A) as shown in the figure below with

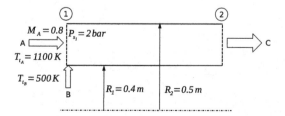

Figure 1.14 Mixing of two compressible flows in an adiabatic annular duct (Problem 1.11).

additional flow and geometric data for a constant-area, adiabatic annular duct. Neglect the axial velocity of flow B. The mass flow rate of flow B is 2 percent of the mass flow rate of flow A, which enters the duct at a Mach number of 0.8. Both flows are fully mixed at section 2. Assume that the total pressure at section 1 corresponds to that of flow A, and neglect any pressure loss due to wall friction.

a) Verify that $P_{s_2} = 177572$ Pa.

b) Compute the loss in total pressure between sections 1 and 2.

1.12 Figure 1.15 shows schematically a 200 MW modern gas turbine engine for power generation. The operating cycle parameters for the baseline design are given as follows:

$P_{t_1} = P_{t_4} = 101.325$ kPa

$T_{t_1} = 15°C$

Compressor pressure ratio $(P_{t_2}/P_{t_1}) = 20$

Compressor pressure ratio at the point of turbine cooling flow extraction $(P_{t_{2'}}/P_{t_1}) = 10$

Turbine cooling flow $(\dot{m}_{cooling}) = 20$ percent of the compressor inlet flow $(\dot{m}_{engine})$

Combustor total pressure loss = 1 percent of compressor exit total pressure

Turbine inlet temperature $(T_3) = 2000$ K

Compressor polytropic efficiency $(\eta_{P_C}) = 92\%$

Turbine polytropic efficiency $(\eta_{P_T}) = 92\%$

Ratio of specific heats across compressor $(\kappa_C) = 1.4$

Ratio of specific heats across turbine $(\kappa_T) = 1.33$

$R_{air} = 287$ J/(kg K)

Calculate engine flow rate $(\dot{m}_{engine})$ and cycle thermodynamic efficiency (η_{th}) for the following cases:

(a) Baseline design

(b) Upgraded baseline design with $\eta_{P_C} = \eta_{P_T} = 93\%$

(c) Upgraded baseline design with $P_{t_{2'}}/P_{t_1} = 9$

(d) Upgraded baseline design with $\dot{m}_{cooling} = 0.18\,\dot{m}_{engine}$

Assume that the turbine cooling flow does not contribute to turbine work. Show details of all your calculations for the baseline design only. Tabulate your calculation results for all four cases.

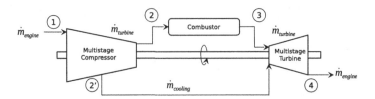

Figure 1.15 An air-cooled gas turbine engine (Problem 1.12).

References

Alexiou, A., and K. Mathioudakis. 2009. Secondary air system component modeling for engine performance simulations. *ASME J. Eng. Gas Turbines Power.* 131(3): 031202.1–031202.9.

Bathie, W. W. 1996. *Fundamentals of Gas Turbines Theory*, 2nd edn. New York: John Wiley & Sons, Inc.

Bergman, B., J. deMare, S. Loren, and T. Svensson (eds.). 2009. *Robust Design Methodology for Reliability: Exploring the Effects of Variation and Uncertainty*, 1st edn. New York: Wiley.

Brack, S., and Y. Muller. 2014. Probabilistic analysis of the secondary air system of a low-pressure turbine. *ASME J. Eng. Gas Turbines Power.* 137(2): 022602.1–022602.8.

Childs, P. R. N. 2011. *Rotating Flow.* New York: Elsevier.

Craney, T. A. 2003. Probabilistic engineering design: reliability review. *R & M Eng. J.* 23(2): 1-7.

Cumpsty, N. A. 2004. *Compressor Aerodynamic*, 2nd edn. Malabar: Krieger Pub Co.

Hartsel, J. E. 1972. Prediction of effects of mass-transfer cooling on blade-row efficiency of turbine airfoils. AIAA 10th Aerospace Sciences Meeting, San Diego. Paper No. AIAA-72-11.

Horlock, J. H., D. T. Watson, and T. V. Jones. 2001. Limitations on gas turbine performance imposed by large turbine cooling flows. *ASME J. Eng. Gas Turbines Power.* 123: 487–494.

Horlock, J. H., and L. Torbidoni. 2006. Calculation of cooled turbine efficiency. ASME Paper No. GT2006-90242.

Johnson, B. V. 2010. Internal air and lubrication systems. In R. Blockley and W. Shyy (eds.), *Encyclopedia of Aerospace Engineering.* Hoboken, NJ: Wiley.

Kutz, K. J., and T. M. Speer. 1994. Simulation of the secondary air system of aero engines. *ASME J. Turbomach.* 116: 306–315.

MacArthur, C. D. 1999. Advanced aero-engine turbine technologies and their application to industrial gas turbines. Proc. 14th Int. Symp. on Air Breathing Engines, Florence. Paper No. 99-7151.

Myers, R. H., D. C. Montgomery, and C. M. Anderson-Cook. 2016. *Response Surface Methodology: Process and Product Optimization Using Designed Experiments*, 4th edn. New York: Wiley.

Olson, E. 2017. GE's new jet engine is a modern engineering marvel. Retrieved September 4, 2017, from http://insights.globalspec.com/article/6373/ge-s-new-jet-engine-is-a-modern-engineering-marvel.

Owen, J. M., and R. H. Rogers. 1989. *Flow and Heat Transfer in Rotating Disc Systems*, Vol. 1: Rotor-Stator Systems. Taunton, UK: Research Studies Press.

Owen, J. M., and R. H. Rogers. 1995. *Flow and Heat Transfer in Rotating Disc Systems*, Vol. 2: Rotating Cavities. Taunton, UK: Research Studies Press.

Saravanamuttoo, H. I. H., G. F. C. Rogers, H. Cohen, P. V. Stranznicky, and A. C. Nix. 2017. *Gas Turbine Theory*, 7th edn. Harlow, UK: Pearson Education Limited.

Schlichting, H. 1979. *Boundary Layer Theory*, 7th edn. New York: McGraw-Hill.

Soares, C. 2014. *Gas Turbines: A Handbook of Air, Land and Sea Applications*, 2nd edn. New York: Elsevier.

Sultanian, B. K. 1980. HOME – a program package on Householder reflection method for linear least-squares data fitting. *J. Inst. Eng. (I).* 60(pt Et. 3): 71–75.

Sultanian, B. K., S. Nagao, and T. Sakamoto. 1999. Experimental and three-dimensional CFD investigation in a gas turbine exhaust system. *ASME J. Eng. Gas Turbines Power.* 121: 364–374.

Timko, L. P. 1980. Energy Efficient Engine: High Pressure Turbine Component Test Performance Report CR-168289.

Vandervort, C., T. Wetzel, and D. Leach. 2017. Engineering and validating a world record gas turbine. Mechanical Engineering. *Mag. ASME.* 12(139): 48–50.

Wilcock, R. C., J. B. Young, and J. H. Horlock. 2005. The effect of turbine blade cooling on the cycle efficiency of gas turbine power plants. *ASME J. Eng. Gas Turbines Power.* 127: 109–120.

Young, J. B., and R. C. Wilcock. 2002a. Modeling the air-cooled gas turbine: part 1 – general thermodynamics. *ASME J. Turbomach.* 124: 2007–213.

Young, J. B., and R. C. Wilcock. 2002b. Modeling the air-cooled gas turbine: part 2 – coolant flows and losses. *ASME J. Turbomach.* 124: 2007–213.

Bibliography

Baskharone, E. A. 2014. *Principles of Turbomachinery in Air-Breathing Engines.* New York: Cambridge University Press.

Box, G. E. P., W. G. Hunter, and J. S. Hunter. 1978. *Statistics for Experimenters: An Introduction to Design, Data Analysis, and Model Building.* New York: John Willey & Sons.

Connors, J. 2010. *The Engines of Pratt & Whitney: A Technical History*, ed. N. Allen. Reston: AIAA.

Cumpsty, N., and A. Heyes. *Jet Propulsion: A Simple Guide to the Aerodynamics and Thermodynamic Design and Performance of Jet Engines*, 3rd edn. New York: Cambridge University Press.

Dixon, S. L., and C. Hall. 2013. *Fluid Mechanics and Thermodynamics of Turbomachinery*, 7th edn. Waltham: Elsevier.

El-Sayed, A. F. 2017. *Aircraft Propulsion and Gas Turbine Engines*, 2nd edn. Boca Rotan: Taylor & Francis.

Flack, R. D. 2010. *Fundamentals of Jet Propulsion with Applications.* New York: Cambridge University Press.

Garvin, R. V. 1998. *Starting Something Big: The Commercial Emergence of GE Aircraft Engines.* Reston: AIAA.

Harman, R. T. C. 1981. *Gas Turbine Engineering: Applications, Cycles and Characteristics.* New York: John Wiley & Sons, Inc.

Korakianitis, T., and D. G. Wilson. 1994. Models for predicting the performance of Brayton-cycle engines. *ASME J. Eng. Gas Turbines & Power.* 161: 381–388.

Kurzke, J. 2002. Performance modeling methodology: Efficiency definitions for cooled single and multistage turbines. ASME Paper No. 2002-GT-30497.

Oates, G. 1997. *Aerothermodynamics of Gas Turbine and Rocket Propulsion*, 3rd edn. Reston: AIAA.

Prabakaran, J., S. Vaidyanathan, and D. Kanagarajan. 2012. Establishing empirical relation to predict temperature difference of vortex tube using response surface methodology. *J. Eng. Sci. Technol.* 7(6): 722–731.

Rangwala, A. S. 2013. *Theory and Practice in Gas Turbines*, 2nd edn. London: New Academic Science Limited.

Rolls-Royce. 2005. *The Jet Engine.* Hoboken, NJ: Wiley.

Smout, P. D., J. H. Chew, and P. R. N. Childs. 2002. ICAS-GT: a European collaborative research program on internal air cooling systems for gas turbines. ASME Paper No. GT2002–30479.

Sultanian, B. K. 2015. *Fluid Mechanics: An Intermediate Approach.* Boca Rotan, FL: Taylor & Francis.

Sultanian, B. K. 2018. *Logan's Turbomachinery: Flowpath Design and Performance Fundamentals*, 3rd edn. Boca Rotan, FL: Taylor & Francis.

Walsh, P. P., and P. Fletcher. 2004. *Gas Turbine Performance*, 2nd edn. Hoboken: Wiley.

Wang, J. X. 2005. *Engineering Robust Design with Six Sigma*. New York: Prentice Hall.

Wilson, D. G., and T. Korakianitis. 2014. *The Design of High-Ef ciency Turbomachinery and Gas Turbines*, 2nd edn. Cambridge, MA: MIT Press.

Nomenclature

BBD	Box-Behnken design
c_p	Specific heat at constant pressure
c_v	Specific heat at constant volume
CC	Combined cycle
CCD	Central composite design
CFD	Computational fluid dynamics
CMC	Ceramic matrix composite
COT	Combustor outlet temperature
DNS	Direct numerical simulation
DOE	Design of experiments
e	Specific total energy
EDF	Electricity of France
FF3	3-Level full factorial design
g_j	Transfer function for y_j
GE9X	Trade mark of GE
GT	Gas turbine
GTCC	Gas turbine combined cycle
IFS	Internal flow system
IAS	Internal air system
LES	Large eddy simulation
k	Number of independent variables in MCS and RSM
$\dot{m}$	Mass flow rate
$\dot{m}_a$	Compressor inlet air mass flow rate
$\dot{m}_c$	Combustor inlet air mass flow rate
$\dot{m}_e$	Turbine exhaust mass flow rate
$\dot{m}_f$	Fuel mass flow rate into combustor
$\dot{m}_g$	Mass flow rate of combustor gases
$\dot{m}_i$	Internal air mass flow rate
$\dot{m}_{i_T}$	Internal air mass flow rate for turbine cooling
M	Mach number
MCS	Monte Carlo simulation
P	Pressure
$\dot{q}_j$	Diffusive energy flux vector
Q_S	Energy source term
R	Gas constant

RSE	Response surface equation
RSM	Response surface modeling
S_i	Source term of the momentum equation in tensor notation
ST	Steam turbine
SAS	Secondary air system
t	Time
T	Temperature
TIT	Turbine inlet temperature
u_i	Velocity in tensor notation
w	Specific work
$\tilde{w}$	Specific work of a cooled gas turbine
$\dot{W}$	Rate of work transfer
x_i	Cartesian coordinates in tensor notation; input (design) variables
y_j	Dependent (response) variables

Subscripts and Superscripts

C	Compressor
ch	Chargeable
CC	Combined cycle
GT	Gas turbine
i	Summation index in RSE
j	Summation index in RSE
nch	Nonchargeable
net	Net (turbine work output minus compressor work input)
s	Static
ST	Steam turbine
t	Total (stagnation)
T	Turbine
$(^-)$	Average

Greek Symbols

β_0	Constant in RSE
β_i	Coefficients of linear terms in RSE
β_{ii}	Coefficients of pure quadratic terms in RSE
β_{ij}	Coefficients of cross-product terms in RSE
η_{C_i}	Compressor isentropic efficiency
η_{P_C}	Compressor polytropic efficiency
η_{P_T}	Turbine polytropic efficiency
η_{T_i}	Turbine isentropic efficiency
$\tilde{\eta}_{T_i}$	Cooled-turbine isentropic efficiency
η_{th}	Cycle thermal efficiency

κ	Ratio of specific heats ($\kappa = c_P/c_v$)
μ_i	Mean of independent variable x_i
π	Pressure ratio
ρ	Density
σ_{ij}	Stress tensor
	Angular velocity

2 Review of Thermodynamics, Fluid Mechanics, and Heat Transfer

2.0 Introduction

The physics-based modeling of various internal flow systems in gas turbines requires a strong foundation in thermodynamics, fluid mechanics, and heat transfer. In this chapter, we review and reinforce a number of key concepts such as enthalpy, rothalpy, entropy, secondary flows, vortex, vorticity, circulation, isentropic efficiency, polytropic efficiency, stream thrust, impulse pressure, adiabatic wall temperature, and so on. Many of these concepts are directly related to the flow or fluid properties that cannot be directly measured by an instrument but are mathematical combinations of different fundamental properties. For example, the specific enthalpy is a combination of the specific internal energy and specific flow work, which is a product of static pressure and specific volume (reciprocal of density). Nevertheless, most of these derived variables play a useful role in modeling of internal flow systems and their physical interpretation. We also present here in most simple terms various laws that govern the flow and heat transfer physics. Central to any thermofluids mathematical modeling is a clear understanding of the control volume analysis of the conservation laws of mass (continuity equation), momentum (linear and angular), energy, and entropy. While large control volume analyses (macro-analyses) are at the heart of one-dimensional modeling of flow networks, small control volume analyses (micro-analyses) are invariably used in detailed predictions by computational fluid dynamics (CFD).

At the end of this chapter we have included a number of worked-out examples, which are designed to further aid in the understanding and application of various concepts and physical laws, a number of problems to help readers gauge their mastery of various topics, and references and bibliography for more detailed discussion of these topics.

2.1 Thermodynamics

The laws of thermodynamics – only the first and second laws are discussed here – are entirely based on the empirical evidence from the physical world. Accordingly, any prediction method that violates any of these laws is unlikely to yield results that are physically realizable. Thermodynamic laws involve state variables such as static pressure, static temperature, and so on, which are fluid properties, and path variables such as work transfer and heat transfer, which are needed to change the fluid properties between any two states. In our discussion in this book, we will assume that the fluid is a simple substance whose state is uniquely defined by any two properties. Additionally, the fluid is assumed to be calorically perfect with constant specific heats c_p and c_v, at constant pressure and constant volume, respectively. Note that these thermophysical properties are generally functions of static temperature. Consistent with our assumption, we can use an average static temperature between two consecutive states to evaluate these properties.

2.1.1   The First Law of Thermodynamics

For a closed system, the first law of thermodynamics simply states that the change in its total energy dE going from state 1 to state 2 results from heat transfer δQ and work transfer δW, the former is considered positive if it occurs into the system from the surroundings and the later is considered positive when the system does work on the surroundings. In equation form we can write

$$dE = \delta Q - \delta W = \delta Q - P_s d\Psi \qquad (2.1)$$

where the total stored energy consists of the internal energy U, kinetic energy $KE = mV^2/2$, and potential energy PE (in the gravitational force field: $PE = mgh$), that is,

$$E = U + \frac{1}{2}mV^2 + mgh \qquad (2.2)$$

For a system doing work on the surrounding, $\Delta\Psi$ is positive. Dividing Equation 2.2 by mass m yields

$$de = \delta q + \delta w = \delta q - P_s d\nu \qquad (2.3)$$

where the specific total energy $e = u + \frac{1}{2}V^2 + gh$, u being the specific internal energy. Neglecting changes in the kinetic and potential energy, Equation 2.3 becomes

$$du = \delta q - P_s d\nu \qquad (2.4)$$

When the gas volume does not change during its state change (isochoric process), Equation 2.4 reduces to $du = \delta q$, indicating that any change in internal energy in this process is entirely as a result of heat transfer. For a calorically perfect gas, which has constant c_v, we can write $u = c_v T_s$ where we have assumed that $u = 0$ at the reference static temperature $T_s = 0$.

2.1.1.1   Enthalpy

As discussed in Section 2.2.5.4, in a flow system, the fluid specific internal energy u appears with the quantity P_s/ρ, which equals specific flow work as a result of pressure force. By combining these two terms, we define the static enthalpy as

$$h_s = u + \frac{P_s}{\rho} \qquad (2.5)$$

Note that the static enthalpy h_s given by Equation 2.5 is a state property, which is a combination of three state properties u, P_s, and ρ. Only for a flow system, P_s/ρ can be interpreted to equal the specific flow work. For a calorically perfect gas, which has constant c_p, we can write $h_s = c_p T_s$ where we have assumed that $h_s = 0$ at the reference temperature $T_s = 0$.

For a perfect gas with $P_s = \rho R T_s$ as its equation of state, we can rewrite Equation 2.5 as

$$h_s = u + R T_s$$

$$c_p T_s = c_v T_s + R T_s$$

giving

$$c_p = c_v + R \tag{2.6}$$

as a simple relation between the two specific heats and gas constant R.

When we add the specific kinetic energy associated with the bulk gas flow to the static enthalpy of the gas, we obtain the total specific enthalpy of the flow as

$$h_t = h_s + \frac{1}{2} V^2 \tag{2.7}$$

For a calorically perfect gas, we can rewrite Equation 2.7 as

$$h_t = c_p T_s + \frac{1}{2} V^2$$

$$= c_p \left(T_s + \frac{V^2}{2 c_p} \right),$$

$$h_t = c_p T_t \tag{2.8}$$

which yields a simple relation between total (stagnation) temperature and total (stagnation) enthalpy.

For a flow system, it is desirable to express the first law of thermodynamics in terms of enthalpy. Accordingly, we express Equation 2.4 as follows:

$$\delta q = du + P_s d\not{v}$$

$$= du + d\left(P_s \not{v} \right) - \not{v} dP_s$$

$$= d\left(u + \frac{P_s}{\rho} \right) - \frac{dP_s}{\rho}$$

where we have used $\not{v} = 1/\rho$. Thus, we finally obtain

$$\delta q = dh_s - \frac{dP_s}{\rho} \tag{2.9}$$

For an isobaric (constant pressure) process Equation 2.9 reduces to

$$\delta q = dh_s, \tag{2.10}$$

which shows that, for a system at constant pressure, the amount of heat transfer appears as a change in fluid static enthalpy.

2.1.2   The Second Law of Thermodynamics

The first law of thermodynamics embodies the energy conservation principle in that the energy is neither created nor destroyed; it may only be converted from one form to another. Empirical evidence shows that, while all of the work energy can be completely converted into heat energy (internal energy), the reverse is not true. The practical restriction placed on an energy conversion process is governed by the second law of thermodynamics with its focus on irreversibility, which entails conversion of organized motion to random molecular motion through the dissipation of mean flow energy by viscosity. As a result of the presence of friction in practical applications, a reversible work transfer is merely an idealization.

One of the tenets of the second law of thermodynamics is that heat transfer always occurs from higher temperature to lower temperature. This fact, for example, is built into the Fourier's law of heat conduction governed by the equation $\dot{q}_x = -k(dT/dx)$, where the negative sign ensures that the one-dimensional heat conduction in the x direction takes place in the direction of negative temperature gradient.

2.1.2.1   Entropy

We can have an infinite number of paths connecting states 1 and 2 of a fluid system, each path with different values of heat transfer and work transfer, and the associated process could be either reversible or irreversible. According to the second law of thermodynamics, for any two reversible paths A and B connecting states 1 and 2, the following equality holds:

$$\left[\int_1^2 \frac{\delta q_{\text{rev}}}{T_{\text{s}}}\right]_A = \left[\int_1^2 \frac{\delta q_{\text{rev}}}{T_{\text{s}}}\right]_B, \tag{2.11}$$

which points to the existence of a state property, which we call entropy. Thus, we can write

$$(s_2 - s_1) = \left[\int_1^2 \frac{\delta q_{\text{rev}}}{T_{\text{s}}}\right]_A = \left[\int_1^2 \frac{\delta q_{\text{rev}}}{T_{\text{s}}}\right]_B \tag{2.12}$$

which is true for any reversible process connecting states 1 and 2. The second law of thermodynamics further states that the following inequality must be true for an irreversible path, say C, connecting states 1 and 2

$$\left[\int_1^2 \frac{\delta q_{\text{irrev}}}{T_{\text{s}}}\right]_C < (s_2 - s_1), \tag{2.13}$$

which means that a part of the entropy increase along path C occurs as a result of irreversible work transfer. Because entropy is a state property, both reversible and irreversible processes connecting states 1 and 2 will have the same change in entropy, that is, $(s_2 - s_1)_A = (s_2 - s_1)_B = (s_2 - s_1)_C$. This fact gives us a means to compute entropy change between any two states. A reversible process that is also adiabatic must, therefore, be isentropic. Thus, the second law of thermodynamics provides us with an important concept of the state property called entropy, which serves to delineate the feasible flow solutions from those that are not physically possible even if they satisfy the remaining conservation equations.

Computing entropy change between two states. Let us consider a reversible process connecting two states of a fluid flow. In this case Equation 2.9 yields

$$\delta q_{\text{rev}} = dh_s - \frac{dP_s}{\rho}$$

Replacing the reversible heat transfer δq_{rev} in terms of entropy in the aforementioned equation by using Equation 2.12, we obtain

$$T_s ds = dh_s - \frac{dP_s}{\rho} \qquad (2.14)$$

Because Equation 2.14 involves only state properties, it holds good for both reversible and irreversible processes connecting any two states. For a calorically perfect gas, we can write Equation 2.14 as

$$ds = c_p \frac{dT_s}{T_s} - \frac{dP_s}{\rho T_s}$$

$$= c_p \frac{dT_s}{T_s} - \left(\frac{P_s}{\rho T_s}\right) \frac{dP_s}{P_s}$$

$$ds = c_p \frac{dT_s}{T_s} - R \frac{dP_s}{P_s} \qquad (2.15)$$

Integrating Equation 2.15 between states 1 and 2 yields

$$\int_1^2 ds = c_p \int_1^2 \frac{dT_s}{T_s} - R \int_1^2 \frac{dP_s}{P_s}$$

$$s_2 - s_1 = c_p \ln\left(\frac{T_{s_2}}{T_{s_1}}\right) - R \ln\left(\frac{P_{s_2}}{P_{s_1}}\right) \qquad (2.16)$$

For an isentropic (adiabatic and reversible) process, Equation 2.16 yields

$$c_p \ln\left(\frac{T_{s_2}}{T_{s_1}}\right) - R \ln\left(\frac{P_{s_2}}{P_{s_1}}\right) = 0 \qquad (2.17)$$

giving

$$\frac{P_{s_2}}{P_{s_1}} = \left(\frac{T_{s_2}}{T_{s_1}}\right)^{\frac{c_p}{R}} = \left(\frac{T_{s_2}}{T_{s_1}}\right)^{\frac{\kappa}{\kappa-1}} \qquad (2.18)$$

where $\kappa = c_p/c_v$. At any point in a gas flow, the total pressure and total temperature are obtained assuming an isentropic stagnation process. Accordingly, at state 1, Equation 2.18 yields

$$c_p \ln\left(\frac{T_{t_1}}{T_{s_1}}\right) - R \ln\left(\frac{P_{t_1}}{P_{s_1}}\right) = 0$$

$$c_p \ln T_{s_1} - R \ln P_{s_1} = c_p \ln T_{t_1} - R \ln P_{t_1} \qquad (2.19)$$

Similarly, we can write at state 2

$$c_p \ln T_{s_2} - R \ln P_{s_2} = c_p \ln T_{t_2} - R \ln P_{t_2} \tag{2.20}$$

Combining Equations 2.19 and 2.20, we obtain

$$c_p \ln \left(\frac{T_{s_2}}{T_{s_1}} \right) - R \ln \left(\frac{P_{s_2}}{P_{s_1}} \right) = c_p \ln \left(\frac{T_{t_2}}{T_{t_1}} \right) - R \ln \left(\frac{P_{t_2}}{P_{t_1}} \right). \tag{2.21}$$

Thus, in terms of total pressure and total temperature, Equation 2.16 becomes

$$s_2 - s_1 = c_p \ln \left(\frac{T_{t_2}}{T_{t_1}} \right) - R \ln \left(\frac{P_{t_2}}{P_{t_1}} \right) \tag{2.22}$$

Note that when we replace all static quantities in Equation 2.16 with the corresponding total quantities, we obtain Equation 2.22. For an isentropic process, Equation 2.22 yields

$$\frac{P_{t_2}}{P_{t_1}} = \left(\frac{T_{t_2}}{T_{t_1}} \right)^{\frac{c_p}{R}} = \left(\frac{T_{t_2}}{T_{t_1}} \right)^{\frac{\kappa}{\kappa-1}} \tag{2.23}$$

If the total temperature in a process connecting states 1 and 2 remains constant; for example, in an adiabatic flow in a stationary duct with wall friction, Equation 2.22 reduces to

$$s_2 - s_1 = -R \ln \left(\frac{P_{t_2}}{P_{t_1}} \right).$$

$$\frac{P_{t_2}}{P_{t_1}} = e^{-\frac{s_2-s_1}{R}} \tag{2.24}$$

Because the wall friction will always increase entropy downstream of an adiabatic pipe flow, Equation 2.24 implies that the total pressure must decrease along this flow.

2.1.3   Thermodynamic Cycles

Figure 2.1a depicts the operation of a simple gas turbine on an open cycle in which both the compressor and turbine are mounted on the same shaft. In this cycle, the ambient air enters the compressor, which increases both air pressure and temperature from the specific work transfer w_C. The compressed air temperature is further increased in the combustor through fuel addition. Hot gases (products of combustion), treated here to have properties same as air (assumed to be a perfect gas), enters the turbine, which converts part of the energy of these gases into the specific work output w_T.

Figure 2.1b uses a fictitious heat exchanger to cool the hot exhaust gases back to the ambient air conditions, making the gas turbine operation a closed cycle, at least thermally. From the steady flow energy equation (derived from the first law of thermodynamics) in this closed cycle we can write

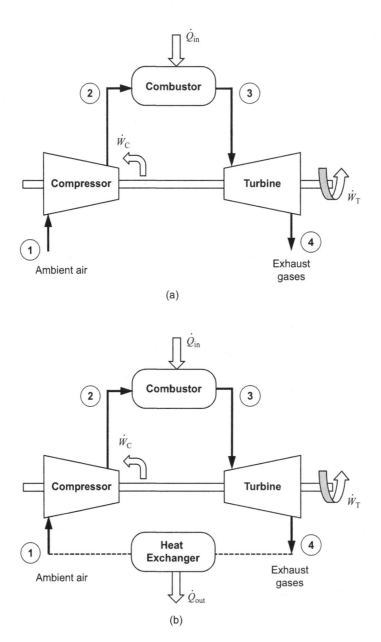

Figure 2.1 Key operational components of a simple gas turbine: (a) open cycle and (b) closed cycle with a fictitious heat exchanger.

$$\dot{Q}_{in} - \dot{Q}_{out} = \dot{W}_T - \dot{W}_C \qquad (2.25)$$

The thermal efficiency of an open-cycle gas turbine shown in Figure 2.1a is defined as the ratio of the net power output to the heat rate (rate of energy transfer from the combustion of fuel in the combustor), giving

$$\eta_{th} = \frac{\dot{W}_{net}}{\dot{Q}_{in}} = \frac{\dot{W}_T - \dot{W}_C}{\dot{Q}_{in}} \qquad (2.26)$$

which also equals the cycle efficiency η_{cycle} for the closed-cycle gas turbine shown in Figure 2.1b.

2.1.4 Isentropic Efficiency

The Brayton cycle of Figure 2.2 maps the operation of the gas turbine engine shown in Figure 2.1 on the $T - S$ diagram. In this cycle, an ideal compressor operates along the isentropic process 1–2i, while the real compressor works along 1–2 with increase in flow entropy as a result of unavoidable friction and other irreversible effects present in the compressor. For the same pressure ratio $\pi_C = P_{t_2}/P_{t_1}$, the actual compressor work input along 1–2 will be higher than the corresponding isentropic (ideal) work input along 1–2i. Accordingly, we define the compressor isentropic efficiency as follows:

$$\eta_{C_i} = \frac{\dot{W}_{C_i}}{\dot{W}_C} = \frac{T_{t_{2i}} - T_{t_1}}{T_{t_2} - T_{t_1}} \qquad (2.27)$$

which further yields

$$\frac{T_{t_2}}{T_{t_1}} = 1 + \frac{T_{t_{2i}}/T_{t_1} - 1}{\eta_{C_i}} \qquad (2.28)$$

Using Equation 2.23 in Equation 2.28 results in

$$\frac{T_{t_2}}{T_{t_1}} = 1 + \frac{\pi_C^{\frac{\kappa-1}{\kappa}} - 1}{\eta_{C_i}} \qquad (2.29)$$

With no drop in total pressure in the combustor ($P_{t_2} = P_{t_3}$), the turbine inlet point 3 lies on the constant-pressure line established by the compressor exit pressure. Let us

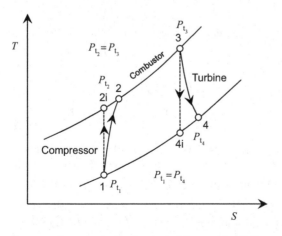

Figure 2.2 Brayton cycle of a gas turbine on $T - S$ diagram.

assume that the expansion in the turbine during work extraction either along the isentropic line 3–4i or the actual line 3–4 with higher entropy occurs to the same total pressure P_{t_4}, which is equal to the compressor inlet pressure P_{t_1}. Because point 4 lies to the right of 4i, the actual work output of the turbine along the process line 3–4 is lower than that along the isentropic (ideal) process line 3–4i. Accordingly, we define the turbine isentropic efficiency as follows:

$$\eta_{T_i} = \frac{\dot{W}_T}{\dot{W}_{T_i}} = \frac{T_{t_3} - T_{t_4}}{T_{t_3} - T_{t_{4i}}}, \tag{2.30}$$

which further yields

$$\frac{T_{t_3}}{T_{t_4}} = \frac{\left(T_{t_3}/T_{t_{4i}}\right)}{\left(1 - \eta_{T_i}\right)\left(T_{t_3}/T_{t_{4i}}\right) + \eta_{T_i}}. \tag{2.31}$$

Substitution of Equation 2.23 in Equation 2.31 gives

$$\frac{T_{t_3}}{T_{t_4}} = \frac{\pi_T^{\frac{\kappa-1}{\kappa}}}{\left(1 - \eta_{T_i}\right)\pi_T^{\frac{\kappa-1}{\kappa}} + \eta_{T_i}} \tag{2.32}$$

where $\pi_T = P_{t_3}/P_{t_4}$ is the pressure ratio across the turbine and equals the pressure ratio $\pi_C = P_{t_2}/P_{t_1}$ across the compressor.

With reference to the gas turbine cycle shown in Figure 2.2, we can express the cycle efficiency or the thermal efficiency in terms of total temperatures as follows:

$$\eta_{th} = \eta_{cycle} = \frac{\dot{W}_{net}}{\dot{Q}_{in}} = \frac{\dot{W}_T - \dot{W}_C}{\dot{Q}_{in}} = \frac{\left(T_{t_3} - T_{t_4}\right) - \left(T_{t_2} - T_{t_1}\right)}{\left(T_{t_3} - T_{t_2}\right)}$$

$$\eta_{th} = \eta_{cycle} = 1 - \frac{\left(T_{t_4} - T_{t_1}\right)}{\left(T_{t_3} - T_{t_2}\right)}$$

$$\eta_{th} = \eta_{cycle} = 1 - \frac{1/\left(T_{t_3}/T_{t_4}\right) - 1/\left(T_{t_3}/T_{t_1}\right)}{1 - \left(T_{t_2}/T_{t_1}\right)/\left(T_{t_3}/T_{t_1}\right)}. \tag{2.33}$$

In Equation 2.33, the temperature ratio T_{t_2}/T_{t_1} across the compressor is given by Equation 2.28 and the temperature ratio T_{t_3}/T_{t_4} across the turbine by Equation 2.32. The equation involves an additional parameter T_{t_3}/T_{t_1}, which is the ratio of turbine inlet temperature to compressor inlet temperature. For an ideal Brayton cycle with $\eta_{C_i} = \eta_{T_i} = 1.0$, which for $\pi_C = \pi_T$ implies that $T_{t_2}/T_{t_1} = T_{t_3}/T_{t_4}$, Equation 2.33 simplifies to

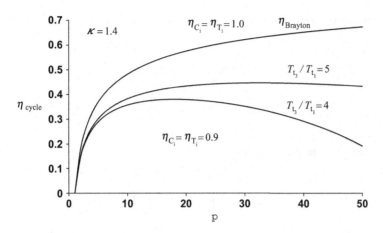

Figure 2.3 Variation of gas turbine cycle efficiency with pressure ratio.

$$\eta_{\text{Brayton}} = 1 - \frac{1}{T_{t_2}/T_{t_1}} = 1 - \frac{1}{\pi^{\frac{\kappa}{\kappa-1}}} \tag{2.34}$$

where $\pi = \pi_{\text{C}} = \pi_{\text{T}}$. It is interestingly to note from Equation 2.34 that η_{Brayton} is independent of the parameter T_{t_3}/T_{t_1}.

Both Equations 2.33 and 2.34 are plotted in Figure 2.3, showing variation of gas turbine cycle efficiency with pressure ratio $\pi = \pi_{\text{C}} = \pi_{\text{T}}$. The ideal Brayton cycle efficiency η_{Brayton} with $\eta_{\text{C}_i} = \eta_{\text{T}_i} = 1.0$ shown in the figure increases monotonically with pressure ratio, initially rather sharply for $\pi < 10$ and then gradually reaching a value of 0.673 for $\pi = 50$. The trend of cycle efficiency variation with pressure ratio with nonideal compressor and turbine depends upon the ratio of turbine inlet temperature to compressor inlet temperature as shown in the figure for $\eta_{\text{C}_i} = \eta_{\text{T}_i} = 0.9$. For $T_{t_3}/T_{t_1} = 4$, cycle efficiency first increases with pressure ratio, reaches a maximum value of 0.380 at $\pi = 18$, and then decreases to $\eta_{\text{cycle}} = 0.191$ at $\pi = 50$. The curve for $T_{t_3}/T_{t_1} = 5$ exhibits a shallower maximum that corresponds to $\eta_{\text{cycle}} = 0.447$ at $\pi = 33$, decreasing to $\eta_{\text{cycle}} = 0.433$ at $\pi = 50$. The figure also shows a well known result that, for a given pressure ratio and compressor inlet temperature, the cycle efficiency increases with turbine inlet temperature.

Optimum pressure ratio for maximum net specific work output. With reference to Figure 2.2, we can express the net specific work output (turbine specific work output minus compressor specific work input, neglecting the fuel mass addition in the combustor) as

$$w_{\text{net}} = c_p \left\{ \left(T_{t_3} - T_{t_4} \right) - \left(T_{t_2} - T_{t_1} \right) \right\} \tag{2.35}$$

Because $T_{t_2}/T_{t_1} = T_{t_3}/T_{t_4}$, we obtain $T_{t_4}/T_{t_1} = T_{t_3}/T_{t_2}$. We now rewrite Equation 2.35 as

$$w_{\text{net}} = c_p \left(T_{t_3} - T_{t_2} + T_{t_1} - \frac{T_{t_3}}{T_{t_2}} T_{t_1} \right)$$

which for an optimum value of w_net with respect to T_{t_2} yields

$$\frac{dw_\text{net}}{dT_{t_2}} = c_p\left(-1 + \frac{T_{t_3}T_{t_1}}{T_{t_2}^{2}}\right) = 0$$

$$T_{t_2} = \sqrt{T_{t_3}T_{t_1}} \qquad (2.36)$$

Equation 2.36 further leads to

$$\pi_\text{optimum} = \left(\frac{P_{t_2}}{P_{t_1}}\right)_\text{optimum} = \left(\frac{T_{t_2}}{T_{t_1}}\right)^{\frac{\kappa-1}{\kappa}} = \left(\frac{T_{t_3}}{T_{t_1}}\right)^{\frac{\kappa-1}{2\kappa}} \qquad (2.37)$$

which, for a given ratio of turbine inlet temperature to compressor inlet temperature, yields the optimum pressure ratio across the compressor needed for maximum net specific work output of the gas turbine. Note that, in the aforementioned derivation, we have assumed a single value of κ, which is typically close to 1.4 for the compressor and 1.33 for the turbine.

2.1.5   Polytropic Efficiency

In a gas turbine, the isentropic efficiencies of compressor and turbine depend upon their operating conditions. For example, if two identical compressors operate in series with equal pressure ratio across each, their isentropic efficiency will be equal. This isentropic efficiency, however, will be higher than if the same compressor were to operate over the higher combined pressure ratio. This results from the fact that the increase in entropy as a result of friction in the upstream compressor requires more work input in the downstream one, generally known as the preheat effect. If we divide the compression process across a compressor or the expansion process across a turbine into a large number of very small, consecutive compression or expansion processes, respectively, then the isentropic efficiency across each is called the polytropic efficiency, or small-stage efficiency, which may be assumed constant for the machine, reflecting its state-of-the-art design engineering.

With reference to Figure 2.2, we assume that the index of polytropic compression along 1–2 is n, which yields

$$\frac{T_{t_2}}{T_{t_1}} = \left(\frac{P_{t_2}}{P_{t_1}}\right)^{\frac{n-1}{n}} = \pi_\text{C}^{\frac{n-1}{n}}. \qquad (2.37)$$

Note that, for $n = \kappa$, this equation yields the isentropic compression along 1–2i.

We define the compressor polytropic efficiency as

$$\eta_{\text{p}_\text{C}} = \frac{(dT_t)_\text{isentropic}}{(dT_t)_\text{actual}} = \frac{(dT_t/T_t)_\text{isentropic}}{(dT_t/T_t)_\text{actual}} \qquad (2.38)$$

Integrating Equation 2.38 between the compressor inlet and exit yields

$$\eta_{P_C} \int_1^2 (dT_t/T_t)_{\text{actual}} = \int_1^{2i} (dT_t/T_t)_{\text{isentropic}}$$

$$\left(\frac{T_{t_2}}{T_{t_1}}\right)^{\eta_{PC}} = \frac{T_{t_{2i}}}{T_{t_1}}$$

$$\pi_C^{\frac{(n-1)\eta_{PC}}{n}} = \pi_C^{\frac{\kappa-1}{\kappa}}$$

$$\frac{n-1}{n} = \frac{\kappa-1}{\eta_{P_C}\kappa}. \tag{2.39}$$

To establish a relation between the compressor isentropic efficiency and its polytropic efficiency, we write

$$\eta_{C_i} = \frac{T_{t_{2i}} - T_{t_1}}{T_{t_2} - T_{t_1}} = \frac{T_{t_{2i}}/T_{t_1} - 1}{T_{t_2}/T_{t_1} - 1}$$

$$\eta_{C_i} = \frac{\pi_C^{\frac{\kappa-1}{\kappa}} - 1}{\pi_C^{\frac{n-1}{n}} - 1}$$

which with the substitution of Equation 2.39 yields

$$\eta_{C_i} = \frac{\pi_C^{\frac{\kappa-1}{\kappa}} - 1}{\pi_C^{\frac{\kappa-1}{\kappa \eta_{PC}}} - 1}. \tag{2.40}$$

Equation 2.40 expresses compressor isentropic efficiency in terms of its polytropic efficiency and pressure ratio. Alternatively, to express compressor polytropic efficiency in terms of its isentropic efficiency and pressure ratio, we can rewrite Equation 2.40 as

$$\eta_{P_C} = \frac{\ln\left(\pi_C^{\frac{\kappa-1}{\kappa}}\right)}{\ln\left(1 + \frac{\pi_C^{\frac{\kappa-1}{\kappa}} - 1}{\eta_{C_i}}\right)}. \tag{2.41}$$

Again, with reference to Figure 2.2, we assume that the index of polytropic expansion in the turbine along 3–4 is n, which yields

$$\frac{T_{t_3}}{T_{t_4}} = \left(\frac{P_{t_3}}{P_{t_4}}\right)^{\frac{n-1}{n}} = \pi_T^{\frac{n-1}{n}} \tag{2.42}$$

Note that for $n = \kappa$, the aforementioned equation yields the isentropic expansion along 3–4i.

We define the turbine polytropic efficiency as

$$\eta_{P_T} = \frac{(dT_t)_{\text{actual}}}{(dT_t)_{\text{isentropic}}} = \frac{(dT_t/T_t)_{\text{actual}}}{(dT_t/T_t)_{\text{isentropic}}}. \tag{2.43}$$

Integrating Equation 2.43 across the turbine yields

$$\eta_{\text{P}_\text{T}} \int_3^{4i} (dT_t/T_t)_{\text{isentropic}} = \int_3^4 (dT_t/T_t)_{\text{actual}}$$

$$\left(\frac{T_{t_3}}{T_{t_4}}\right) = \left(\frac{T_{t_3}}{T_{t_{4i}}}\right)^{\eta_{\text{P}_\text{T}}}$$

$$\pi_T^{\frac{n-1}{n}} = \pi_T^{\frac{(\kappa-1)\eta_{\text{P}_\text{T}}}{\kappa}}$$

$$\frac{n-1}{n} = \frac{(\kappa-1)\eta_{\text{P}_\text{T}}}{\kappa}. \tag{2.44}$$

To establish a relation between the turbine isentropic efficiency and its polytropic efficiency, we write

$$\eta_{\text{T}_i} = \frac{T_{t_3} - T_{t_4}}{T_{t_3} - T_{t_{4i}}} = \frac{1 - T_{t_4}/T_{t_3}}{1 - T_{t_{4i}}/T_{t_3}}$$

$$\eta_{\text{T}_i} = \frac{1 - (1/\pi_T)^{\frac{n-1}{n}}}{1 - (1/\pi_T)^{\frac{\kappa-1}{\kappa}}}$$

which using Equation 2.44 yields

$$\eta_{\text{T}_i} = \frac{1 - (1/\pi_T)^{\frac{\eta_{\text{P}_\text{T}}(\kappa-1)}{\kappa}}}{1 - (1/\pi_T)^{\frac{\kappa-1}{\kappa}}}. \tag{2.45}$$

Equation 2.45 expresses turbine isentropic efficiency in terms its polytropic efficiency and pressure ratio. Alternatively, to express turbine polytropic efficiency in terms of its isentropic efficiency and pressure ratio, we rewrite Equation 2.45 as

$$\eta_{\text{P}_\text{T}} = \frac{\ln\left(1 - \eta_{\text{T}_i}\left\{1 - (1/\pi_T)^{\frac{\kappa-1}{\kappa}}\right\}\right)}{\left(\frac{\kappa-1}{\kappa}\right)\ln(1/\pi_T)}. \tag{2.46}$$

Equations 2.40 and 2.45 are plotted in Figure 2.4 for $\eta_{\text{P}_\text{C}} = \eta_{\text{P}_\text{T}} = 0.9$. In this figure, we have assumed $\kappa = 1.4$ for the compressor and $\kappa = 1.33$ for the turbine. For the pressure ratio very close to 1.0, the compressor and the turbine each has equal isentropic and polytropic efficiencies. The figure further shows that, for the compressor, the isentropic efficiency decreases with pressure ratio and, for the turbine, it increases with the pressure ratio. The rate of efficiency decrease for the compressor, however, is higher than the corresponding rate of efficiency increase for the turbine.

2.2   Fluid Mechanics

2.2.1   Secondary Flows

In a gas turbine operation, flows that directly partake in the energy conversion along the turbine and compressor flow path (core flow) are generally referred to as the primary

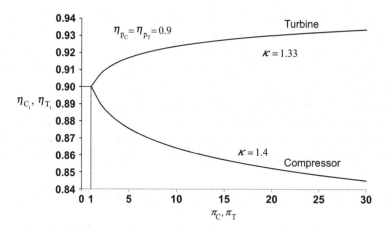

Figure 2.4 Variation of compressor and turbine isentropic efficiencies with pressure ratio for a constant value of their polytropic efficiency.

flows. The flows that are used for cooling and sealing of gas turbine components are often called secondary flows. These flows should not be confused with the secondary flows, which are normal to the main flow direction, found in many internal flows. For example, a turbulent flow in a noncircular duct features secondary flows that have velocities in the plane of the duct cross-section, which is perpendicular to the main flow direction. These are driven by the anisotropy in turbulence intensity and gradients in normal turbulent stresses and are called secondary flow of the second kind. Note that these secondary flows do not exist in a constant-viscosity laminar flow in a noncircular duct or the turbulent flow in a circular duct.

Because the local gradient in static pressure is the common primary mechanism to generate the velocity field in a flow, when a uniform flow enters a bend in a circular duct it exits with secondary flows. In the bend region, as a result of the centrifugal force, the static pressure at the outer radius of curvature is higher than at the inner radius of curvature. This radial pressure gradient in the pipe cross-section generates secondary flows that we observe at the exit plane. These are known as the secondary flows of the first kind.

2.2.2   Vorticity and Vortex

Vorticity is a kinematic vector property of a flow field and equals the curl of its velocity vector ($\vec{\zeta} = \nabla \times \vec{V}$). At any point in the flow, the vorticity represents the local rigid-body rotation of fluid particles and equals twice the rotation vector ($\vec{\zeta} = 2\vec{\omega}$). A vortex, on the other hand, refers to the bulk circular motion of the entire flow field. Streamlines in a vortex flow are essentially concentric circles, and so are the isobars (lines of constant static pressure). The local pressure gradient along the radius of curvature is given by the local equilibrium equation $dP_s/dr = \rho V_\theta^2/r$. This characteristic of a vortex flow is so different from a nonrotating flow where the static pressure typically varies along a streamline, as dictated by the Bernoulli equation for an ideal incompressible flow.

A forced vortex features bulk rigid-body rotation with constant angular velocity (in radians per second) for all streamlines. As a result, the outer streamline features higher tangential, or swirl, velocity than the inner ones. In a free vortex, which is assumed to be free from any external torque, the angular momentum (the product of radius and tangential velocity), not the angular velocity, remains constant everywhere in the flow. In this vortex, the outer streamline has lower tangential velocity than the inner one. As a result of the presence of centrifugal force, all vortex motions feature positive radial gradient in the static pressure distribution. Lugt (1995) presents an excellent discussion of vortex flows in both nature and technology.

2.2.3   Sudden Expansion Pipe Flow With and Without Swirl

In gas turbine design applications, flows often feature changes in pipe diameter, going from a smaller-diameter pipe to a larger-diameter pipe (sudden expansion) or from a larger-diameter pipe to a smaller-diameter pipe (sudden contraction). Such flows, with their undesirable excess loss in total pressure, generally become necessary as a result of spatial constraints.

Following a sudden expansion, shown in Figure 2.5a for the case without swirl, the flow enters the larger pipe in the form of a circular jet whose shear layer grows both inward toward the pipe axis and outward toward the pipe wall. While the inward growth of the shear layer occurs at the expense of the potential (inviscid) core flow, which decays downstream until the shear layers merge at the pipe axis, its outward growth happens through the process of entrainment until it finally reattaches to the pipe wall. The fluid entrained into this shear layer is continuously replenished from downstream through a favorable reverse pressure gradient. This gives rise to a primary recirculation region such that the flow in its forward branch is driven by the central jet momentum, and in its reverse branch, by an adverse pressure gradient. Sultanian, Neitzel, and Metzger (1986) present accurate CFD predictions of this flow field using an advanced algebraic stress transport model of turbulence. They also demonstrate that the standard high-Reynolds number $\kappa - \varepsilon$ turbulence model significantly under-predicts the experimentally found reattachment lengths.

Figure 2.5b shows the main features of a swirling flow in a sudden pipe expansion. The wall-bounded primary recirculation zone in this flow field is found to shrink with the increasing incoming swirl velocity. When the swirl exceeds a critical value (greater than 45-degree swirl angle), an on-axis recirculation region appears. Sultanian (1984) discusses at length the overwhelming complexity of this flow field from a computational viewpoint. The major difficulties associated with predicting this three-dimensional flow field lies in accurately modeling its highly anisotropic turbulence field.

A distinguishing feature of a critically swirling flow in a sudden pipe expansion is the region of on-axis recirculation caused by vortex breakdown. Such a feature is widely used in modern gas turbine combustors as an aerodynamic flame holder, which in early designs corresponded to the wake region of a bluff body, to ignite the fresh charge entering the combustor. Here is a simple explanation behind the vortex breakdown in this flow field. The swirl in a free or forced vortex causes positive radial pressure gradient, that is, the static pressure always increases radially outward – the higher the

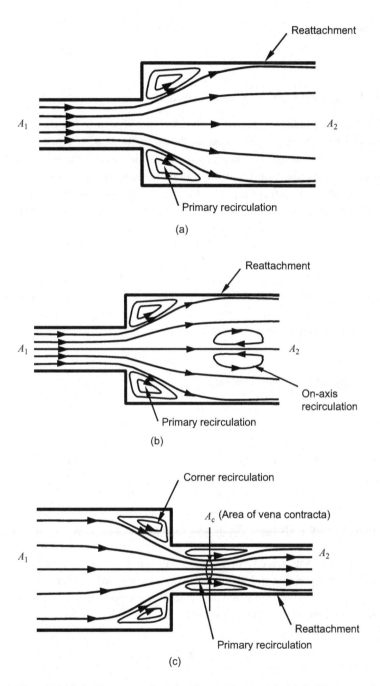

Figure 2.5 (a) Sudden expansion pipe flow without swirl, (b) Sudden expansion pipe flow with swirl, and (c) Sudden contraction pipe flow without swirl.

swirl, the higher is this pressure gradient. Near the sudden expansion, as a result of high swirl, the static pressure at the pipe axis is much lower than the section-average static pressure. Further downstream, swirl decays as a result of wall shear stress, and the static pressure distribution tends to become more uniform over the pipe cross section. As a result, the static pressure at the pipe axis becomes higher downstream than upstream, causing on-axis flow reversal.

2.2.4   Sudden Contraction Pipe Flow: Vena Contracta

Key features of a sudden contraction pipe flow are shown in Figure 2.5c. When the flow from the large pipe of area A_1 abruptly enters the smaller pipe of area $A_2 < A_1$, the outer streamline detaches from the pipe wall near the area change, creating a short region of corner recirculation and the minimum flow cross-section, called the vena contracta, in the downstream pipe. At the vena contracta, having area $A_c < A_2$ and surrounded by the primary recirculation zone in the downstream pipe, the flow has the maximum velocity and minimum static pressure. Downstream of the vena contracta, the flow behaves very similar to the one in a sudden pipe expansion, discussed in Section 2.2.2, expanding from A_c to A_2. Experimental data in a sudden contraction pipe flow indicate that A_c/A_2 is a function of A_1/A_2. For example, $A_c/A_2 = 0.68$ for $A_1/A_2 = 2$. It is worth noting here that, for given A_1/A_2, A_c does not depend on the total-to-static pressure ratio at the vena contracta for incompressible flows but increases slightly for compressible flows.

2.2.5   Stream Thrust and Impulse Pressure

The concept of stream thrust is grounded in the linear momentum equation. The stream thrust at any section of an internal flow, be it incompressible or compressible, is defined as the sum of the pressure force and inertia force (momentum flow rate). In any particular direction, the stream thrust thus represents the total force of the flow stream – we need this much force in the opposite direction to bring the flow velocity to zero (stagnation). For example, the force needed to hold a plate in place to an incoming fluid jet equals the stream thrust associated with the jet, and not the jet area times the jet stagnation pressure at the plate. As another example, let us consider a constant-area duct with zero wall friction, say a stream tube. The stream thrust associated with an incoming flow with nonuniform velocity distribution will remain constant throughout the duct, although its static pressure and total pressure at each section will change downstream. If the velocity profile is uniform at the duct exit, its static pressure will be higher and total pressure lower than their respective values at the duct inlet.

Given by the equation

$$S_T = P_s A + \dot{m} V \tag{2.47}$$

the stream thrust is a vector quantity having the same direction as the momentum velocity (see Section 2.2.5) in the chosen coordinate system. Use of stream thrust at inlets and outlets of a control volume for the momentum equation in each coordinate direction makes it analogous to the continuity and energy equations, which involve only scalar quantities. The net change in stream thrust summed over all inflows and outflows of a control volume must equal all body and surface forces (except the pressure forces at inlets and outlets) acting on the control volume.

Impulse pressure equals the stream thrust per unit area. It is given by the equation

$$P_i = P_s + \rho V^2, \tag{2.48}$$

which is valid for both incompressible and compressible flows. The aforementioned equation should not be confused with the equation $P_t = P_s + \rho V^2/2$, which is used to compute total pressure in an incompressible flow and approximately in a compressible flow at low Mach numbers ($M \leq 0.3$). Note that the second term on the right-hand side of Equation 2.48 equals twice the dynamic pressure of an incompressible flow.

2.2.6 Control Volume Analyses of Conservation Laws

Control volume analyses of the conservation laws of mass, momentum, energy, and entropy form the core foundation for the mathematical modeling of all fluid flows and their quantitative predictions for practical applications. In Newtonian mechanics, conservation laws are written in an inertial reference frame for a control system of identified mass – the Lagrangian viewpoint. If Φ_{CS} stands for the mass, linear momentum, energy, or entropy of a control system with the corresponding intensive (per unit mass) properties represented by ϕ, the substantial, total, or material derivative of Φ_{CS} in a flow field, where the control system instantaneously assumes the local flow properties, can be written as

$$\frac{D\Phi_{CS}}{Dt} = \Pi_S + \Pi_B. \tag{2.49}$$

In Equation 2.49, the terms on the right-hand side are the change agents, Π_S working on the surface of the control system and Π_B working on the body of the control system. Their net effect is to cause $D\Phi_{CS}/Dt$ on the left-hand side of the equation. Note that Φ_{CS} is the instantaneous value of the extensive property Φ for the entire control system.

In most engineering applications, a control volume analysis consistent with the Eulerian viewpoint is used. Facilitated by the Reynolds transport theorem, see Sultanian (2015), the left-hand side of Equation 2.49 can be replaced by

$$\frac{D\Phi_{CS}}{Dt} = \frac{\partial}{\partial t} \iiint\limits_{CV} \rho\phi\, d\Psi + \iint\limits_{CVS} \phi\left(\rho \vec{V} \bullet d\vec{A}\right) \tag{2.50}$$

in which the first term on the right-hand side represents the time rate of change of Φ within the control volume and the second term represents the net outflow (total outflow minus total inflow) of Φ through the control volume surface.

Combining Equations 2.49 and 2.50, we obtain

$$\Pi_S + \Pi_B = \frac{\partial}{\partial t} \iiint\limits_{CV} \rho\phi\, d\Psi + \iint\limits_{CVS} \phi\left(\rho \vec{V} \bullet d\vec{A}\right). \tag{2.51}$$

Invoking Gauss divergence theorem, we can replace the surface integral in Equation 2.51 by volume integral, giving

$$\Pi_S + \Pi_B = \iiint\limits_{CV} \left[\frac{\partial(\rho\phi)}{\partial t} + \nabla \bullet \left(\rho\phi\vec{V} \right) \right] d\Psi \qquad (2.52)$$

where we have assumed that $\partial\Psi/\partial t = 0$, allowing us to pull $\partial/\partial t$ inside the volume integral. Note that Equation 2.51 is convenient to use for an integral control volume analysis, and Equation 2.52 forms the basis for a point-wise differential control volume analysis.

2.2.6.1   Mass Conservation (Continuity Equation)

According to the law of mass conservation we must have $\Pi_S = \Pi_B = 0$ as there can be no agent that will change the mass of a system. In this case, we have $\phi = 1$, and Equation 2.51 reduces to

$$\frac{\partial}{\partial t} \iiint\limits_{CV} \rho d\Psi + \iint\limits_{CVS} \rho\,\vec{V} \bullet d\vec{A} = 0, \qquad (2.53)$$

which simply states that, for a given control volume with multiple inlets and outlets, the imbalance between mass inflows and outflows equals the rate of accumulation or reduction of mass in the control volume.

For $\phi = 1$, Equation 2.52 reduces to the volume integral

$$\iiint\limits_{CV} \left[\frac{\partial\rho}{\partial t} + \nabla \bullet \left(\rho\vec{V} \right) \right] d\Psi = 0$$

whose integrand must be zero, giving

$$\frac{\partial\rho}{\partial t} + \nabla \bullet \left(\rho\vec{V} \right) = 0. \qquad (2.54)$$

Note that Equation 2.54, which is a partial differential equation, must be satisfied at every point in a flow field. For a steady compressible flow, the continuity equation becomes $\nabla \bullet \left(\rho\vec{V} \right) = 0$, and for an incompressible flow, be it steady or unsteady, we obtain $\nabla \bullet \vec{V} = 0$, which states that the velocity field of an incompressible flow must be divergence free.

Expanding terms within the volume integral on the right-hand side of Equation 2.52, using the vector identity for the divergence term involving scalar coefficients, and regrouping them, we obtain

$$\Pi_S + \Pi_B = \iiint\limits_{CV} \left[\rho\left\{ \frac{\partial\phi}{\partial t} + \vec{V} \bullet \nabla\phi \right\} + \phi\left\{ \frac{\partial\rho}{\partial t} + \nabla \bullet \left(\rho\vec{V} \right) \right\} \right] d\Psi.$$

Using Equation 2.54, the expression within the second curly brackets in the right-hand side of the aforementioned equation becomes zero. We finally obtain a simpler form of Equation 2.52

$$\Pi_S + \Pi_B = \iiint\limits_{CV} \left[\rho\left\{ \frac{\partial\phi}{\partial t} + \vec{V} \bullet \nabla\phi \right\} \right] d\Psi. \qquad (2.55)$$

2.2.6.2 Linear Momentum Equation

The linear momentum equation for a system is grounded in the Newton's second law of motion, which simply states that the force acting on a system equals the mass of the system times its acceleration in the direction of the force. So the forces acting on a system are the change agent for its linear momentum. Accordingly, Equation 2.52 for the integral form of the linear momentum equation in an inertial (nonaccelerating) reference frame becomes

$$\vec{F}_S + \vec{F}_B = \frac{\partial}{\partial t} \iiint_{CV} \vec{V}\rho \, d\forall + \iint_{CVS} \vec{V} \left(\rho\vec{V} \cdot d\vec{A} \right) \qquad (2.56)$$

where $\Pi_S = \vec{F}_S$ and $\Pi_B = \vec{F}_B$, which are surface and body forces, respectively, and $\phi = \vec{V}$ is the intensive property corresponding to the linear momentum $\vec{M}$. Surface forces arising from the static pressure at inlets, outlets, and walls of a control volume are compressive in nature, being always normal to the surface. Wall shear stresses, which also contribute to the surface forces, are locally parallel to the surface. In this text, we consider only two kinds of body forces, one as a result of gravity (weight) and the other as a result of rotation (centrifugal force).

It is instructive to analyze Equation 2.56 as three independent equations, one for each coordinate direction of a Cartesian coordinate system. This is possible because the forces acting along one coordinate direction will not change the momentum in the other coordinate direction, which is orthogonal. Accordingly, for the following presentation, let us only consider the x component of Equation 2.56 in an assigned Cartesian coordinate system:

$$F_{S_x} + F_{B_x} = \frac{\partial}{\partial t} \iiint_{CV} V_x\rho \, d\forall + \iint_{CVS} V_x \left(\rho \vec{V} \cdot d\vec{A} \right). \qquad (2.57)$$

If the velocity, static pressure, and density are uniform at each inlet and outlet of a control volume, we can replace the surface integral on the right-hand of Equation 2.57 by algebraic summations. For a control volume with N_i inlets and N_j outlets having M_x as its total instantaneous value of x momentum, Equation 2.57 simplifies to

$$F_{S_x} + F_{B_x} = \left(\frac{\partial M_x}{\partial t} \right)_{CV} + \left(\sum_{j=1}^{j=N_{out}} V_{x_j}\dot{m}_j \right)_{outlets} - \left(\sum_{i=1}^{i=N_{in}} V_{x_i}\dot{m}_i \right)_{inlets} \qquad (2.58)$$

where

$V_{x_i} \equiv x$-momentum velocity at inlet i
$V_{x_j} \equiv x$-momentum velocity at outlet j
$\dot{m}_i \equiv$ Mass flow rate at inlet i ($\dot{m}_i = \rho_i A_i V_i$)
$\dot{m}_j \equiv$ Mass flow rate at outlet j ($\dot{m}_j = \rho_j A_j V_j$)

For an inertial (nonaccelerating) control volume, Equation 2.58 reads the following:

(Sum of surface forces: pressure force and shear force)
+ (Sum of body forces: gravitational force and centrifugal force under rotation)
= (Time-rate of change of x-momentum within the control volume)
+ (Total x-momentum flow rate leaving the control volume)
− (Total x-momentum flow rate entering the control volume)

In Equation 2.58, V_i and V_j are the mass velocity at each inlet and outlet, respectively. They are always positive and independent of the chosen coordinate system. By contrast, V_{x_i} and V_{x_j} are the x-momentum velocity at inlet i and outlet j, respectively. Whether they are positive or negative is determined relative to the positive direction of the chosen coordinate system. Identifying mass velocity (the velocity that produces mass flow rate) and momentum velocity (the velocity component for which the linear momentum equation is being considered) greatly simplifies the evaluation of the momentum flow rate with the correct sign at each inlet and outlet of a control volume.

To demonstrate the use of mass velocity and momentum velocity in the control volume analysis of linear momentum, let us consider a steady flow through the control volume shown in Figure 2.6. Based on the direction of each flow velocity, it is easy to see that sections 2 and 3 are inlets and sections 1, 4, and 5 are outlets. The mass velocity and x-momentum velocity for each inlet and outlet are summarized in Table 2.1. Note that, with reference to the Cartesian coordinate system shown in the figure, the x-momentum velocities for inlet 3 and outlet 1 are negative. For outlet 5, whereas the mass velocity V_5 in the y direction is nonzero, the corresponding x-momentum velocity is zero. With various quantities given in Table 2.1 for a steady flow through the control volume, the right-hand side of Equation 2.58 can be easily evaluated.

If we separate out the pressure force at each inlet and outlet of the control volume from the total surface forces $\vec{F_s}$ on the left-hand side of Equation 2.58, and combine it with the corresponding momentum flow on the right-hand side of the equation, we can express the momentum equation in terms of stream thrust in the x direction. In Table 2.1, note that the stream thrust at each inlet and outlet has the same sign as the momentum velocity.

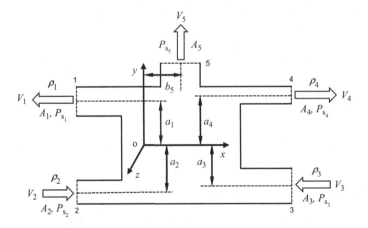

Figure 2.6 A control volume showing flow properties at its two inlets and three outlets.

Table 2.1 x-Momentum Flow Rate and Stream Thrust at Each Inlet and Outlet of the Control Volume of Figure 2.6

Inlet/ Outlet	Mass Velocity	Mass Flow Rate	x-Momentum Velocity	x-Momentum Flow Rate	Stream Thrust
1 (outlet)	V_1	$\dot{m}_1 = \rho_1 A_1 V_1$	$-V_1$	$-\dot{m}_1 V_1$	$-\left(\dot{m}_1 V_1 + P_{s_1} A_1\right)$
2 (inlet)	V_2	$\dot{m}_2 = \rho_2 A_2 V_2$	V_2	$\dot{m}_2 V_2$	$\left(\dot{m}_2 V_2 + P_{s_2} A_2\right)$
3 (inlet)	V_3	$\dot{m}_3 = \rho_3 A_3 V_3$	$-V_3$	$-\dot{m}_3 V_3$	$-\left(\dot{m}_3 V_3 + P_{s_3} A_3\right)$
4 (outlet)	V_4	$\dot{m}_4 = \rho_4 A_4 V_4$	V_4	$\dot{m}_4 V_4$	$\left(\dot{m}_4 V_4 + P_{s_4} A_4\right)$
5 (outlet)	V_5	$\dot{m}_5 = \rho_5 A_5 V_5$	0	0	0

How to handle nonuniform flow properties at inlets and outlets and to express the linear momentum equation in a noninertial (accelerating or decelerating) reference frame attached to the control volume are discussed in comprehensive details by Sultanian (2015).

2.2.6.3 Angular Momentum Equation

Because angular momentum is expressed as $\vec{H} = \vec{r} \times \vec{M}$, the angular momentum equation is not a new conservation law, but simply a moment of the linear momentum equation. In word form, the angular momentum equation can be expressed as

(Sum of the torque from surface forces: pressure force and shear force)
+ (Sum of the torque from body forces: gravitational force and centrifugal force)
= (Time-rate of change of angular momentum within the control volume)
+ (Total angular momentum flow rate leaving the control volume)
− (Total angular momentum flow rate entering the control volume)

In equation form, we can write the aforementioned statement in an inertial coordinate system as

$$\vec{\Gamma}_S + \vec{\Gamma}_B = \left(\frac{\partial \vec{H}}{\partial t}\right)_{CV} + \left(\sum_{j=1}^{j=N_{out}} \dot{\vec{H}}_j\right)_{outlets} - \left(\sum_{i=1}^{i=N_{in}} \dot{\vec{H}}_i\right)_{inlets}. \qquad (2.58)$$

Similar to the approach presented in the previous section for the evaluation of linear momentum flow at an inlet or outlet of a control volume by using the concept of mass velocity and momentum velocity, we can evaluate each angular momentum flow rate by taking the product of mass velocity and angular momentum, which is considered positive in the counterclockwise direction and negative in the clockwise direction. Table 2.2 summarizes the flow rate of z-component of the angular momentum for each inlet and outlet of the control volume shown in Figure 2.6, demonstrating the application of the approach proposed here.

Table 2.2 z-Component Angular Momentum Flow Rates at Inlets and Outlets of the Control Volume of Figure 2.6

Inlet / Outlet	Mass velocity	Specific z-component angular momentum	z-component angular momentum flow rate
1 (outlet)	V_1	$a_1 V_1$	$a_1 V_1 (\rho_1 A_1 V_1)$
2 (inlet)	V_2	$a_2 V_2$	$a_2 V_2 (\rho_2 A_2 V_2)$
3 (inlet)	V_3	$-a_3 V_3$	$-a_3 V_3 (\rho_3 A_3 V_3)$
4 (outlet)	V_4	$-a_4 V_4$	$-a_4 V_4 (\rho_4 A_4 V_4)$
5 (outlet)	V_5	$b_5 V_5$	$b_5 V_5 (\rho_5 A_5 V_5)$

A coordinate system rotating at constant angular velocity is considered noninertial. In this noninertial coordinate system with no linear and rotational accelerations, we can write, see Sultanian (2015), the integral angular momentum equation for a control volume as

$$\vec{\Gamma}_S + \vec{\Gamma}_B - \iiint_{CV} \vec{r} \times \left(2\vec{\Omega} \times \vec{W} + \vec{\Omega} \times \left(\vec{\Omega} \times \vec{r} \right) \right) \rho \, d\Psi = \iint_{CVS} \vec{r} \times \vec{W} \left(\rho \, \vec{W} \bullet d\vec{A} \right)$$

(2.59)

where $\vec{\Omega}$ is the constant speed of rotation and the relative velocity $\vec{W}$ is the flow velocity in this coordinate system with its corresponding value in the stationary (inertial) coordinate system represented by the absolute $\vec{V}$. Note that expressing the linear momentum equation in this noninertial coordinate system gives rise two new terms: the Coriolis acceleration term $2\vec{\Omega} \times \vec{W}$ and the centrifugal acceleration term $\vec{\Omega} \times (\vec{\Omega} \times \vec{r})$, whose moments appear in Equation 2.59.

For turbomachinery flows, including those in a gas turbine with a single axis of rotation, a cylindrical coordinate system shown in Figure 2.7 is a natural choice. In this coordinate system, the axial coordinate direction is aligned with the axis of rotation. As shown in the figure, the tangential direction is perpendicular to the meridional plane, which is formed by the axial and radial directions. For a constant rotational speed Ω and steady flow, Sultanian (2015) presents a mathematically rigorous simplification of Equation 2.59 to yield

$$\Gamma_S + \Gamma_B = \iint_{CVS} r V_\theta \left(\rho \, \vec{W} \bullet d\vec{A} \right).$$

(2.60)

In the derivation of Equation 2.60 from Equation 2.59, it is seen that the moment of the centrifugal force term makes no contribution and the volume integral of the moment of the Coriolis term contributes to the surface integral and converts the flow of angular momentum in the noninertial reference into that in the inertial reference frame at each inlet and outlet of the control volume. Equation 2.60 involves only the tangential velocity V_θ in the inertial reference frame, forming the basis for the Euler's turbomachinery equation discussed in Section 2.2.8.

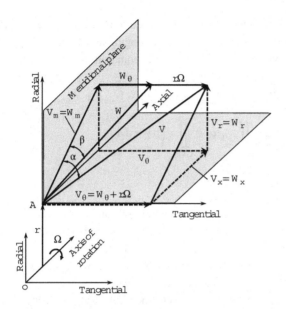

Figure 2.7 Cylindrical coordinate system used in turbomachinery flows.

With reference to Figure 2.7, we see that the angular momentum vector at point A in the flow about the turbomachinery axis of rotation is the cross-product of the radial vector $\vec{r}$ and the velocity vector $\vec{V}$. Because the radial velocity V_r is collinear with the radial vector, its contribution to angular momentum will be zero. The contribution of axial velocity V_x to angular momentum will be in the tangential direction and not along the axis of rotation. Thus, only the tangential velocity V_θ contributes to the angular momentum vector along the rotation axis, giving the angular momentum flow rate $\dot{H}_x = \dot{m}\,rV_\theta$.

For uniform properties at control volume inlets and outlets, we can replace the surface integral in Equation 2.60 by algebraic summations, giving

$$\Gamma_S + \Gamma_B = \left(\sum_{j=1}^{j=N_{\text{out}}} r_j V_{\theta_j} \dot{m}_j\right)_{\text{outlets}} - \left(\sum_{i=1}^{i=N_{\text{in}}} r_i V_{\theta_i} \dot{m}_i\right)_{\text{inlets}}. \qquad (2.61)$$

2.2.6.4 Energy Equation

The energy equation embodies the first law of thermodynamics. According to Equation 2.1, heat transfer and work transfer are the agents responsible for change in the total energy of a system. Thus, with $\phi = e = u + V^2/2 + gz$ as the intensive property corresponding to the total energy E, Equation 2.51 for the energy conservation becomes

$$\dot{Q} - \dot{W} = \frac{\partial}{\partial t} \iiint_{\text{CV}} \rho\, e\, \mathrm{d}\Psi + \iint_{\text{CVS}} e\left(\rho\, \vec{V} \bullet \mathrm{d}\vec{A}\right). \qquad (2.62)$$

In Equation 2.62, according to the convention used in Equation 2.1, the work transfer into the control volume becomes negative. Recognizing different types of work transfer into the control volume, we can write this equation as

$$\dot{Q} + \dot{W}_{\text{pressure}} + \dot{W}_{\text{shear}} + \dot{W}_{\text{rotation}} + \dot{W}_{\text{shaft}} + \dot{W}_{\text{other}}$$

$$= \frac{\partial}{\partial t} \iiint_{\text{CV}} e\rho \, d\mathcal{V} + \iint_{\text{CVS}} e\left(\rho \vec{V} \bullet d\vec{A}\right) \tag{2.63}$$

where

$\dot{Q} \equiv$ Rate of heat transfer into the control volume through its control surface
$\dot{W}_{\text{pressure}} \equiv$ Rate of work transfer into the control volume by pressure force
$\dot{W}_{\text{shear}} \equiv$ Rate of work transfer into the control volume by shear force
$\dot{W}_{\text{rotation}} \equiv$ Rate of Euler work transfer into the control volume by rotation
$\dot{W}_{\text{shaft}} \equiv$ Rate of shaft work into the control volume
$\dot{W}_{\text{other}} \equiv$ Rate of other energy transfers into the control volume: chemical energy, electrical energy, electromagnetic energy, and so on.

In a nondeformable control volume, the pressure force can do work only at its inflows and outflows. Within the control volume itself, the pressure force at any point has an equal and opposite pressure force, and does not contribute to the work transfer term $\dot{W}_{\text{pressure}}$. At an inlet to the control volume, because $\vec{V}$ and $\vec{A}$ are in opposite direction, $\rho \vec{V} \cdot d\vec{A}$ will be negative, and the pressure force will do work on the flow entering the control volume.

According to our convention, we express the work transfer by the pressure force at inlets as

$$\left(\dot{W}_{\text{pressure}}\right)_{\text{inlets}} = -\left\{ \iint_{\text{CVS}} \frac{P_s}{\rho}\left(\rho \vec{V} \cdot d\vec{A}\right) \right\}_{\text{inlets}}.$$

At control volume outlets, $\rho \vec{V} \cdot d\vec{A}$ is positive, and the pressure force opposes the flow exiting the control volume. We can thus write

$$\left(\dot{W}_{\text{pressure}}\right)_{\text{outlets}} = -\left\{ \iint_{\text{CVS}} \frac{P_s}{\rho}\left(\rho \vec{V} \cdot d\vec{A}\right) \right\}_{\text{outlets}}.$$

Combining the aforementioned two equations, we can express $\dot{W}_{\text{pressure}}$ by the following surface integral, which is true at all inlets and outlets:

$$\dot{W}_{\text{pressure}} = -\iint_{\text{CVS}} \frac{P_s}{\rho}\left(\rho \vec{V} \cdot d\vec{A}\right)$$

where P_s/ρ is the specific flow work.

Transferring $\dot{W}_{\text{pressure}}$ to the right-hand side of Equation 2.63 yields

$$\dot{Q} + \dot{W}_{\text{shear}} + \dot{W}_{\text{rotation}} + \dot{W}_{\text{shaft}} + \dot{W}_{\text{other}}$$

$$= \frac{\partial}{\partial t} \iiint_{\text{CV}} e\rho \, \mathrm{d}V + \iint_{\text{CVS}} \left(u + \frac{P_{\text{s}}}{\rho} + \frac{V^2}{2} + gz \right) \left(\rho \vec{V} \cdot \mathrm{d}\vec{A} \right). \quad (2.64)$$

Replacing $u + P_{\text{s}}/\rho$ by the specific static enthalpy h_{s} in Equation 2.64, we obtain

$$\dot{Q} + \dot{W}_{\text{shear}} + \dot{W}_{\text{rotation}} + \dot{W}_{\text{shaft}} + \dot{W}_{\text{other}}$$

$$= \frac{\partial}{\partial t} \iiint_{\text{CV}} e\rho \, \mathrm{d}V + \iint_{\text{CVS}} \left(h_{\text{s}} + \frac{V^2}{2} + gz \right) \left(\rho \vec{V} \cdot \mathrm{d}\vec{A} \right). \quad (2.65)$$

For uniform properties at control volume inlets and outlets, we can replace the surface integral in Equation 2.65 by algebraic summations, giving

$$\dot{Q} + \dot{W}_{\text{shear}} + \dot{W}_{\text{rotation}} + \dot{W}_{\text{shaft}} + \dot{W}_{\text{other}} = \left(\frac{\partial E}{\partial t} \right)_{\text{CV}}$$

$$+ \left(\sum_{j=1}^{j=N_{\text{out}}} \left(h_{\text{s}} + \frac{V^2}{2} + gz \right)_j \dot{m}_j \right)_{\text{outlets}} \quad (2.66)$$

$$- \left(\sum_{i=1}^{i=N_{\text{in}}} \left(h_{\text{s}} + \frac{V^2}{2} + gz \right)_i \dot{m}_i \right)_{\text{inlets}}.$$

Note that, in Equation 2.66, we can further combine the specific static enthalpy h_{s} and specific kinetic energy $V^2/2$ to yield the specific total enthalpy $h_{\text{t}} = h_{\text{s}} + V^2/2$.

2.2.6.5 Entropy Equation

The entropy equation is founded in the second law of thermodynamics, which states that the entropy of an isolated system increases if the processes within the system are irreversible; the entropy remains constant when the processes are reversible. By introducing the concept of entropy production as a change agent, in addition to the change in entropy as a result of heat transfer, we can write Equation 2.51 for entropy as

$$\dot{P}_{\text{entropy}} + \frac{\dot{Q}}{T} = \frac{\partial}{\partial t} \iiint_{\text{CV}} \rho s \, \mathrm{d}V + \iint_{\text{CVS}} s \left(\rho \vec{V} \cdot \mathrm{d}\vec{A} \right) \quad (2.67)$$

where $\phi = s$ is the entropy per unit mass. In this equation, the rate of entropy production $\dot{P}_{\text{entropy}}$ accounts for entropy generation within the control volume from all irreversible processes and $\dot{Q}$ is the rate of heat transfer into the control volume through its surface at temperature T.

The unsteady integral term on the right-hand side of Equation 2.67 is the rate of storage of entropy within the control volume and should not be confused with $\dot{P}_{\text{entropy}}$, which represents the rate of entropy production associated with the irreversibility of energy transfer into the control volume. For the case with no heat transfer, $\dot{P}_{\text{entropy}}$ equals the rate of increase in entropy stored within the control volume plus the excess of the entropy outflow rate over its inflow rate through the control-volume surface. Further note that there is no counterpart to $\dot{P}_{\text{entropy}}$ in the equations of mass, momentum, and energy whose conservation laws dictate zero production within the control volume.

In a steady flow, the unsteady integral term in Equation 2.67 vanishes, yielding Equation 2.68 in which $\dot{P}_{\text{entropy}}$ and other terms are considered time-independent.

$$\dot{P}_{\text{entropy}} + \frac{\dot{Q}}{T} = \iint\limits_{\text{CVS}} s\left(\rho \vec{V} \bullet d\vec{A}\right) \tag{2.68}$$

For uniform properties at control volume inlets and outlets, we can again replace the surface integral in Equation 2.68 by algebraic summations, giving

$$\dot{P}_{\text{entropy}} + \frac{\dot{Q}}{T} = \left(\sum_{j=1}^{j=N_{\text{out}}} s_j \dot{m}_j\right)_{\text{outlets}} - \left(\sum_{i=1}^{i=N_{\text{in}}} s_i \dot{m}_i\right)_{\text{inlets}} \tag{2.69}$$

2.2.7   Euler's Turbomachinery Equation

Euler's turbomachinery equation is widely used in all types of turbomachinery. This equation, derived from the angular momentum equation discussed in Section 2.2.5.3, determines power transfer between the fluid and the rotor blades. In pumps, fans, and compressors, the power transfer occurs into the fluid to increase its outflow rate of angular momentum over the inflow rate; in turbines, the power transfer occurs to the rotor from the fluid to decrease its outflow rate of angular momentum over the inflow rate. The power transfer to or from the fluid is simply the product of the torque and rotor angular velocity in radians per second.

Let's consider a steady adiabatic flow in a rotating passage between two blades of a mixed axial-radial-flow turbomachinery, as shown in Figure 2.8. Each of the velocity vectors, $\vec{V}_1$ at the inlet 1 and $\vec{V}_2$ at outlet 2, has components in axial, radial, and tangential directions. The meridional velocity, which is the resultant of the axial and the radial velocities, is the mass velocity in the rotating passage flow at sections 1 and 2. Let V_{θ_1} be the tangential velocity at inlet 1 and V_{θ_2} at outlet 2. For a constant mass flow rate $\dot{m}$ through the rotating passage, the angular momentum equation, Equation 2.61, yields

$$\Gamma = \dot{m}\left(r_2 V_{\theta_2} - r_1 V_{\theta_1}\right) \tag{2.70}$$

The aerodynamic power transfer, which is the rate of work transfer as a result of the aerodynamic torque acting on the fluid control volume, is given by

$$\dot{W} = \Gamma\Omega = \dot{m}\left(r_2 V_{\theta_2} - r_1 V_{\theta_1}\right)\Omega = \dot{m}\left(U_2 V_{\theta_2} - U_1 V_{\theta_1}\right) \tag{2.71}$$

where U_1 and U_2 are rotor tangential velocities at inlet 1 and outlet 2, respectively. Using steady flow energy equation in terms of total enthalpy at sections 1 and 2, we can also write

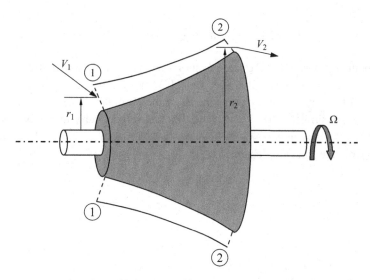

Figure 2.8 Flow through an axial-radial-flow turbomachinery passage between adjacent blades.

$$\dot{W} = \dot{m}\left(h_{t_2} - h_{t_1}\right). \qquad (2.72)$$

Combining Equations 2.71 and 2.72 yields

$$\left(h_{t_2} - h_{t_1}\right) = \left(U_2 V_{\theta_2} - U_1 V_{\theta_1}\right). \qquad (2.73)$$

Equation (2.73) is known as the Euler's turbomachinery equation. This equation simply states that, under adiabatic conditions, the change in specific total enthalpy of the fluid flow between any two sections of a rotor equals the difference in the products of the rotor tangential velocity and flow tangential velocity at these sections. This equation further reveals that for turbines, where the work is done by the fluid, we have $h_{t_2} < h_{t_1}$ and $U_2 V_{\theta_2} < U_1 V_{\theta_1}$; and for compressors, fans, and pumps, where the work is done on the fluid, we have $h_{t_2} > h_{t_1}$ and $U_2 V_{\theta_2} > U_1 V_{\theta_1}$.

For axial flow machines, where $r_1 \approx r_2$ and $U_1 \approx U_2$, Equation 2.73 reveals that the change in total enthalpy is entirely as a result of the change in the flow tangential velocity, that is, $\Delta h_t = \overline{U}(\Delta V_\theta)$, requiring blades with camber (bow). By contrast, for radial machines (radial turbines and centrifugal compressors), the change in total enthalpy results largely as a result of the change in rotor tangential velocity as a result of the change in radius, that is, $\Delta h_t = (\Delta U)\overline{V}_\theta$.

2.2.7.1 Rothalpy

The concept of rothalpy is grounded in Euler's turbomachinery equation presented in the preceding section. Rearranging Equation 2.73 yields

$$h_{t_1} - U_1 V_{\theta_1} = h_{t_2} - U_2 V_{\theta_2} \qquad (2.74)$$

which reveals that the quantity $\left(h_t - U V_\theta\right)$ at any point in a rotor flow remains constant under adiabatic conditions (no heat transfer). This quantity is called rothalpy expressed as

$$I = h_t - UV_\theta = h_s + \frac{V^2}{2} - UV_\theta = c_p T_{t_s} - UV_\theta \qquad (2.75)$$

where both h_t and V_θ are in the inertial (stationary) reference frame. Let us now convert Equation 2.75 into the rotor reference frame. If V_x, V_r, and V_θ are, respectively, the axial, radial, and tangential velocity components of the absolute velocity $\vec{V}$ at any point in the flow; and similarly, if W_x, W_r, and W_θ are, respectively, the axial, radial, and tangential velocity components of the corresponding relative velocity $\vec{W}$ at that point, we can write the following equations:

$$V^2 = V_x^2 + V_r^2 + V_\theta^2 \qquad (2.76)$$

$$W^2 = W_x^2 + W_r^2 + W_\theta^2 \qquad (2.77)$$

$$V_\theta = W_\theta + U \qquad (2.78)$$

where U is the local tangential velocity of the rotor. Substituting for V_θ from Equation 2.78 into Equation 2.76 and noting that $W_x = V_x$ and $W_r = V_r$, we obtain

$$V^2 = W_x^2 + W_r^2 + W_\theta^2 + 2W_\theta U + U^2. \qquad (2.79)$$

Using Equations 2.78 and 2.79, we rewrite Equation 2.75 as

$$I = h_s + \frac{W_x^2 + W_r^2 + W_\theta^2 + 2W_\theta U + U^2}{2} - U(W_\theta + U),$$

which simplifies to

$$I = h_s + \frac{W^2}{2} - \frac{U^2}{2} = h_{t_R} - \frac{U^2}{2} \qquad (2.80)$$

where h_{t_R} is the specific total enthalpy in the rotor reference frame. Equation 2.80 yields an interesting interpretation of rothalpy in a rotating system in that the change in fluid relative total enthalpy between any two radii in this system equals, under adiabatic conditions, the change in dynamic enthalpy associated with the solid-body rotation over these radii.

For a calorically perfect gas (constant c_p), we can also write Equation 2.80 as

$$I = c_p T_{t_R} - \frac{U^2}{2} \qquad (2.81)$$

where T_{t_R} is the fluid total temperature in the rotor reference frame.

2.2.7.2   An Alternate Form of the Euler's Turbomachinery Equation

For an adiabatic flow in a rotor, the rothalpy remains constant, that is, $I_1 = I_2$. Using Equation 2.80 we can write

$$h_{s_1} + \frac{W_1^2}{2} - \frac{U_1^2}{2} = h_{s_2} + \frac{W_2^2}{2} - \frac{U_2^2}{2}$$

$$h_{s_2} - h_{s_1} = \left(\frac{W_1^2}{2} - \frac{W_2^2}{2}\right) + \left(\frac{U_2^2}{2} - \frac{U_1^2}{2}\right). \qquad (2.82)$$

Equation 2.82 expresses the change in fluid static enthalpy in a rotor in terms of changes in the flow relative velocity and rotor tangential velocity. From the definition of total enthalpy, we can write its change between locations 1 and 2 as

$$h_{t_2} - h_{t_1} = \left(h_{s_2} - h_{s_1} \right) + \left(\frac{V_2^2}{2} - \frac{V_1^2}{2} \right). \tag{2.83}$$

Using Equation 2.82 to substitute for $\left(h_{s_2} - h_{s_1} \right)$ in Equation 2.83 yields the following alternate form of Euler's turbomachinery equation earlier derived as Equation 2.73:

$$h_{t_2} - h_{t_1} = \left(\frac{W_1^2}{2} - \frac{W_2^2}{2} \right) + \left(\frac{U_2^2}{2} - \frac{U_1^2}{2} \right) + \left(\frac{V_2^2}{2} - \frac{V_1^2}{2} \right) \tag{2.84}$$

where the terms on the right-hand side have the following physical interpretations:

$\left(\frac{W_1^2}{2} - \frac{W_2^2}{2} \right) \equiv$ Static enthalpy change as a result of change in rotor passage geometry

$\left(\frac{U_2^2}{2} - \frac{U_1^2}{2} \right) \equiv$ Static enthalpy change as a result of rotation of the rotor

$\left(\frac{V_2^2}{2} - \frac{V_1^2}{2} \right) \equiv$ Dynamic enthalpy change as a result of change in flow absolute velocity

2.2.7.3 Total Pressure and Total Temperature in Stator and Rotor Reference Frames

In turbomachinery, we need to consider flow properties in both stator (absolute, stationary, or inertial) and rotor (relative, rotating, or noninertial) reference frames. Because turbomachinery generally has a single axis of rotation, tangential (swirl) velocity is the only velocity component that is impacted by the change of reference frames from stator to rotor or vice versa. Both the axial and radial velocity components do not change. Further, all static properties of the fluid; for example, the static temperature and static pressure, remain the same in both stator and rotor reference frames.

Total temperature in both incompressible and compressible flows is computed by the equations:

$$\text{Stator reference frame: } T_{t_S} = T_s + \frac{V^2}{2c_p} \tag{2.85}$$

$$\text{Rotor reference frame: } T_{t_R} = T_s + \frac{W^2}{2c_p} \tag{2.86}$$

In an incompressible flow (constant density), the total pressure is evaluated by the equations:

$$\text{Stator reference frame: } P_{t_S} = P_s + \frac{\rho V^2}{2} \tag{2.87}$$

$$\text{Rotor reference frame: } P_{t_R} = P_s + \frac{\rho W^2}{2} \tag{2.88}$$

In a compressible flow, an isentropic stagnation process is assumed to obtain total properties. Accordingly, we invoke the isentropic relations to compute the total pressure using the following equations in each reference frame:

$$\text{Stator reference frame}: \frac{P_{t_S}}{P_S} = \left(\frac{T_{t_S}}{T_S}\right)^{\frac{\kappa}{\kappa-1}} = \left(1 + \frac{\kappa-1}{2}M_S^2\right)^{\frac{\kappa}{\kappa-1}} \qquad (2.89)$$

$$\text{Rotor reference frame}: \frac{P_{t_R}}{P_S} = \left(\frac{T_{t_R}}{T_S}\right)^{\frac{\kappa}{\kappa-1}} = \left(1 + \frac{\kappa-1}{2}M_R^2\right)^{\frac{\kappa}{\kappa-1}} \qquad (2.90)$$

Conversion from stator reference frame to rotor reference frame. Tangential velocities in the two reference frames are related by Equation 2.78, giving $W_\theta = V_\theta - U$. Because the fluid static temperature remains invariant in both reference frames, we can write

$$T_{t_R} - \frac{W^2}{2c_p} = T_{t_S} - \frac{V^2}{2c_p}. \qquad (2.91)$$

Because $W_x = V_x$ and $W_r = V_r$, Equation 2.91 reduces to

$$T_{t_R} = T_{t_S} + \frac{W_\theta^2}{2c_p} - \frac{V_\theta^2}{2c_p}. \qquad (2.92)$$

Substituting $W_\theta = V_\theta - U$ in Equation 2.92, we obtain

$$T_{t_R} = T_{t_S} + \frac{(V_\theta - U)^2}{2c_p} - \frac{V_\theta^2}{2c_p}$$

$$T_{t_R} = T_{t_S} + \frac{U(U - 2V_\theta)^2}{2c_p}, \qquad (2.93)$$

which yields $T_{t_R} = T_{t_S}$ for $V_\theta = U/2$.

For an isentropic compressible flow, we can use Equation 2.93 to obtain the pressure ratio as

$$\frac{P_{t_R}}{P_{t_S}} = \left(\frac{T_{t_R}}{T_{t_S}}\right)^{\frac{\kappa}{\kappa-1}} = \left\{1 + \frac{U(U - 2V_\theta)}{2c_p T_{t_S}}\right\}^{\frac{\kappa}{\kappa-1}}$$

$$P_{t_R} = P_{t_S}\left\{1 + \frac{U(U - 2V_\theta)}{2c_p T_{t_S}}\right\}^{\frac{\kappa}{\kappa-1}}. \qquad (2.94)$$

Conversion from rotor reference frame to stator reference frame. Using Equation 2.78 to substitute for V_θ in Equation 2.93, and rearranging terms, we obtain

$$T_{t_S} = T_{t_R} - \frac{U(U - 2W_\theta - 2U)^2}{2c_p} = T_{t_R} - \frac{U(U + 2W_\theta)^2}{2c_p} \qquad (2.95)$$

$$\frac{P_{t_S}}{P_{t_R}} = \left(\frac{T_{t_S}}{T_{t_R}}\right)^{\frac{\kappa}{\kappa-1}} = \left\{1 - \frac{U(U + 2W_\theta)}{2c_p T_{t_R}}\right\}^{\frac{\kappa}{\kappa-1}}, \qquad (2.96)$$

which uses the fact that both the static pressure and static temperature, being fluid state properties, are independent of the reference frame. Equation 2.96 thus gives

$$P_{t_s} = P_{t_R} \left\{ 1 - \frac{U(U + 2W_\theta)}{2c_p T_{t_R}} \right\}^{\frac{\kappa}{\kappa-1}}.$$ (2.97)

An easier approach to first converting total temperatures, and then total pressures, between stator and rotor reference frames is by equating the expressions of rothalpy in these reference frames, as given by Equations 2.75 ($I = c_p T_{t_s} - UV_\theta$) and 2.81 ($I = c_p T_{t_R} - U^2/2$) and converting tangential velocities in these reference frames by using Equation 2.78.

2.2.8 Compressible Flow with Area Change, Friction, Heat Transfer, Rotation, and Normal Shocks

Compressible flows are ubiquitous in gas turbines. The key parameter that characterizes a compressible flow is its Mach number ($M = V/C$), which is the ratio of the flow velocity to speed of sound ($C = \sqrt{\kappa R T_s}$). For $M \leq 0.3$, the gas flow may be treated as incompressible for most practical engineering calculations using a constant density, which is obtained using the perfect gas law $P_s/\rho = RT_s$. With several examples from gas turbine design applications, Sultanian (2015) presents comprehensive details of compressible flows with area change, friction, heat transfer, rotation, normal, and oblique shocks. In our presentation of internal compressible flows in this section, all the flow properties are uniform to be uniform at each section.

Let us briefly reflect on the fundamental difference between a compressible flow and an incompressible flow with constant density. The total energy of a flow, as measured by its total temperature, is the sum of the fluid static temperature, which is the average kinetic energy associated with the gas molecules, and the dynamic temperature based on the flow kinetic energy. In an incompressible flow, there is negligible coupling between the external flow energy and its internal one. For example, when water flows through a converging-diverging (C-D) nozzle with constant entropy (constant total pressure and total temperature), the throat section with its minimum area features the maximum flow velocity, or the flow kinetic energy, with negligible decrease in its static temperature, a measure of the internal energy of the flow. Note that the specific heat of water is more than four times that of air. Further, in an incompressible flow, total pressure ($P_t = P_s + \rho V^2/2$) may be interpreted as the total mechanical energy per unit volume of the flow. The constancy of total pressure in an ideal incompressible flow leads to the famous Bernoulli equation with its tenet of "high velocity, low pressure" at any point in the flow.

A compressible flow, by contrast, exhibits quite different characteristics, particularly at high Mach numbers. For an isentropic airflow in a choked C-D nozzle with exit pressure ratio (ratio of total-to-static pressure) of more than two, both the static pressure and static temperature continuously decrease in the flow direction with continuous increase in Mach number. In the diverging section with increasing flow area, Mach number increases mainly as a result of the increase in the flow velocity needed to maintain a constant mass flow rate through the nozzle under significantly decreasing density. This behavior is somewhat counterintuitive based on our understanding of an incompressible internal flow where, for a given mass flow rate, the flow velocity always decreases with increasing area.

Note that the square of Mach number indicates the ratio of the external energy to internal energy of the flow. This provides a useful physical insight into Mach number. When the airflow expands in a C-D nozzle, increasing Mach number entails increase in the external flow energy and decrease in the internal one, the total flow energy remaining constant (constant total temperature). In other words, the change in Mach number is associated of the exchange between the internal flow energy and external flow energy. This two-way coupling between the internal and external flow energies in a compressible flow, also known as the compressibility effect, makes it fundamentally different from an incompressible flow. Note that the compressibility effect increases with Mach number. At low Mach numbers ($M \leq 0.3$), as stated before, the compressibility effect may be neglected, and the compressible flow may be assumed to behave like an incompressible flow. On the other hand, an incompressible flow cannot be modeled to behave like a compressible flow. Based on this discussion, one should therefore avoid the use of Bernoulli equation or its extended form, also called the mechanical energy equation, in a compressible flow.

The phenomenon of choking is again unique to a compressible flow. At any section of fixed area, when the Mach number is unity ($M = 1$), the flow is considered to be choked at that section, meaning the downstream changes in the pressure boundary condition will have no effect on the mass flow rate through the section. Note, however, that this maximum mass flow rate can still be changed by changing the upstream boundary conditions where the flow is subsonic ($M < 1$). Choking has a simple physical interpretation. Because small pressure changes in a gas travel at the speed of sound in all directions, for $M = 1$ at a section, the flow velocity equals the local speed of sound and prevents any sound wave to travel upstream of this section. As a result, the lowering of the C-D nozzle exit static pressure, after it is choked at the throat, will not flow more regardless of how much we decrease this pressure as long as the upstream total pressure and total temperature remain constant.

2.2.8.1   Isentropic Flow

Isentropic gas flows are hard to find in a real word, including gas turbines. Nevertheless, these flows serve as ideal reference flows in many design applications. From Equation 2.17, we obtain the following relation between the pressure ratio and temperature ratio at two points of an isentropic compressible flow:

$$\frac{P_{s_2}}{P_{s_1}} = \left(\frac{T_{s_2}}{T_{s_1}}\right)^{\frac{c_p}{R}} = \left(\frac{T_{s_2}}{T_{s_1}}\right)^{\frac{\kappa}{\kappa-1}} \tag{2.98}$$

Because the stagnation process in a compressible flow is assumed isentropic, a practical application of Equation 2.98 is to compute the maximum local stagnation pressure using the equation

Table 2.3 Isentropic Compressible Flow Equations

$$\frac{T_t}{T_s} = 1 + \frac{\kappa - 1}{2}M^2$$

$$\frac{P_t}{P_s} = \left(\frac{T_t}{T_s}\right)^{\frac{c_p}{R}} = \left(\frac{T_t}{T_s}\right)^{\frac{\kappa}{\kappa-1}} = \left(1 + \frac{\kappa - 1}{2}M^2\right)^{\frac{\kappa}{\kappa-1}} \qquad \frac{\rho_t}{\rho_s} = \left(\frac{T_t}{T_s}\right)^{\frac{1}{\kappa-1}} = \left(1 + \frac{\kappa - 1}{2}M^2\right)^{\frac{1}{\kappa-1}}$$

With starred quantities corresponding to the sonic condition ($M = 1$):

$$\frac{T_t}{T_s^*} = \frac{\kappa + 1}{2} \qquad \frac{P_t}{P_s^*} = \left(\frac{\kappa + 1}{2}\right)^{\frac{\kappa}{\kappa-1}} \qquad \frac{\rho_t}{\rho_s^*} = \left(\frac{\kappa + 1}{2}\right)^{\frac{1}{\kappa-1}}$$

$$\frac{A}{A^*} = \frac{1}{M}\sqrt{\left\{\frac{2 + (\kappa - 1)M^2}{\kappa + 1}\right\}^{\frac{\kappa+1}{\kappa-1}}}$$

$$\frac{P_t}{P_s} = \left(\frac{T_t}{T_s}\right)^{\frac{c_p}{R}} = \left(\frac{T_t}{T_s}\right)^{\frac{\kappa}{\kappa-1}} \tag{2.99}$$

where the total temperature is computed as $T_t = T_s + 0.5V^2/c_p$. It is often convenient to express compressible flow equations in terms of Mach number. These equations for an isentropic compressible flow are summarized in Table 2.3.

2.2.8.2 Mass Flow Functions

In an internal flow, mass flow rate is simply calculated as $\dot{m} = \rho A V$, where the velocity V is normal to the flow area A. In steady state, regardless of changes in density, area, and velocity, the mass flow rate remains constant at each section. Further, for an incompressible flow and subsonic compressible flow in a duct, the exit static pressure must always equal the ambient static pressure, which acts as the discharge pressure boundary condition. In our approach in this textbook, we use only the properties at a section to compute mass flow rate at that section. For the flow to occur in either direction at a section, the total pressure at the section must be greater than the static pressure. Discounting the change in entropy as a result of heat transfer, an internal flow at any section occurs in the direction of increasing entropy.

In a compressible flow, the mass flow functions provide a convenient means of computing mass flow rate at a section without explicitly using the fluid density, which does not remain constant in such a flow. This approach is widely used by design engineers in the gas turbine industry. Starting with the equation $\dot{m} = \rho A V$ and using the isentropic flow equations along with the equation of state ($P_s/\rho = RT_s$), we can easily express the mass flow rate at a section as

$$\dot{m} = \frac{A\hat{F}_{f_s}P_s}{\sqrt{RT_t}} = \frac{AF_{f_s}P_s}{\sqrt{T_t}} \tag{2.100}$$

where $\hat{F}_{f_s}$ is the dimensionless static-pressure mass flow function given by

$$\hat{F}_{f_s} = M\sqrt{\kappa\left(1 + \frac{\kappa - 1}{2}M^2\right)} \tag{2.101}$$

and F_{f_s}, having the dimensions of $1/\sqrt{R}$, is the static-pressure mass flow function given by

$$F_{f_s} = M\sqrt{\frac{\kappa}{R}\left(1 + \frac{\kappa - 1}{2}M^2\right)}. \tag{2.102}$$

Note that, for a calorically perfect gas, both $\hat{F}_{f_s}$ and F_{f_s} are functions of Mach number only.

Equation 2.100 shows that, for a given Mach number, the mass flow rate at a section is proportional to the static pressure and inversely proportional to the square root of the total temperature. Both the static pressure and total temperature are assumed to be uniform over that section.

Using P_t/P_s in terms of Mach number from Table 2.3, we can replace P_s in Equation 2.100 by P_t and alternatively express the mass flow rate as

$$\dot{m} = \frac{A\hat{F}_{f_t}P_t}{\sqrt{RT_t}} = \frac{AF_{f_t}P_t}{\sqrt{T_t}} \tag{2.103}$$

where $\hat{F}_{f_t}$ is the dimensionless total-pressure mass flow function given by

$$\hat{F}_{f_t} = M\sqrt{\frac{\kappa}{\left(1 + \frac{\kappa-1}{2}M^2\right)^{\frac{\kappa+1}{\kappa-1}}}} \tag{2.104}$$

and F_{f_t}, having the dimensions of $\sqrt{R}$, is the total-pressure mass flow function given by

$$F_{f_t} = M\sqrt{\frac{\kappa}{R\left(1 + \frac{\kappa-1}{2}M^2\right)^{\frac{\kappa+1}{\kappa-1}}}}. \tag{2.105}$$

For a calorically perfect gas both $\hat{F}_{f_t}$ and F_{f_t} are functions of Mach number only.

Equation 2.103 shows that, for a given Mach number, the mass flow rate at a section is proportional to the total pressure and inversely proportional to the square root of the total temperature. Both the total pressure and total temperature are assumed to be uniform over that section.

Figure 2.9 shows the variations of mass flow functions $\hat{F}_{f_t}$ and $\hat{F}_{f_s}$ with Mach number for $\kappa = 1.4$. As shown in the figure, $\hat{F}_{f_t}$ increases as Mach number increases for $M < 1$ (subsonic flow) and decreases as Mach number increases for $M > 1$ (supersonic flow), and it has a maximum value of 0.6847 at $M = 1$(sonic flow), which corresponds to the choked flow conditions at the flow area. For example, for the ambient air at $20°$C and 1.013 bar, Equation 2.103 computes 239.2 kg/s as the maximum possible mass flow rate through an area of 1.0 m^2. Furthermore, the figure shows that each value of Mach number yields a single value of mass flow function $\hat{F}_{f_t}$, but each value of flow function

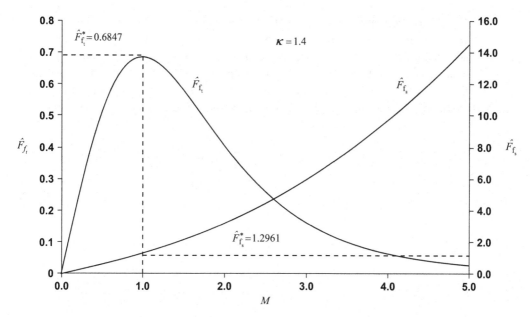

Figure 2.9 Variations of mass flow functions $\hat{F}_{f_t}$ and $\hat{F}_{f_s}$ with Mach number for $\kappa = 1.4$.

$\hat{F}_{f_t}$ corresponds to two values (except at $M = 1$) of Mach number – one subsonic and the other supersonic. While the calculation of $\hat{F}_{f_t}$ by Equation 2.103 for a given value of Mach number is direct, we need to use an iterative method (e.g., "Goal Seek" in MS Excel) to compute subsonic and supersonic Mach numbers for a given value of $\hat{F}_{f_t}$.

As shown in Figure 2.9, unlike $\hat{F}_{f_t}$, $\hat{F}_{f_s}$ increases monotonically with Mach number without exhibiting a maximum value at $M = 1$, implying one-to-one correspondence between $\hat{F}_{f_s}$ and M. Note that the ratio $\hat{F}_{f_s}/\hat{F}_{f_t}$ equals the total-to-static pressure ratio P_t/P_s, which increases monotonically with Mach number, more rapidly for $M > 1$. For a given value of $\hat{F}_{f_s}$, we present here a direct method to calculate the corresponding Mach number.

Finding Mach number for a given static-pressure mass flow function. Squaring both sides of Equation 2.101, we obtain

$$\hat{F}_{f_s}^2 = M^2\kappa\left(1 + \frac{\kappa-1}{2}M^2\right) = \kappa M^2 + 0.5\kappa(\kappa-1)M^4$$

$$0.5\kappa(\kappa-1)M^4 + \kappa M^2 - \hat{F}_{f_s}^2 = 0,$$

which is a quadratic equation in M^2 with the positive root given by

$$M^2 = \frac{-\kappa + \sqrt{\kappa^2 + 2\kappa(\kappa-1)\hat{F}_{f_s}^2}}{\kappa(\kappa-1)}$$

$$M = \sqrt{\frac{-\kappa + \sqrt{\kappa^2 + 2\kappa(\kappa-1)\hat{F}_{f_s}^2}}{\kappa(\kappa-1)}}. \tag{2.106}$$

Use of isentropic flow tables to compute mass flow functions. Not all isentropic flow tables list mass flow functions; however, they all list quantities like pressure ratio P_t/P_s and area ratio A/A^* against Mach number. For an isentropic flow, we know that the total pressure and total temperature remain constant in a variable-area duct. In a duct flow, we can equate the mass flow rate at a give section of area A to that at the sonic throat area A^*(imaginary if it is outside the duct), giving

$$\dot{m} = \frac{A\hat{F}_{f_t}P_t}{\sqrt{RT_t}} = \frac{A^*\hat{F}_{f_t}^*P_t}{\sqrt{RT_t}}$$

$$\hat{F}_{f_t} = \frac{\hat{F}_{f_t}^*}{\left(\dfrac{A}{A^*}\right)} = \frac{0.6847}{\left(\dfrac{A}{A^*}\right)} \tag{2.107}$$

and

$$\hat{F}_{f_s} = \hat{F}_{f_t}\left(\frac{P_t}{P_s}\right). \tag{2.108}$$

Knowing the Mach number at a section, we can look up P_t/P_s and A/A^* in isentropic flow tables; then use Equation 2.107 to compute $\hat{F}_{f_t}$, followed by Equation 2.108 to compute $\hat{F}_{f_s}$ at that section.

2.2.8.3   Impulse Functions

As discussed in Section 2.2.5, the concept of stream thrust is grounded in the linear momentum equation. The change in stream thrust (the total stream thrust at outflows minus the total stream thrust at inflows) across a duct represents the total thrust on a duct as a result of fluid flow.

Using impulse functions, we can express the stream thrust at any section of a compressible duct flow as

$$S_T = P_iA = P_sAI_{f_s} = P_tAI_{f_t} \tag{2.109}$$

where I_{f_s} is the static-pressure impulse function given by

$$I_{f_s} = \left(1 + \kappa M^2\right) \tag{2.110}$$

and I_{f_t} is the total-pressure impulse function given by

$$I_{f_t} = \frac{1 + \kappa M^2}{\left(1 + \frac{\kappa-1}{2}M^2\right)^{\frac{\kappa}{\kappa-1}}}. \tag{2.111}$$

Note that the ratio I_{f_s}/I_{f_t} equals the total-to-static pressure ratio P_t/P_s. For a given gas flow, both I_{f_s} and I_{f_t} are functions of Mach number only.

Figure 2.10 shows the monotonically increasing parabolic variation of I_{f_s} with Mach number for $\kappa = 1.4$. As shown in the figure, I_{f_s} increases with M for $M < 1$, has a maximum value of 1.2679 at $M = 1$, and decreases with M for $M > 1$.

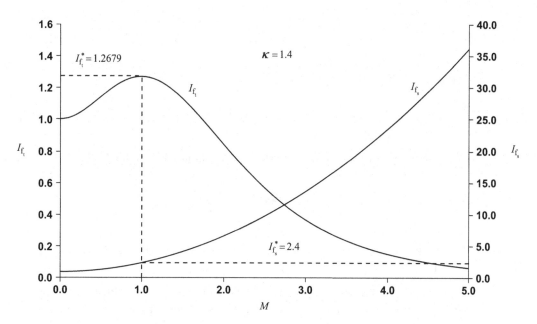

Figure 2.10 Variations of impulse functions I_{f_t} and I_{f_s} with Mach number for $\kappa = 1.4$.

2.2.8.4   Normal Shock Function

When a supersonic compressible flow needs to adjust to a sudden increase in downstream static pressure, it undergoes a normal shock, which is a very thin nonequilibrium region featuring an abrupt increase in entropy, while its total temperature and mass flow rate remain constant. The flow upstream of a normal shock is always supersonic and the downstream flow always subsonic. Thus, the static temperature across a normal shock increases, and the flow velocity decreases with simultaneous increase in density to satisfy continuity.

Let us consider a constant-area control volume simulating a normal shock under the boundary conditions of no heat transfer and surface force. In this case, therefore, the mass flow rate, the stream thrust, and the total temperature over the control volume all remain constant. Thus, the ratio of mass flow rate to stream thrust at the inlet (upstream of the normal shock) and outlet (downstream of the normal shock) of our control volume must be equal. Evaluating this ratio using Equations 2.103 and 2.109 yields

$$\frac{\dot{m}}{S_T} = \frac{\left(\dfrac{A\hat{F}_{f_t}P_t}{\sqrt{RT_t}}\right)}{P_t A I_{f_t}} = \left(\frac{\hat{F}_{f_t}}{I_{f_t}}\right)\frac{1}{\sqrt{RT_t}}.$$

The aforementioned equation reveals that the ratio $\hat{F}_{f_t}/I_{f_t}$ must remain constant across a normal shock. We call this ratio, which is the ratio of total-pressure mass flow function to the total-pressure impulse function, the normal shock function N expressed as the following function of M and κ:

$$N = \frac{\hat{F}_{f_t}}{I_{f_t}} = \frac{\hat{F}_{f_s}}{I_{f_s}} = \frac{M}{1 + \kappa M^2} \sqrt{\kappa \left(1 + \frac{\kappa - 1}{2} M^2 \right)} \qquad (2.112)$$

For $M = 1$, Equation 2.112 yields

$$N^*(1, \kappa) = \sqrt{\frac{\kappa}{2(1 + \kappa)}}, \qquad (2.113)$$

which for $\kappa = 1.4$ equals 0.540. For $M \to \infty$, Equation 2.112 yields N_∞ as

$$N_\infty(\infty, \kappa) = \sqrt{\frac{\kappa - 1}{2\kappa}}. \qquad (2.114)$$

For $\kappa = 1.4$ we obtain $N_\infty = 0.378$. It is interesting to note that the corresponding value of the subsonic Mach number is also equal to 0.378. This can be easily demonstrated by substituting $M = \sqrt{(\kappa - 1)/(2\kappa)}$ in Equation 2.112 and obtaining $N = \sqrt{(\kappa - 1)/(2\kappa)}$ after some simplification of the right-hand of the equation. From this we can conclude that, even for the strongest normal shock, the downstream subsonic Mach number will not be lower than $\sqrt{(\kappa - 1)/(2\kappa)}$.

The behavior of the normal shock function N as a function of Mach number is depicted in Figure 2.11 for $\kappa = 1.4$. Like the total-pressure mass flow function $\hat{F}_{f_t}$ and the total-pressure impulse function I_{f_t}, the figure shows that N also increases with M for $M < 1$, reaches a maximum value of $N^* = 0.540$ at $M = 1$, and decreases with M for $M > 1$.

In addition to simplifying the computations of compressible flows without the explicit use of density, the mass flow functions, impulse functions, and normal shock function presented in the foregoing sections are very helpful in explaining many seemingly nonintuitive behaviors these flows, as we will see in the following sections.

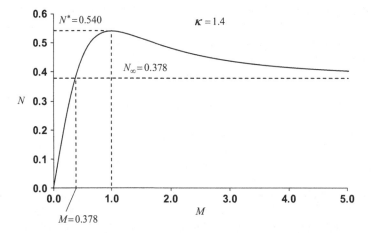

Figure 2.11 Variation of normal shock function N with Mach number for $\kappa = 1.4$.

2.2.8.5   Isentropic Flow with Area Change

For the differential control volumes for mass conservation, force-momentum balance, and energy conservation shown in Figure 2.12, Sultanian (2015) presents a detailed derivation of the following differential equation for flow velocity change under the influence of area change, friction, heat transfer, and rotation:

$$\left(1 - M^2\right)\frac{dV}{V} = -\frac{dA}{A} - M_\theta^2\frac{dr}{r} + \left(\frac{\kappa M^2}{\rho A V^2}\right)\delta F_{\text{sh}} + \frac{\delta \dot{Q}}{\dot{m} c_p T_s} \qquad (2.115)$$

where the rotational Mach number $M_\theta = r\Omega/\sqrt{\kappa R T_s}$. Being nonlinear, Equation 2.115 generally requires a numerical solution when all effects are presents, as in a serpentine passage used in internal cooling of gas turbine blades (with rotation) and vanes (without rotation).

To develop a good intuitive understanding of a compressible duct flow, let us consider one effect at a time. For an isentropic compressible flow in a variable-area duct, Equation 2.115 reduces to

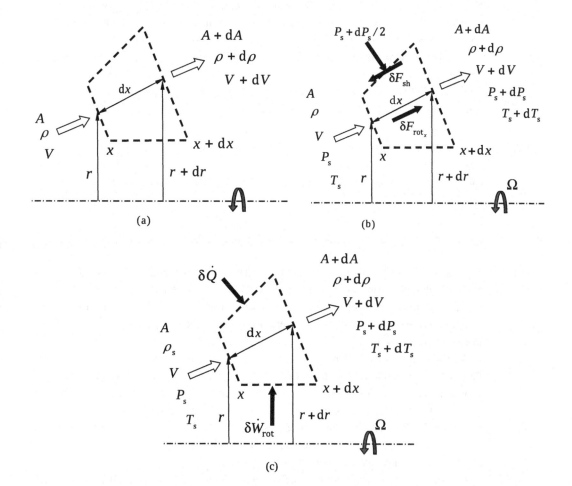

Figure 2.12 Differential control volumes: (a) mass, (b) linear momentum, and (c) energy.

$$\left(1 - M^2\right) \frac{\mathrm{d}V}{V} = -\frac{\mathrm{d}A}{A}, \qquad (2.116)$$

which clearly shows that, for subsonic ($M < 1$) flows with the coefficient $(1 - M^2)$ being positive, the flow velocity increases as the duct area decreases and vice versa. For supersonic ($M > 1$) flows; however, the coefficient $(1 - M^2)$ being negative, the flow velocity increases with duct area and vice versa. As discussed earlier, a supersonic C-D nozzle used in rocket engines features this counterintuitive compressible flow behavior.

2.2.8.6   Isentropic Flow with Rotation

For an isentropic flow in a constant-area duct rotating around an axis other than along the flow direction, Equation 2.115 reduces to

$$\left(1 - M^2\right) \frac{\mathrm{d}V}{V} = -M_\theta^2 \frac{\mathrm{d}r}{r}. \qquad (2.117)$$

Note that the duct rotation essentially effect the flow in two ways; first, as a body force (centrifugal force) in the momentum equation, and second, as rotational work transfer in the energy equation. For the velocity change in the flow, both these effects are combined into the term on the right-hand side of Equation 2.117.

Because M_θ^2 is positive definite, Equation 2.117 reveals that, for a subsonic flow, the velocity will increase when the flow is radially inward ($\mathrm{d}r < 0$) and decrease when the flow is radially outward ($\mathrm{d}r > 0$). When the flow is supersonic in a constant-area rotating duct flow, it will exhibit the opposite behavior.

2.2.8.7   Nonisentropic Flow with Friction: Fanno Flow

The nonisentropic compressible flow with friction in a constant-area duct with no rotation and no heat transfer is known as Fanno flow. In this case, Equation 2.115 reduces to

$$\left(1 - M^2\right) \frac{\mathrm{d}V}{V} = \left(\frac{\kappa M^2}{\rho A V^2}\right) \delta F_{\mathrm{sh}}. \qquad (2.118)$$

Because the right-hand side of Equation 2.118 is always positive, we infer from this equation that the friction will always accelerate a subsonic flow and decelerate a supersonic flow. The fact that the friction accelerates a subsonic flow in a constant-area duct is counterintuitive to our experience on the effects of friction, which is known to slow things down.

A simple physics-based reasoning behind why the flow accelerates in a subsonic Fanno flow, in which the total temperature, flow area, and mass flow rate remain constant, runs as follows. For the flow to overcome wall shear force as a result of friction, which always opposes the flow, we need a net pressure force in the flow direction. This entails that the static pressure and density (from the equation of state) decrease downstream. Because the duct area remains constant, the flow velocity must increase to maintain the constant mass flow rate.

Another way to explain the flow behavior in a Fanno flow is through the continuity equation expressed in terms of the total-pressure mass flow function, Equation 2.103.

Table 2.4 Fanno Flow Equations

$$\frac{T_s}{T_s^*} = \frac{\kappa+1}{2+(\kappa-1)M^2} \qquad \frac{\rho}{\rho^*} = \frac{1}{M}\sqrt{\frac{2+(\kappa-1)M^2}{\kappa+1}} \qquad \frac{P_s}{P_s^*} = \frac{1}{M}\sqrt{\frac{\kappa+1}{2+(\kappa-1)M^2}}$$

$$\frac{P_t}{P_t^*} = \frac{1}{M}\left(\frac{2+(\kappa-1)M^2}{\kappa+1}\right)^{\frac{\kappa+1}{2(\kappa-1)}} \qquad \frac{P_i}{P_i^*} = \frac{P_s(1+\kappa M^2)}{P_s^*(\kappa+1)}$$

$$\frac{fL_{max}}{D_h} = \frac{\kappa+1}{2\kappa}\ln\left\{\frac{(\kappa+1)M^2}{2+(\kappa-1)M^2}\right\} + \frac{1}{\kappa}\left(\frac{1}{M^2}-1\right)$$

The effect of friction in an adiabatic flow is always to increase entropy and decrease total pressure in the flow direction. Because the mass flow rate at each section of a steady internal flow remains constant, Equation 2.103 dictates that the downstream total-pressure mass flow function ($\hat{F}_{f_t}$) must increase in the flow direction. According to the variation of $\hat{F}_{f_t}$ with Mach number shown in Figure 2.9, it is clear that the downstream Mach number should move closer toward the sonic condition ($M=1$). This explains why in a Fanno flow the downstream Mach number increases if the flow is subsonic and decreases if it is supersonic. This reasoning can also be extended to conclude that the exit Mach number of a Fanno flow is limited to $M=1$(frictional choking) for both subsonic and supersonic flows at the inlet.

Table 2.4 summarizes a set of equations to evaluate changes in various properties in a Fanno flow. Detailed derivations of these equations are given by Sultanian (2015).

2.2.8.8 Nonisentropic Flow with Heat Transfer: Rayleigh Flow
The nonisentropic compressible flow in a nonrotating, constant-area, frictionless duct with heat transfer (both heating and cooling of the flow) is known as Rayleigh flow. Because friction is always present in real duct flows, Rayleigh flows are seldom found in engineering applications. Nevertheless, it is instructive to study these flows to develop a better understanding of the effects of heat transfer. The following reduced form of Equation 2.115 shows that the heating will increase flow velocity for a subsonic Rayleigh flow ($M<1$) and will decrease it for a supersonic Rayleigh flow ($M>1$). In addition, this equation also shows that cooling the flow will have opposite trends in velocity changes in the flow direction.

$$\left(1-M^2\right)\frac{dV}{V} = \frac{\delta\dot{Q}}{\dot{m}c_p T_s}. \qquad (2.119)$$

In a Rayleigh flow, in addition to the area and mass flow rate, the stream thrust (zero shear force from the wall) remains constant at each section. Combining Equations 2.103 and 2.109, we can express the stream thrust as

$$S_T = \frac{\dot{m}\sqrt{RT_t}}{N}. \qquad (2.120)$$

If the upstream and downstream sections of a Rayleigh flow are designated by subscripts 1 and 2, respectively, we have $S_{T_1} = S_{T_2}$, which according to Equation 2.120 yields

$$\frac{N_2}{N_1} = \sqrt{\frac{T_{t_2}}{T_{t_1}}}.$$

For heating, we have $T_{t_2} > T_{t_1}$, which means $N_2 > N_1$. According to the variation of N with Mach number shown in Figure 2.11, we further conclude that M_2 should be closer to $M = 1$ than to M_1. In other words, if the upstream flow is subsonic, the Mach number will increase downstream. By contrast, if the flow is supersonic, the Mach number will decrease downstream. Thus, heating in a Rayleigh flow raises a subsonic Mach number and lowers a supersonic Mach number toward the sonic condition (thermal choking).

From the constancy of stream thrust between any two sections of a Rayleigh flow, we can write

$$P_{t_1} I_{f_{t_1}} = P_{t_2} I_{f_{t_2}}$$

$$\frac{P_{t_2}}{P_{t_1}} = \frac{I_{f_{t_1}}}{I_{f_{t_2}}}. \tag{2.121}$$

Based on the dependence of I_{f_t} on Mach number shown in Figure 2.10, because M_2 is on the same side as M_1 and closer to $M = 1$, we can infer that $I_{f_{t_1}} < I_{f_{t_2}}$. Thus, Equation 2.121 yields $P_{t_2} < P_{t_1}$ for both subsonic and supersonic Rayleigh flows. This result explains another nonintuitive behavior of a Rayleigh flow in that the increase in entropy as a result of heating leads to decrease in total pressure in the flow direction. Conversely, cooling in a Rayleigh flow increases the total pressure downstream.

A subsonic Rayleigh flow exhibits yet another nonintuitive feature. In the Mach number range $1/\sqrt{\kappa} < M < 1$, the static temperature decreases with heating and increases with cooling. For the case of heating in this Mach number range, although the total temperature increases, the flow accelerates very rapidly to satisfy the continuity equation with concomitant reduction in the static temperature. The opposite happens for the case of cooling.

Table 2.5 summarizes a set of equations to evaluate changes in various properties in a Rayleigh flow. Detailed derivations of these equations are given by Sultanian (2015).

Table 2.5 Rayleigh Flow Equations

$$\frac{T_s}{T_s^*} = M^2 \left(\frac{\kappa + 1}{2 + \kappa M^2}\right)^2 \qquad \frac{\rho}{\rho^*} = \frac{1}{M^2}\left(\frac{1 + \kappa M^2}{\kappa + 1}\right) \qquad \frac{P_s}{P_s^*} = \frac{\kappa + 1}{1 + \kappa M^2}$$

$$\frac{P_t}{P_t^*} = \left(\frac{\kappa + 1}{1 + \kappa M^2}\right)\left(\frac{2 + (\kappa - 1)M^2}{\kappa + 1}\right)^{\frac{\kappa}{\kappa - 1}} \qquad \frac{T_t}{T_t^*} = M^2\left(\frac{\kappa + 1}{1 + \kappa M^2}\right)^2\left(\frac{2 + (\kappa - 1)M^2}{\kappa + 1}\right)$$

2.2.8.9 Nonisentropic Flow: Normal Shock

The normal shock, which occurs only in a supersonic compressible flow, is a compression wave normal to the flow direction. With no area-change, friction, heat transfer, and rotation, the right-hand side of Equation 2.115 vanishes, yielding

$$\left(1 - M^2\right)\frac{dV}{V} = 0. \tag{2.122}$$

A trivial solution of Equation 2.122 corresponds to $dV = 0$, which implies equal velocities upstream and downstream of a normal shock. The second solution corresponds to abrupt changes in flow properties across the normal shock. These changes are primarily driven by the nonequilibrium and nonisentropic processes, allowing the flow to adjust itself to a higher downstream static pressure boundary condition by becoming subsonic. Because the changes of properties across a normal shock are discontinuous, the governing differential equations cannot be integrated across it. However, we can use a control volume analysis of the governing conservation laws over a normal shock to develop simple algebraic equations, summarized in Table 2.6, which relate flow properties before and after the shock.

As we noted in Section 2.2.8.4, the normal shock function N remains constant across a normal shock. Figure 2.11 shows that, for a given value of N above the asymptotic value of 0.378, we obtain two values of M, one subsonic and the other supersonic. Let us go through a simple physical reasoning to show that the Mach number upstream of a normal shock must be supersonic. Let the sections upstream and downstream of a normal shock be denoted by subscripts 1 and 2, respectively. As discussed earlier, the flow across a normal shock is adiabatic ($T_{t_1} = T_{t_2}$). For entropy to increase across a normal shock, Equation 2.22 yields $P_{t_2} < P_{t_1}$. Because the static pressure increases across a normal shock (compression wave), we obtain $P_{t_2}/P_{s_2} < P_{t_1}/P_{s_1}$, which from isentropic relations leads to the result $M_2 < M_1$. Thus, for N remaining constant across a normal shock, we conclude that the flow before a normal shock must be supersonic, resulting in a subsonic flow after the shock.

Table 2.6 Normal Shock Equations

$$M_2^2 = \frac{2 + (\kappa - 1)M_1^2}{2\kappa M_1^2 - (\kappa - 1)} \qquad \frac{P_{s_2}}{P_{s_1}} = \frac{2\kappa M_1^2 - (\kappa - 1)}{(\kappa + 1)} \qquad \frac{\rho_2}{\rho_1} = \frac{(\kappa + 1)M_1^2}{2 + (\kappa - 1)M_1^2}$$

$$\frac{T_{s_2}}{T_{s_1}} = \left\{\frac{2\kappa M_1^2 - (\kappa - 1)}{(\kappa + 1)}\right\}\left\{\frac{2 + (\kappa - 1)M_1^2}{(\kappa + 1)M_1^2}\right\}$$

$$\frac{P_{t_2}}{P_{t_1}} = \left\{\frac{(\kappa + 1)}{2\kappa M_1^2 - (\kappa - 1)}\right\}^{\frac{1}{\kappa - 1}}\left[\frac{(\kappa + 1)M_1^2}{\{2 + (\kappa - 1)M_1^2\}}\right]^{\frac{\kappa}{\kappa - 1}}$$

$$\frac{P_{t_2}}{P_{s_1}} = \left\{\frac{2\kappa M_1^2 - (\kappa - 1)}{(\kappa + 1)}\right\}^{\frac{-1}{\kappa - 1}}\left\{\frac{(\kappa + 1)}{2}M_1^2\right\}^{\frac{\kappa}{\kappa - 1}}$$

2.2.9   Navier-Stokes Equations

In fluid mechanics, time-dependent Navier-Stokes (N-S) equations are the partial differential equations that govern the linear momentum conservation in a Newtonian fluid flow, be it laminar or turbulent, incompressible or compressible. The major difficulty associated with solving the N-S equations lies in the nonlinear inertia (convection) terms that involve the product of velocity with its spatial gradient. Only for a limited number of laminar flow problems, the N-S equations are known have analytical solutions; some of these solutions are presented by Sultanian (2015). For all other laminar flows, numerical solutions are the only way forward. For accurately predicting a turbulent flow, the method of direct numerical simulation (DNS), which requires enormous computing power to resolve very small length and time scales, is still beyond the reach of most engineering applications. In such cases, approximate numerical predictions are carried out using a number of statistical turbulence models within the framework of CFD technology, which has made tremendous progress over the last fifty years.

The N-S equations are invariably solved along with the continuity equation (Equation 2.54). For a compressible flow, be it laminar or turbulent, we must additionally solve the energy equation. In such a flow, the governing momentum and energy equations are coupled through density, which depends on both static pressure and static temperature via the equation of state $\left(\rho = P_s/(RT_s)\right)$.

Without going through a detailed derivation, we simply state here the Navier-Stokes equations in their most general form as follows:

Vector notation:

$$\rho\frac{D\vec{V}}{Dt} = \rho\left[\frac{\partial \vec{V}}{\partial t} + \left(\vec{V}\bullet\nabla\right)\vec{V}\right] = -\nabla P_s + \vec{F} - \nabla\times\left[\mu\left(\nabla\times\vec{V}\right)\right] + \nabla\left[(\lambda+2\mu)\nabla\bullet\vec{V}\right]$$

(2.123)

where λ is the second coefficient of viscosity ($\lambda = -2\mu/3$ for a monatomic gas).

Tensor notation:

$$\rho\frac{DU_i}{Dt} = \rho\left[\frac{\partial U_i}{\partial t} + U_j\frac{\partial U_i}{\partial x_j}\right] = -\frac{\partial P_s}{\partial x_i} + F_i + \frac{\partial}{\partial x_j}\left[\mu\left(\frac{\partial U_i}{\partial x_j} + \frac{\partial U_j}{\partial x_i}\right)\right] + \frac{\partial}{\partial x_i}\left(\lambda\frac{\partial U_m}{\partial x_m}\right).$$

(2.124)

As discussed in the foregoing, Equation 2.213 or 2.214 can only be solved numerically using the state-of the-art CFD methods where variable density and viscosity pose little additional difficulty. Although most practical flows are turbulent, closed-form analytical solutions of these equations are possible only for laminar flows with constant density and viscosity. These flows, however, provide useful limiting cases and are worth pursuing for the following reasons: (a) key flow features under rotation are sufficiently captured, enhancing the intuitive understanding of such flows, which are not a part of our day-to-day experience, (b) these solutions provide a basis for what to expect for variable density and viscosity or when the flow turns turbulent, and (c) these solutions are ideal for CFD validation for numerical accuracy; any mismatch of the corresponding turbulent flow prediction with the bench-mark quality measurements can then be fully attributed to deficiencies in physical models, including turbulence models.

For constant density and viscosity, Equations 2.123 and 124 reduce to the following equations:

Vector notation:

$$\rho\frac{D\vec{V}}{Dt} = \rho\left[\frac{\partial\vec{V}}{\partial t} + \left(\vec{V}\bullet\nabla\right)\vec{V}\right] = -\nabla P_s + \vec{F} + \mu\nabla^2\vec{V}. \qquad (2.125)$$

Tensor notation:

$$\rho\frac{DU_i}{Dt} = \rho\left[\frac{\partial U_i}{\partial t} + U_j\frac{\partial U_i}{\partial x_j}\right] = -\frac{\partial P_s}{\partial x_i} + F_i + \mu\frac{\partial^2 U_i}{\partial x_j\partial x_j}. \qquad (2.126)$$

Note that, in conformity with the Newton's second law of motion, Equations 2.123, 2.124, 2.125, and 2.126 are valid only in an inertial (stationary, nonaccelerating, or nonrotating) reference frame.

2.2.9.1 Inertial Reference Frame: Cartesian Coordinates

In this coordinate system, the velocity $\vec{V}$ (also called absolute velocity) has components $V_x, V_y,$ and V_z in $x, y,$ and z coordinate directions, respectively. Equation 2.215 in each coordinate direction and the continuity equation are written as follows:

x-coordinate direction:

$$\rho\left[\frac{\partial V_x}{\partial t} + V_x\frac{\partial V_x}{\partial x} + V_y\frac{\partial V_x}{\partial y} + V_z\frac{\partial V_x}{\partial z}\right] = -\frac{\partial P_s}{\partial x} + F_x + \mu\left[\frac{\partial^2 V_x}{\partial x^2} + \frac{\partial^2 V_x}{\partial y^2} + \frac{\partial^2 V_x}{\partial z^2}\right] \qquad (2.127)$$

y-coordinate direction:

$$\rho\left[\frac{\partial V_y}{\partial t} + V_x\frac{\partial V_y}{\partial x} + V_y\frac{\partial V_y}{\partial y} + V_z\frac{\partial V_y}{\partial z}\right] = -\frac{\partial P_s}{\partial y} + F_y + \mu\left[\frac{\partial^2 V_y}{\partial x^2} + \frac{\partial^2 V_y}{\partial y^2} + \frac{\partial^2 V_y}{\partial z^2}\right] \qquad (2.128)$$

z-coordinate direction:

$$\rho\left[\frac{\partial V_z}{\partial t} + V_x\frac{\partial V_z}{\partial x} + V_y\frac{\partial V_z}{\partial y} + V_z\frac{\partial V_z}{\partial z}\right] = -\frac{\partial P_s}{\partial z} + F_z + \mu\left[\frac{\partial^2 V_z}{\partial x^2} + \frac{\partial^2 V_z}{\partial y^2} + \frac{\partial^2 V_z}{\partial z^2}\right] \qquad (2.129)$$

Continuity equation:

$$\frac{\partial V_x}{\partial x} + \frac{\partial V_y}{\partial y} + \frac{\partial V_z}{\partial z} = 0 \qquad (2.130)$$

2.2.9.2 Inertial Reference Frame: Cylindrical Coordinates

In this coordinate system, the velocity $\vec{V}$ (also called absolute velocity) has components $V_r, V_\theta,$ and V_x in $r, \theta,$ and x(axial) coordinate directions respectively, as shown in

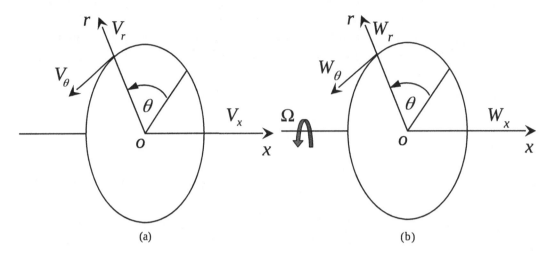

Figure 2.13 (a) Inertial cylindrical coordinate system and (b) rotating (noninertial) cylindrical coordinate system.

Figure 2.13a. Equation 2.215 in each coordinate direction and the continuity equation are written as follows:

r-coordinate direction:

$$\rho\left[\frac{\partial V_r}{\partial t} + V_r\frac{\partial V_r}{\partial r} + \frac{V_\theta}{r}\frac{\partial V_r}{\partial \theta} + V_x\frac{\partial V_r}{\partial x} - \frac{V_\theta^2}{r}\right]$$

$$= -\frac{\partial P_s}{\partial r} + F_r + \mu\left[\frac{\partial^2 V_r}{\partial r^2} + \frac{1}{r}\frac{\partial V_r}{\partial r} + \frac{1}{r^2}\frac{\partial^2 V_r}{\partial \theta^2} - \frac{2}{r^2}\frac{\partial V_\theta}{\partial \theta} + \frac{\partial^2 V_r}{\partial x^2} - \frac{V_r}{r^2}\right] \qquad (2.131)$$

θ-coordinate direction:

$$\rho\left[\frac{\partial V_\theta}{\partial t} + V_r\frac{\partial V_\theta}{\partial r} + \frac{V_\theta}{r}\frac{\partial V_\theta}{\partial \theta} + V_x\frac{\partial V_\theta}{\partial x} + \frac{V_r V_\theta}{r}\right]$$

$$= -\frac{1}{r}\frac{\partial P_s}{\partial \theta} + F_\theta + \mu\left[\frac{\partial^2 V_\theta}{\partial r^2} + \frac{1}{r}\frac{\partial V_\theta}{\partial r} + \frac{1}{r^2}\frac{\partial^2 V_r}{\partial \theta^2} + \frac{2}{r^2}\frac{\partial V_r}{\partial \theta} + \frac{\partial^2 V_\theta}{\partial x^2} - \frac{V_\theta}{r^2}\right] \qquad (2.132)$$

x-coordinate direction:

$$\rho\left[\frac{\partial V_x}{\partial t} + V_r\frac{\partial V_x}{\partial r} + \frac{V_\theta}{r}\frac{\partial V_x}{\partial \theta} + V_x\frac{\partial V_x}{\partial x}\right]$$

$$= -\frac{\partial P_s}{\partial x} + F_x + \mu\left[\frac{\partial^2 V_x}{\partial r^2} + \frac{1}{r}\frac{\partial V_x}{\partial r} + \frac{1}{r^2}\frac{\partial^2 V_x}{\partial \theta^2} + \frac{\partial^2 V_x}{\partial x^2}\right] \qquad (2.133)$$

Continuity equation:

$$\frac{\partial V_r}{\partial r} + \frac{V_r}{r} + \frac{1}{r}\frac{\partial V_\theta}{\partial \theta} + \frac{\partial V_x}{\partial x} = 0 \qquad (2.134)$$

Note that the last term $\left(-V_\theta^2/r\right)$ within brackets on the left-hand side of Equation 2.131 is the centripetal acceleration (negative of the centrifugal acceleration), which results from the flow tangential velocity. For $V_r \ll V_\theta, V_x \ll V_\theta, F_r = 0$, and away from the viscosity dominated region, Equation 2.131 reduces to $\partial P_s / \partial r = \rho V_\theta^2 / r$, which is known as the radial equilibrium equation for a rotating flow.

2.2.9.3 Noninertial Reference Frame: Cylindrical Coordinates with Constant Rotation

Let us now consider a cylindrical coordinate system with its coordinate axes rotating with constant $\vec{\Omega}$ along an arbitrary axis of rotation, as shown in Figure 2.13b for constant rotation around the x axis. In this coordinate system, which is noninertial, the flow velocity $\vec{W}$ (also called relative velocity) has components W_r, W_θ, and W_x in r, θ, and x(axial) coordinate directions, respectively. The relative velocity $\vec{W}$ is related to the absolute velocity $\vec{V}$ by the following equation:

$$\vec{V} = \vec{W} + \vec{\Omega} \times \vec{r} \qquad (2.135)$$

where $\vec{r}$ is the position vector in the rotating reference frame.

Centrifugal and Coriolis forces. Using Equation 2.135, the N–S equations (Equation 2.125) in absolute velocity can be easily converted (see Greitzer, Tan, and Graf (2004) for details) into the corresponding equations in relative velocity of the rotating coordinate system

$$\rho \frac{D\vec{W}}{Dt} + \rho \, \vec{\Omega} \times \left(\vec{\Omega} \times \vec{r} \right) + 2\rho \, \vec{\Omega} \times \vec{W}$$

$$= \rho \left[\frac{\partial \vec{W}}{\partial t} + \left(\vec{W} \bullet \nabla \right) \vec{W} \right] + \rho \, \vec{\Omega} \times \left(\vec{\Omega} \times \vec{r} \right) + 2\rho \, \vec{\Omega} \times \vec{W}$$

$$= -\nabla P_s + \vec{F} + \mu \nabla^2 \vec{W} . \qquad (2.136)$$

Transferring $\rho \, \vec{\Omega} \times \left(\vec{\Omega} \times \vec{r} \right)$ and $2\rho \, \vec{\Omega} \times \vec{W}$ to the right-hand side of Equation 2.136, we obtain

$$\rho \frac{D\vec{W}}{Dt} = \rho \left[\frac{\partial \vec{W}}{\partial t} + \left(\vec{W} \bullet \nabla \right) \vec{W} \right]$$

$$= -\nabla P_s - \rho \, \vec{\Omega} \times \left(\vec{\Omega} \times \vec{r} \right) - 2\rho \vec{\Omega} \times \vec{W} + \vec{F} + \mu \nabla^2 \vec{W} . \qquad (2.137)$$

Note that the left-hand side of Equation 2.137, representing the substantial derivative of the relative velocity $\vec{W}$, is identical in form to the left-hand side of Equation 2.125, which uses the absolute velocity $\vec{V}$. Further note that the pressure forces, viscous forces, and $\vec{F}$ on the right-hand side of Equation 2.137 also take the same form as in Equation 2.125.

On the right-hand side of Equation 2.137, $\rho \vec{\Omega} \times (\vec{\Omega} \times \vec{r})$ and $2\rho \vec{\Omega} \times \vec{W}$ are the centrifugal force and Coriolis force per unit volume, respectively. Because these forces have no resemblance to the forces that appear in this equation, they are sometimes referred to as the "fictitious forces" for an observer in the inertial reference frame. These forces, however, are real for an observer in the rotating (noninertial) reference frame.

They simply arise from the kinematics of going from rotating reference frame, which is noninertial, to an inertial reference frame in which the Newton's laws of motion are valid. Note that the vector that results from $\rho \vec{\Omega} \times (\vec{\Omega} \times \vec{r})$ is co-linear with $\vec{r}$ and points toward the origin (centripetal). Any relative velocity component in the direction of $\vec{\Omega}$ will have no contribution to the Coriolis force ($2\rho \vec{\Omega} \times \vec{W}$). Only the relative velocities in a plane normal to the direction of $\vec{\Omega}$ will contribute to this force.

Let us now see how we can express the centrifugal force $\rho \vec{\Omega} \times (\vec{\Omega} \times \vec{r})$ as the gradient of a scalar quantity, just as the conservative gravitation force can be expressed as the gradient of a potential function. Without loss of generality, let the x axis in our rotating coordinate system coincide with the rotational velocity vector $\vec{\Omega}$. Then the radius of rotation will be measured off this axis, and $(\vec{\Omega} \times \vec{r})$ can be written as $(\vec{\Omega} \times r\hat{e}_r)$, where $\hat{e}_r$ is the unit vector in the radial direction, because the component of the displacement vector $\vec{r}$ along the direction of $\vec{\Omega}$ will have no contribution to the cross product. Thus, we obtain $\rho \vec{\Omega} \times (\vec{\Omega} \times \vec{r}) = \rho \vec{\Omega} \times (\vec{\Omega} \times r\hat{e}_r)$. Using the vector identity $\vec{A} \times (\vec{B} \times \vec{C}) = \vec{B}(\vec{A} \bullet \vec{C}) - \vec{C}(\vec{A} \bullet \vec{B})$, we can write $\rho \vec{\Omega} \times (\vec{\Omega} \times \vec{r}) = \rho \vec{\Omega} \times (\vec{\Omega} \times r\hat{e}_r) = -\rho r\Omega^2 \hat{e}_r$, which can be further expressed as $\nabla(-\rho r^2 \Omega^2/2)$. Using this result in Equation 2.137 and defining a reduced static pressure $\tilde{P}_s = P_s - \rho r^2 \Omega^2/2$, we can alternatively write this equation as

$$\rho \frac{D\vec{W}}{Dt} = \rho \left[\frac{\partial \vec{W}}{\partial t} + \left(\vec{W} \bullet \nabla \right) \vec{W} \right] = -\nabla \tilde{P}_s - 2\rho \vec{\Omega} \times \vec{W} + \vec{F} + \mu \nabla^2 \vec{W} \qquad (2.138)$$

which reveals that the net effect of the centrifugal force as a result of rotation of the coordinate system is to alter the static pressure field to the extent needed to balance this centrifugal force.

A cylindrical coordinate system rotating with constant Ω about the x axis, which could be thought of as the gas turbine axis of rotation, is shown in Figure 2.13b. For this case, Equation 2.137 in each coordinate direction and the continuity equation can be written as follows:

r-coordinate direction:

$$\rho \left[\frac{\partial W_r}{\partial t} + W_r \frac{\partial W_r}{\partial r} + \frac{W_\theta}{r} \frac{\partial W_r}{\partial \theta} + W_x \frac{\partial W_r}{\partial x} - \frac{W_\theta^2}{r} \right]$$

$$= -\frac{\partial P_s}{\partial r} + \rho \Omega^2 r + 2\rho \Omega W_\theta + F_r$$

$$+ \mu \left[\frac{\partial^2 W_r}{\partial r^2} + \frac{1}{r} \frac{\partial W_r}{\partial r} + \frac{1}{r^2} \frac{\partial^2 W_r}{\partial \theta^2} - \frac{2}{r^2} \frac{\partial W_\theta}{\partial \theta} + \frac{\partial^2 W_r}{\partial x^2} - \frac{W_r}{r^2} \right] \qquad (2.139)$$

θ-coordinate direction:

$$\rho \left[\frac{\partial W_\theta}{\partial t} + W_r \frac{\partial W_\theta}{\partial r} + \frac{W_\theta}{r} \frac{\partial W_\theta}{\partial \theta} + W_x \frac{\partial W_\theta}{\partial x} + \frac{W_r W_\theta}{r} \right]$$

$$= -\frac{1}{r} \frac{\partial P_s}{\partial \theta} - 2\rho \Omega W_r + F_\theta$$

$$+ \mu \left[\frac{\partial^2 W_\theta}{\partial r^2} + \frac{1}{r} \frac{\partial W_\theta}{\partial r} + \frac{1}{r^2} \frac{\partial^2 W_r}{\partial \theta^2} + \frac{2}{r^2} \frac{\partial W_r}{\partial \theta} + \frac{\partial^2 W_\theta}{\partial x^2} - \frac{W_\theta}{r^2} \right] \qquad (2.140)$$

x-coordinate direction:

$$\rho\left[\frac{\partial W_x}{\partial t}+W_r\frac{\partial W_x}{\partial r}+\frac{W_\theta}{r}\frac{\partial W_x}{\partial \theta}+W_x\frac{\partial W_x}{\partial x}\right]$$

$$=-\frac{\partial P_s}{\partial x}+F_x+\mu\left[\frac{\partial^2 W_x}{\partial r^2}+\frac{1}{r}\frac{\partial W_x}{\partial r}+\frac{1}{r^2}\frac{\partial^2 W_x}{\partial \theta^2}+\frac{\partial^2 W_x}{\partial x^2}\right] \qquad (2.141)$$

Continuity equation:

$$\frac{\partial W_r}{\partial r}+\frac{W_r}{r}+\frac{1}{r}\frac{\partial W_\theta}{\partial \theta}+\frac{\partial W_x}{\partial x}=0 \qquad (2.142)$$

In the radial direction, according to Equation 2.139, we have two centrifugal force terms $\rho W_\theta^2/r$ and $\rho \Omega^2 r$, the first results from the relative tangential velocity W_θ of the flow, and the second from the fact that the coordinate system is rotating with constant Ω. Note that the Coriolis force has two components. The component $2\rho\Omega W_\theta$ is in the radial direction and $(-2\rho\Omega W_r)$ in the tangential direction (Equation 2.139).

The centrifugal and Coriolis forces are real forces in a rotating noninertial reference frame. We often experience centrifugal forces while driving on a curve or when a car skids on an icy road with a sharp turn. If you are standing on a counterclockwise rotating platform, such as a merry-go-round in a theme park, you will experience a centrifugal force trying to push you radially outward. If the platform is slippery or icy, you will soon slip to the outer edge of the platform, but not exactly along the radial direction from where you started. As you are slipping radially outward, you will gradually turn to your right as a result of the Coriolis force.

Consider a steady, two-dimensional, incompressible, inviscid flow from a source located at the origin ($r=0$). For an observer in an inertial reference frame, the flow will have radial streamlines emanating from the origin with the absolute radial velocity V_r, decaying inversely with radius to satisfy continuity. The static pressure gradient $\partial P_s/\partial r$ in this flow will equal $\left(-\rho V_r \partial V_r/\partial r\right)$. This pressure gradient is independent of the reference frame. The kinematics of this source flow will appear different to an observer in a reference frame rotating counterclockwise at an angular velocity Ω. This observer will see the flow having two velocities in the plane, one radial $W_r=V_r$ and the other tangential $W_\theta=-r\Omega$, spiraling away from the origin at lower radii and assuming circular streamlines for $V_r \ll W_\theta$. Because the pressure gradient at each radial location also equals $\left(-\rho W_r \partial W_r/\partial r\right)$, we find that the additional centrifugal force term due to the tangential velocity (W_θ), $\rho W_\theta^2/r=\rho r\Omega^2$, together with the centrifugal force $\rho r\Omega^2$ as a result of coordinate system rotation cancels the Coriolis force ($2\rho\Omega W_\theta=-2\rho r\Omega^2$).

The existence of the Coriolis and centripetal accelerations as a result of coordinate system rotation is easily demonstrated for a vortex flow with circular streamlines and circular isobars. For a streamline of radius r with velocity V_θ in this flow, the centripetal acceleration V_θ^2/r in the inertial reference frame is driven by the static pressure gradient $\partial P_s/\partial r$; recall the local equilibrium equation $\partial P_s/\partial r=\rho V_\theta^2/r$. Substituting $V_\theta=W_\theta+r\Omega$ for the rotating reference frame, we obtain $V_\theta^2/r=W_\theta^2/r+2\Omega W_\theta+r\Omega^2$, which shows that the centripetal acceleration W_θ^2/r in the rotating reference frame is not

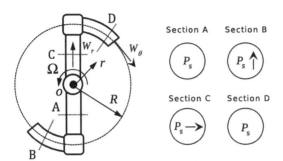

Figure 2.14 A sprinkler system featuring centrifugal and Coriolis forces.

sufficient to yield the required static pressure gradient, which is independent of the reference frame. The Coriolis acceleration $2\Omega W_\theta$ and the centripetal acceleration $r\Omega^2$ as a result of coordinate system rotation are needed to make up for the difference.

To develop further insight into the centrifugal and Coriolis forces, let us consider a frictionless water sprinkler shown in Figure 2.14. Only one sprinkler arm is shown flowing, the other is blanked off. Each arm is made of a straight pipe in the radial direction and a curved pipe in the form of a circular arc, having the radius of curvature equal to that of the circle traversed by the sprinkler. For the sprinkler, the maximum angular velocity is given by $\Omega = W_\theta/R$, where W_θ is the constant relative tangential velocity in the curved pipe. The radial pipe has constant relative radial flow velocity W_r and has no tangential and axial velocities.

In the blanked-off arm of the sprinkler, water is in a rigid-body rotation. Because all relative velocity components are zero in this arm, there are no Coriolis forces. However, the centrifugal force in this arm corresponds to $\rho\Omega^2 r$, increasing linearly with radius. As a result, we have uniform static pressure at each section of the straight pipe, but radially increasing static pressure distribution in each section of the curved pipe, as shown in Figure 2.14 for sections A and B.

In the rotating sprinkler arm that is flowing, the straight pipe features a radial velocity W_r and will generate a uniform Coriolis force $(-2\rho\Omega W_r)$ in the negative tangential direction, causing a static pressure gradient in the direction shown for section C in Figure 2.14. The flow features in the curved pipe with flow are quite interesting. In this rotating pipe, we have no radial static pressure gradient at any section, as shown for section D in Figure 2.14. How do we explain this apparently nonintuitive flow behavior? The answer lies in finding the centrifugal and Coriolis forces in this pipe. Because the absolute flow velocity in this pipe is zero ($V_\theta = R\Omega - W_\theta = 0$), the centrifugal force $\rho V_\theta^2/R = 0$, resulting in a radially uniform static pressure distribution. Now, let us look at the centrifugal and Coriolis forces in the rotating reference frame attached to the sprinkler. The centrifugal force from W_θ equals $\rho W_\theta^2/R$ and that from Ω equals $\rho\Omega^2 R$. Because $\Omega = W_\theta/R$, both these forces are equal, giving $2\rho\Omega^2 R$ as the total centrifugal force. The Coriolis force in this curve pipe can be computed as $2\rho\Omega(-W_\theta) = -2\rho\Omega^2 R$, which is opposite and equal to the centrifugal force, yielding a net zero force, which we concluded from using the absolute velocity (inertial reference frame).

2.2.9.4   Rotating Couette Flow

In this section, we demonstrate a procedure to obtain exact solutions of the Navier-Stokes equations governing a steady incompressible laminar flow between concentric cylinders of radii R_1 and R_2, as shown in Figure 2.15. Among other applications, like CFD validation, the solutions offer a basis for measuring fluid viscosity by a device. In this case, the inner cylinder is rotating at a constant angular velocity Ω_1, and the outer cylinder with Ω_2, with no imposed pressure gradient and zero flow in the axial direction ($V_x = 0$).

The solutions developed here are based on the following assumptions for the flow being: (i) axisymmetric ($\partial/\partial\theta = 0$), (ii) fully developed ($\partial/\partial x = 0$), and (iii) with zero body force as a result of gravity.

Continuity equation in cylindrical coordinates. The continuity equation, Equation 2.134, in the inertial cylindrical polar coordinates is expressed as

$$\frac{\partial V_r}{\partial r} + \frac{V_r}{r} + \frac{1}{r}\frac{\partial V_\theta}{\partial \theta} + \frac{\partial V_x}{\partial x} = 0$$

$$\frac{1}{r}\frac{\partial(rV_r)}{\partial r} + \frac{1}{r}\frac{\partial V_\theta}{\partial \theta} + \frac{\partial V_x}{\partial x} = 0$$

Using $V_x = \partial V_\theta/\partial\theta = 0$ from the assumptions, this equation reduces to

$$\frac{\partial V_r}{\partial r} + \frac{V_r}{r} = \frac{\partial(rV_r)}{\partial r} = 0$$

whose integration yields $rV_r = C$ in r direction, where C is the integration constant. Because $V_r = 0$ at the inner cylinder wall at $r = R_1$, it must be zero everywhere in the flow, giving $C = 0$.

Reduced Navier-Stokes Equations in cylindrical coordinates. With $V_x = 0$, the x-momentum equation, Equation 2.133, is identically satisfied. Under the assumptions made in this problem and $V_r = 0$ from the continuity equation, the r-momentum equation, Equation 2.131, reduces to

$$\frac{dP_s}{dr} = \frac{\rho V_\theta^2}{r}, \tag{2.143}$$

which governs the radial distribution of the static pressure between two cylinders.

The θ-momentum equation, Equation 2.132, reduces to

$$\frac{d^2 V_\theta}{dr^2} + \frac{1}{r}\frac{dV_\theta}{dr} - \frac{V_\theta}{r^2} = 0,$$

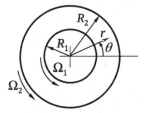

Figure 2.15 Couette flow between two concentric, rotating cylinders.

which can be alternatively written as

$$\frac{d}{dr}\left[\frac{1}{r}\frac{d}{dr}(rV_\theta)\right] = 0. \tag{2.144}$$

Equation 2.144 is a homogeneous second-order ordinary differential equation and requires two boundary conditions for a unique solution. These two boundary conditions in this case correspond to no-slip condition at the walls of the inner and outer cylinders, namely, $V_\theta = R_1\Omega_1 @ r = R_1$ and $V_\theta = R_2\Omega_2 @ r = R_2$.

Tangential velocity distribution. Integrating Equation 2.144 once, we obtain

$$\frac{1}{r}\frac{d}{dr}(rV_\theta) = \tilde{C}$$

$$\frac{d}{dr}(rV_\theta) = r\tilde{C}.$$

Integrating this equation once more yields

$$rV_\theta = r^2\tilde{C}/2 + C_2 = C_1 r^2 + C_2$$

where $C_1 = \tilde{C}/2$. Thus, the general solution for the distribution of V_θ between the cylinders is given by

$$V_\theta = C_1 r + C_2/r \tag{2.145}$$

where C_1 and C_2 are the integration constants, which are to be determined from the boundary conditions at the inner and outer cylinders, yielding the following two equation:

$$C_1 R_1 + C_2/R_1 = R_1\Omega_1$$

$$C_1 R_2 + C_2/R_2 = R_2\Omega_2$$

Solving the aforementioned two simultaneous algebraic equations yields:

$$C_1 = \frac{R_2^2\Omega_2 - R_1^2\Omega_1}{R_2^2 - R_1^2} \tag{2.146}$$

and

$$C_2 = \frac{R_1^2 R_2^2(\Omega_1 - \Omega_2)}{R_2^2 - R_1^2} \tag{2.147}$$

Substituting Equations 2.146 and 2.147 into Equation 2.145, we finally obtain

$$V_\theta = \frac{1}{R_2^2 - R_1^2}\left[(R_2^2\Omega_2 - R_1^2\Omega_1)r + \frac{R_1^2 R_2^2}{r}(\Omega_1 - \Omega_2)\right] \tag{2.148}$$

Static pressure distribution. Substituting for V_θ from Equation 2.145 into Equation 2.143, we obtain

$$\frac{dP_s}{dr} = \rho\left(C_1^2 r + \frac{2C_1 C_2}{r} + \frac{C_2^2}{r^3}\right),$$

which upon integration and substitution for C_1 and C_2 from Equations 2.146 and 2.147, respectively, finally yields

$$P_s - P_{s_1} = \frac{\rho}{\left(R_2^2 - R_1^2\right)^2} \left[\frac{\left(R_2^2\Omega_2 - R_1^2\Omega_1\right)^2 \left(r^2 - R_1^2\right)}{2} \right.$$
$$\left. +2R_1^2 R_2^2 \left(R_2^2\Omega_2 - R_1^2\Omega_1\right)\left(\Omega_1 - \Omega_2\right) \ln \frac{r}{R_1} - \frac{R_1^4 R_2^4}{2}\left(\Omega_1 - \Omega_2\right)^2 \left(\frac{1}{r^2} - \frac{1}{r_1^2}\right) \right]$$

$$(2.149)$$

where P_{s_1} is the static pressure at the surface of the inner cylinder.

Torques on inner and outer cylinders. The torque on each cylinder, per unit length in the axial direction, can be computed as the product of the wall shear stress, cylinder circumference, and radius. Knowing the distribution of velocity V_θ from Equation 2.148, we can evaluate the tangential shear stress distribution in the flow between the two cylinders by the equation

$$\tau_{r\theta} = \mu r \frac{\mathrm{d}}{\mathrm{d}r}\left(\frac{V_\theta}{r}\right)$$

Substituting for V_θ yields

$$\tau_{r\theta} = \frac{2\mu R_1^2 R_2^2 \left(\Omega_2 - \Omega_1\right)}{r^2 \left(R_2^2 - R_1^2\right)}$$

$$(2.150)$$

For $r = R_1$, Equation 2.150 yields the torque on the inner cylinder

$$\Gamma_1 = \frac{4\pi\mu R_1^2 R_2^2 \left(\Omega_2 - \Omega_1\right)}{\left(R_2^2 - R_1^2\right)}.$$

Similarly, for $r = R_2$, Equation 2.150 yields the torque on the outer cylinder

$$\Gamma_2 = -\frac{4\pi\mu R_1^2 R_2^2 \left(\Omega_2 - \Omega_1\right)}{\left(R_2^2 - R_1^2\right)}.$$

Note that the torques Γ_1 and Γ_2 on the two cylinders are equal and opposite, implying zero net torque acting on the fluid; as a result, its angular momentum remains constant.

2.2.9.5 Taylor-Proudman Theorem

Taylor-Proudman theorem embodies the behavior of a steady, incompressible, laminar flow that is dominated by rotation (Coriolis forces). To understand this theorem, let us consider the Navier-Stokes equation, Equation 2.138. When we use the velocity scale W_{ref} and length scale L_{ref} to nondimensionalize this equation, two dimensionless numbers emerge. The first is called the Rossby number $Ro = W_{\mathrm{ref}}/\Omega L_{\mathrm{ref}}$ and the second, the Ekman number $Ek = \mu/\rho\Omega L_{\mathrm{ref}}^2$. While the Rossby number is a measure of the inertia force associated with the relative velocity in comparison to the Coriolis forces, the Ekman number is the ratio of viscous forces to Coriolis forces. For a steady incompressible flow outside a wall boundary layer and with $Ro \ll 1$ and $Ek \ll 1$, that is, the Coriolis forces dominate over the inertia and viscous forces, Equation 2.138 reduces to

$$2\rho \vec{\Omega} \times \vec{W} = -\nabla \tilde{P}_s. \tag{2.151}$$

The flows that are governed by Equation 2.151 are known as geostrophic flows in which the modified (by including the potential function associated with the centrifugal forces as a result of the rotation of the coordinate axes) static pressure gradients are balanced by the Coriolis forces.

Taking the curl of both sides of Equation 2.151 and noting that the curl of the gradient of a scalar function is identically zero, we obtain

$$\nabla \times (\vec{\Omega} \times \vec{W}) = 0. \tag{2.152}$$

Using the vector identity $\nabla \times (\vec{A} \times \vec{B}) = (\vec{B} \bullet \nabla)\vec{A} - \vec{B}(\nabla \bullet \vec{A}) - (\vec{A} \bullet \nabla)\vec{B} + \vec{A}(\nabla \bullet \vec{B})$ and invoking the continuity ($\nabla \bullet \vec{W} = 0$) and the fact that $\vec{\Omega}$ is constant, we finally obtain from Equation 2.152

$$(\vec{\Omega} \bullet \nabla)\vec{W} = 0. \tag{2.153}$$

Assuming the direction of $\vec{\Omega}$ as the axis of rotation along the x direction, Equation 2.153 yields $\partial \vec{W}/\partial x = 0$, which indicates that all three components of the relative velocity do not change in the direction parallel to the axis of rotation – the statement of the Taylor-Proudman theorem. In a flow system with solid boundaries perpendicular to the axis of rotation, we have $W_x = 0$ at some specified value of x. The Taylor-Proudman theorem implies that this flow system will remain entirely two-dimensional in planes perpendicular to the axis of rotation.

Taylor columns. Taylor columns are a unique and interesting manifestation of the Taylor-Proudman theorem. These columns occur when there is relative motion between an obstacle and fluid in a rotating system dominated by rotation. To illustrate a Taylor column, let us consider a rotating cylinder filled with liquid. We have a short cylinder attached to the bottom surface of the cylinder. If you rotate this system at some angular velocity Ω, eventually the liquid will attain solid-body rotation with all the solid surfaces. Now, if we reduce the cylinder angular velocity slightly, the liquid will continue to rotate at the old angular velocity and develop a relative velocity ($\Omega_{\text{liquid}} > \Omega_{\text{cylinder}}$) such that the short cylinder at the bottom of the rotating cylinder is rendered into cross-flow with the surrounding fluid, as shown in Figure 2.16.

In the flow system shown in Figure 2.16, as a result of strong rotation, there is no flow on the top of the short cylinder. According to the Taylor-Proudman theorem, the velocity components will not change in the direction of the axis of rotation, the x direction in this case. As a result, a solid column of liquid develops on the top of the short cylinder, as if the cylinder is extended up to the free surface of the liquid in the rotating cylinder. In planes perpendicular to the x axis, we have identical two-dimensional velocity distributions, as shown in the figure.

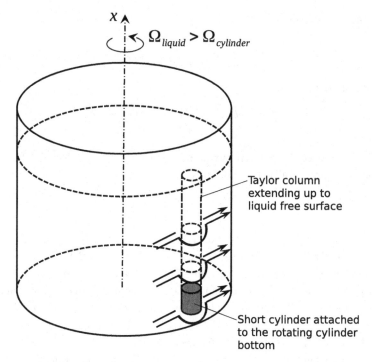

$\Omega_{liquid} > \Omega_{cylinder}$

Taylor column
extending up to
liquid free surface

Short cylinder attached
to the rotating cylinder
bottom

Figure 2.16 Taylor column.

2.2.9.6 Ekman Boundary Layers

Rotor surfaces associated with rotor disks and rotor cavities abound in gas turbines. Boundary layers formed on these surfaces are generally known as Ekman layers. These layers feature some unique characteristics not found in a typical boundary layer such as the one developed on a flat plate under zero pressure gradient where wall shear stress reduces momentum in the flow direction, resulting in a continuously (growing) thickening boundary layer. In contrast, the Ekman layer does not entrain fluid from the region outside the boundary layer and consequently maintains a constant thickness. For low Rossby numbers, where the Ekman boundary layer is sandwiched between the geostrophic flow and the rotating wall, we obtain a set of linear boundary layer equations with closed-form analytical solutions for the distributions of velocities in the boundary layer. These solutions, systematically developed here, provide further insight into an Ekman layer.

Let us consider a rotating disk whose axis of rotation is along the x-axis. For an axisymmetric ($\partial/\partial\theta = 0$), steady, incompressible, laminar flow with constant viscosity, no body forces, and $Ro \ll 1$, the Navier-Stokes equations (Equations 2.139, 2.140, and 2.141) reduce to:

r-momentum equation:

$$-2\rho\Omega W_\theta = -\frac{\partial \tilde{P}_s}{\partial r} + \mu \frac{\partial}{\partial r}\left[\frac{1}{r}\frac{\partial}{\partial r}\left(rW_r\right)\right] + \mu \frac{\partial^2 W_r}{\partial x^2} \qquad (2.154)$$

θ-momentum equation:

$$2\rho\Omega W_r = \mu\frac{\partial}{\partial r}\left[\frac{1}{r}\frac{\partial}{\partial r}(rW_\theta)\right] + \mu\frac{\partial^2 W_\theta}{\partial x^2} \tag{2.155}$$

x-momentum equation:

$$0 = -\frac{\partial\tilde{P}_s}{\partial x} + \mu\left[\frac{\partial^2 W_x}{\partial r^2} + \frac{1}{r}\frac{\partial W_x}{\partial r} + \frac{\partial^2 W_x}{\partial x^2}\right] \tag{2.156}$$

where we have used the reduced pressure given by $\tilde{P}_s = P_s - \rho r^2\Omega^2/2$. The continuity equation (Equation 2.142) also simplifies to

$$\frac{1}{r}\frac{\partial}{\partial r}(rW_r) + \frac{\partial W_x}{\partial x} = 0 \tag{2.157}$$

The flow field outside the disk boundary layer (Ekman layer) is assumed geostrophic ($Ro \ll 1$ and $Ek \ll 1$) with relative velocities $W_{r_\infty}, W_{\theta_\infty}$, and W_{x_∞}. From the Taylor-Proudman theorem, discussed in Section 2.2.9.5, we know that these velocities do not vary along the x direction. From Equation 2.155, without the viscous terms on the right-hand side, we obtain $W_{r_\infty} = 0$ which should prevail at the outer edge of the Ekman layer. Substituting $W_{r_\infty} = 0$ in Equation 2.157 simply yields $\partial W_x/\partial x = 0$, which is also obtained from the Taylor-Proudman theorem.

Under the assumption $\partial(rW_r)/\partial r = 0$ within the Ekman layer, Equation 2.157 yields $\partial W_x/\partial x = 0$, that is, W_x is constant across the boundary layer. Because, as a result of no slip condition, $W_x = 0$ at the wall, we obtain $W_x = 0$ everywhere in the boundary layer, which implies that the Ekman boundary layer is nonentraining. We further conclude that $W_{x_\infty} = 0$. Substituting $W_x = 0$ in Equation 2.156, we obtain $\partial\tilde{P}_s/\partial x = 0$, that is, the reduced static pressure also does not vary along the rotation axis, both in the Ekman layer and the outside geostrophic flow.

In the geostrophic flow region, Equation 2.154 yields

$$-2\rho\Omega W_{\theta_\infty} = -\frac{\partial\tilde{P}_s}{\partial r}. \tag{2.158}$$

Combining Equations 2.154 and 2.158 and noting that $rW_r = 0$, we obtain the following r-momentum equation in the Ekman boundary layer

$$-2\Omega\left(W_\theta - W_{\theta_\infty}\right) = v\frac{\partial^2 W_r}{\partial x^2} \tag{2.159}$$

where $v = \mu/\rho$ is the kinematic viscosity.

Because

$$\mu\frac{\partial}{\partial r}\left[\frac{1}{r}\frac{\partial}{\partial r}(rW_\theta)\right] \ll \mu\frac{\partial^2 W_\theta}{\partial x^2}$$

in Equation 2.155, we can finally write the θ-momentum equation in the Ekman boundary layer as

$$2\Omega W_r = v\frac{\partial^2 W_\theta}{\partial x^2} \tag{2.160}$$

Equations 2.159 and 2.160 indicate that the Coriolis forces are in balance with the shear forces in an Ekman boundary layer. The relevant boundary conditions to solve these coupled second-order differential equations are as follows:

$$W_r = W_\theta = 0 \text{ at } x = 0 \tag{2.161}$$

$$W_r \to 0 \text{ and } W_\theta \to W_{\theta_\infty} \text{ as } x \to \infty. \tag{2.162}$$

To facilitate the solution of the system of Ekman boundary layer equations and boundary conditions, let us define the following dimensionless variables:

$$\tilde{W}_r = \frac{W_r}{W_{\theta_\infty}} \qquad \tilde{W}_\theta = \frac{W_\theta - W_{\theta_\infty}}{W_{\theta_\infty}} \qquad \zeta = \frac{x}{\sqrt{\nu/\Omega}}.$$

Using these variables, Equations 2.159 and 2.160 can be re-written as

$$\frac{d^2\tilde{W}_r}{d\zeta^2} = -2\tilde{W}_\theta \tag{2.163}$$

and

$$\frac{d^2\tilde{W}_\theta}{d\zeta^2} = 2\tilde{W}_r. \tag{2.164}$$

Let us now define a complex velocity $Z = \tilde{W}_\theta + i\tilde{W}_r$. Multiplying Equation 2.164 by i and adding it to Equation 2.163, we obtain the second-order differential equation

$$\frac{d^2Z}{d\zeta^2} = -2iZ \tag{2.165}$$

with the boundary conditions, which correspond to Equations 2.161 and 2.162, given by

$$Z = -1 \text{ at } \zeta = 0 \tag{2.166}$$

and

$$Z \to 0 \text{ as } x \to \infty \tag{2.167}$$

We can write the general solution of Equation 2.165 as

$$Z = Ae^{-(1-i)\zeta} + Be^{(1-i)\zeta} \tag{2.168}$$

where the coefficients A and B need to be determined from the boundary conditions. From the second boundary condition (Equation 2.167), we obtain $B = 0$, and the first boundary condition (Equation 2.166) yields $A = -1$. Thus, the closed-form analytical solution of Equation 2.165 finally becomes

$$Z = -e^{-(1-i)\zeta} = -e^{-\zeta}(\cos\zeta + i\sin\zeta) \tag{2.169}$$

Equating the real and imaginary parts on both sides of Equation 2.169, we finally obtain

$$\tilde{W}_r = -e^{-\zeta} \sin \zeta$$
$$W_r = -W_{\theta_\infty} e^{-x/\sqrt{v/\Omega}} \sin \left(x/\sqrt{v/\Omega} \right) \tag{2.170}$$

and

$$\tilde{W}_\theta = -e^{-\zeta} \cos \zeta$$
$$W_\theta = W_{\theta_\infty} \left\{ 1 - e^{-x/\sqrt{v/\Omega}} \cos \left(x/\sqrt{v/\Omega} \right) \right\}. \tag{2.171}$$

It is interesting to note from Equation 2.170 that for $W_{\theta_\infty} > 0$, which means that the fluid in the geostrophic region is rotating faster than the wall, the flow in the Ekman boundary layer is radially inward. Conversely, for $W_{\theta_\infty} < 0$, we will have radially outward flow in the Ekman layer.

Distributions of radial and tangential velocities, based on Equations 2.170 and 2.171, are depicted in Figure 2.17 for the case with $W_{\theta_\infty} < 0$. The undershoot of the radial velocity for $\pi < \zeta < 2\pi$, representing a slight flow reversal in the Ekman layer, and the overshoot of the tangential velocity for $\pi/2 < \zeta < 3\pi/2$ are interesting features not found in boundary layers in nonrotating flow systems. In Figure 2.17, the plot of $\tilde{\theta}$, which is the angle the resultant velocity vector in the Ekman boundary layer makes with the tangential velocity, shows that it varies from zero at the edge of the boundary layer to 45° near the wall, exhibiting a spiraling fluid motion, which is known as Ekman spiral.

2.3   Internal Flow

Many of the general concepts of fluid mechanics are discussed in Section 2.2. Some of the special concepts, which are founded in the conservation laws of mass, momentum,

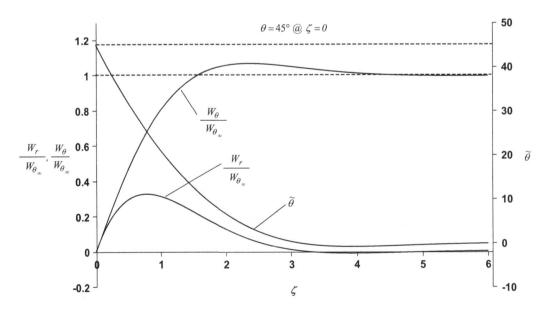

Figure 2.17 Velocity distributions in an Ekman boundary layer.

energy, and entropy and widely used in gas turbine design engineering, are presented in this section.

These concepts include isentropic and nonisentropic vortices, windage, and the relation between the loss coefficient and discharge coefficient for both incompressible and compressible flows. This section also includes the quantitative evaluation of static and total pressures and temperatures in a nonisentropic generalized vortex, which is an important feature of gas turbine internal flow systems.

2.3.1 Isentropic and Nonisentropic Vortices: Pressure and Temperature Changes

2.3.1.1 Isentropic Free Vortex

A free vortex is characterized by constant angular momentum, and it must be free from any applied torque. As such it is seldom found in a gas turbine internal flow system, but the tendency toward a free vortex behavior exits in radial inflows in a rotor cavity. Often, the flowpath designs of turbine blades are based on a free vortex assumption. It is, however, instructive to model a free vortex and understand how pressure and temperature change in this vortex flow.

Like the Euler's turbomachinery equation, discussed in Section 2.2.7, we will formulate here an isentropic free vortex in the inertial reference frame. To see how such a formulation is simpler than the alternative formulation in the rotor reference frame, let us examine the radial equilibrium equation in both reference frames. Converting Equation 2.143 in the rotor reference frame, we obtain

$$\frac{\mathrm{d}P_\mathrm{s}}{\mathrm{d}r} = \frac{\rho V_\theta^2}{r} = \frac{\rho \left(W_\theta + r\Omega \right)^2}{r}$$

$$\frac{\mathrm{d}P_\mathrm{s}}{\mathrm{d}r} = \frac{\rho W_\theta^2}{r} + 2W_\theta \Omega + \rho r \Omega^2. \tag{2.172}$$

Compared to one term, representing the centrifugal force per unit volume on the right-hand side of Equation 2.143, the right-hand side of Equation 2.172 consists of three terms: (1) the centrifugal force as a result of the relative tangential velocity, (2) the Coriolis force, and (3) the centrifugal force as a result of rotation of the rotor reference frame.

It is customary to express the fluid absolute tangential velocity as a fraction of the local tangential velocity of the rotor. Thus, we define the local swirl factor of a vortex flow as

$$S_\mathrm{f} = \frac{V_\theta}{r\Omega}. \tag{2.173}$$

For a free vortex, we can express the swirl factor and tangential velocity, shown in Figure 2.18, by the following equations:

$$V_\theta = \frac{C_1}{r} \tag{2.174}$$

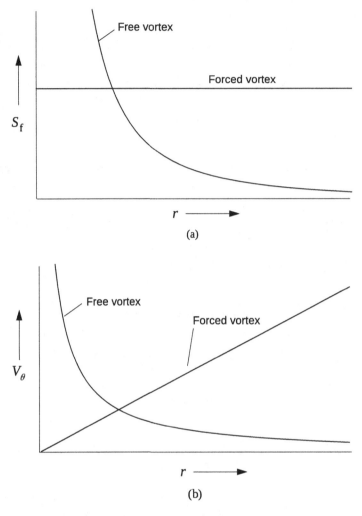

Figure 2.18 (a) Radial variation of swirl factor in free and forced vortices and (b) Radial variation of tangential velocity in free and forced vortices.

and

$$S_f = \frac{C_1}{r^2\Omega} \tag{2.175}$$

where C_1 is a constant related to the vortex strength.

    Static pressure and static temperature variations in an isentropic free vortex. Using the equation of state to replace ρ in the radial equilibrium equation, we obtain

$$\frac{\mathrm{d}P_s}{\mathrm{d}r} = \frac{\rho V_\theta^2}{r} = \frac{P_s V_\theta^2}{RT_s r}$$

$$\frac{\mathrm{d}P_s}{P_s} = \frac{V_\theta^2}{RT_s} \frac{\mathrm{d}r}{r} \tag{2.176}$$

For an isentropic ($ds = 0$) vortex flow, substituting

$$\frac{c_p}{R}\frac{dT_s}{T_s} = \frac{dP_s}{P_s}$$

from Equation 2.15 into Equation 2.176 yields

$$\frac{c_p}{R}\frac{dT_s}{T_s} = \frac{V_\theta^2}{RT_s}\frac{dr}{r}$$

$$dT_s = \frac{V_\theta^2}{c_p}\frac{dr}{r}. \tag{2.177}$$

Substituting for V_θ from Equation 2.174 yields

$$dT_s = \frac{C_1^2}{c_p}\frac{dr}{r^3},$$

which upon integration between states 1 and 2 further yields

$$T_{s_2} - T_{s_1} = \frac{C_1^2}{2c_p}\left(\frac{1}{r_1^2} - \frac{1}{r_2^2}\right) \tag{2.178}$$

and can be easily expressed in the form

$$\frac{T_{s_2}}{T_{s_1}} = 1 + \left(\frac{\kappa - 1}{2}\right)M_{\theta_1}^2\left(\frac{r_2^2}{r_1^2} - 1\right)\frac{r_1^2}{r_2^2} \tag{2.179}$$

where

$$M_{\theta_1} = \frac{V_{\theta_1}}{\sqrt{\kappa R T_{s_1}}} = \frac{C_1}{r\left(\sqrt{\kappa R T_{s_1}}\right)} \tag{2.180}$$

Using the relation between pressure ratio and temperature ratio in an isentropic flow, we can write

$$\frac{P_{s_2}}{P_{s_1}} = \left(\frac{T_{s_2}}{T_{s_1}}\right)^{\frac{\kappa}{\kappa-1}} = \left\{1 + \left(\frac{\kappa-1}{2}\right)M_{\theta_1}^2\left(\frac{r_2^2}{r_1^2} - 1\right)\frac{r_1^2}{r_2^2}\right\}^{\frac{\kappa}{\kappa-1}}. \tag{2.181}$$

Total pressure and total temperature variations in an isentropic free vortex. Knowing the static temperature and velocity components at any point of a free vortex, we can compute the total temperature by the equation

$$T_t = T_s + \frac{V_x^2}{2c_p} + \frac{V_r^2}{2c_p} + \frac{V_\theta^2}{2c_p}$$

If the axial and radial velocities do not change between two points, the aforementioned equation yields

$$T_{t_2} - T_{t_1} = \left(T_{s_2} - T_{s_1} \right) + \left(\frac{V_{\theta_2}^2}{2c_p} - \frac{V_{\theta_1}^2}{2c_p} \right), \tag{2.182}$$

which shows that the change in total temperature between two points consists of two parts: the change in static temperature and the change in dynamic temperature.

Using Equation 2.174, Equation 2.182 becomes

$$T_{t_2} - T_{t_1} = \left(T_{s_2} - T_{s_1} \right) + \frac{C_1^2}{2c_p} \left(\frac{1}{r_2^2} - \frac{1}{r_1^2} \right),$$

which upon substitution from Equation 2.178 reduces to

$$T_{t_2} - T_{t_1} = \left(T_{s_2} - T_{s_1} \right) - \left(T_{s_2} - T_{s_1} \right) = 0. \tag{2.183}$$

Equation 2.183 reveals an interesting, but expected, result that there is no change in total temperature in an isentropic free vortex. In this case, the change in static temperature is opposite and equal to the change in the dynamic temperature resulting from the change in the tangential velocity. Thus, for an isentropic free vortex we obtain $T_{t_2}/T_{t_1} = 1$ and $P_{t_2}/P_{t_1} = 1$.

2.3.1.2   Isentropic Forced Vortex

Like every particle in a solid-body rotation, the fluid in a forced vortex rotates at a constant angular velocity. Thus, unlike a free vortex, the tangential velocity in a forced vortex increases linearly with radius. Figure 2.18 shows the radial variations of the tangential velocity and swirl factor in a forced vortex. The flow in a rotating duct behaves like a forced vortex. With the constant swirl factor ($S_f = C_2$) characterizing a forced vortex, the linear variation of the tangential velocity with radius can be expressed as

$$V_\theta = rS_f\Omega = rC_2\Omega \tag{2.184}$$

where C_2 is a constant representing the vortex strength as a fraction of the rotor angular velocity Ω.

Static pressure and static temperature variations in an isentropic forced vortex. Equation 2.177 is valid for all isentropic vortex flows. For a forced vortex, we use Equation 2.184 to replace V_θ in this equation, giving

$$dT_s = \left(\frac{C_2^2\Omega^2}{c_p} \right) rdr. \tag{2.185}$$

Integration of Equation 2.185 between states 1 and 2 yields

$$T_{s_2} - T_{s_1} = \frac{C_2^2\Omega^2}{2c_p} (r_2^2 - r_1^2), \tag{2.186}$$

which shows that the change in static temperature between any two points of an isentropic forced vortex equals the change in the dynamic temperature at these points. Upon further rearrangement, Equation 2.186 becomes

$$\frac{T_{s_2}}{T_{s_1}} = 1 + \left(\frac{\kappa - 1}{2}\right) M_{\theta_1}^2 \left(\frac{r_2^2}{r_1^2} - 1\right) \tag{2.187}$$

where

$$M_{\theta_1} = \frac{V_{\theta_1}}{\sqrt{\kappa R T_{s_1}}} = \frac{r C_2 \Omega}{\sqrt{\kappa R T_{s_1}}}. \tag{2.188}$$

Using the relation between pressure ratio and temperature ratio in an isentropic flow, we finally obtain

$$\frac{P_{s_2}}{P_{s_1}} = \left(\frac{T_{s_2}}{T_{s_1}}\right)^{\frac{\kappa}{\kappa-1}} = \left\{1 + \left(\frac{\kappa - 1}{2}\right) M_{\theta_1}^2 \left(\frac{r_2^2}{r_1^2} - 1\right)\right\}^{\frac{\kappa}{\kappa-1}}. \tag{2.189}$$

Comparing Equations 2.179 and 2.187 for static temperature change in a free vortex and a forced vortex, respectively, we can write

$$\left(\frac{T_{s_2} - T_{s_1}}{T_{s_1}}\right)_{\text{Forced vortex}} = \left(\frac{T_{s_2} - T_{s_1}}{T_{s_1}}\right)_{\text{Free vortex}} \left(\frac{r_2^2}{r_1^2}\right). \tag{2.190}$$

Equation 2.190 indicates that, for both radially outward ($r_2 > r_1$) and radially inward ($r_1 > r_2$) flows with identical conditions at r_1 (inlet), the static temperature at r_2 (outlet) for a forced vortex is always higher than that for a free vortex. Similarly, comparing Equations 2.181 and 2.189, we conclude that, with the identical conditions at r_1 (inlet), the static pressure at r_2 (outlet) for a forced vortex is always higher than that for a free vortex. These observations have important repercussions in the design of internal flow systems of gas turbines.

Total pressure and total temperature variations in an isentropic forced vortex. Equations 2.182 and 2.184 yield

$$T_{t_2} - T_{t_1} = \left(T_{s_2} - T_{s_1}\right) + \frac{C_2^2 \Omega^2}{2 c_p} \left(r_2^2 - r_1^2\right) \tag{2.191}$$

on the right-hand side of which the first expression within the parentheses represents the change in static temperature and the remaining terms represent the change in dynamic temperature based on the tangential velocity.

Using Equation 2.186, we can express Equation 2.191 as

$$T_{t_2} - T_{t_1} = \left(T_{s_2} - T_{s_1}\right) + \left(T_{s_2} - T_{s_1}\right) = 2\left(T_{s_2} - T_{s_1}\right), \tag{2.192}$$

which shows that, between any two points, the change in total temperature in an isentropic forced vortex equals twice the change in its static temperature, and can be alternatively expressed as

$$\frac{T_{t_2}}{T_{t_1}} = 1 + \left(\frac{S_f^2 \Omega^2 \left(r_2^2 - r_1^2\right)}{c_p T_{t_1}}\right). \tag{2.193}$$

Using the relation between the ratio of total pressures and ratio of total temperatures between any two points in an isentropic flow, we obtain

$$\frac{P_{t_2}}{P_{t_1}} = \left(\frac{T_{t_2}}{T_{t_1}}\right)^{\frac{\kappa}{\kappa-1}} = \left[1 + \left(\frac{S_f^2 \Omega^2 \left(r_2^2 - r_1^2\right)}{c_p T_{t_1}}\right)\right]^{\frac{\kappa}{\kappa-1}}. \quad (2.194)$$

2.3.1.3   Isothermal Forced Vortex

An accurate prediction of the static pressure distribution on a rotor disk forming a rotor-rotor or rotor-stator cavity of a gas turbine is critical to the evaluation axial rotor thrust. The current design practice by and large is based on the static pressure distribution in a series of isentropic force vortices, neglecting the effect of temperature change as a result of heat transfer on this distribution. In this section, we will examine such an effect by finding the analytical solution of static pressure distribution in an isothermal forced vortex and comparing it to the one for the isentropic case derived in the foregoing section.

In an isothermal forced vortex, the static temperature is assumed to remain constant between any two points. This implies cooling of a radially outward flowing forced vortex and heating of a radially inward flowing forced vortex, which are otherwise isentropic. Using Equations 2.176 and 2.184 we can write

$$\frac{dP_s}{P_s} = \left(\frac{C_2^2 \Omega^2}{R T_s}\right) r \, dr, \quad (2.195)$$

which can be integrated between two points to yield

$$\ln\left(\frac{P_{s_2}}{P_{s_1}}\right) = \frac{\kappa M_{\theta_1}^2}{2}\left(\frac{r_2^2}{r_1^2} - 1\right). \quad (2.196)$$

Variations of static pressure in a forced vortex under isentropic and isothermal flow conditions, based on Equations 2.189 and 2.196, are shown in Figure 2.19 for $\kappa = 1.4$ and $M_{\theta_1} = 0.5$. Beyond a small change in radius, radial variation of static pressure in an isothermal forced vortex is higher than that in an isentropic forced vortex. The

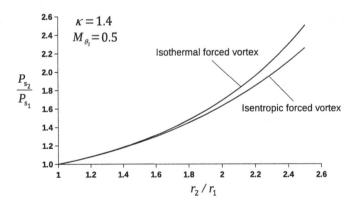

Figure 2.19 Comparison of static pressure variations in isentropic and isothermal forced vortices.

comparison shown in this figure indicates that, for an accurate prediction of static pressure in a forced vortex, we must account for temperature variation as a result of heat transfer, as discussed in Section 2.3.1.4 for a nonisentropic generalized vortex.

2.3.1.4   Nonisentropic Generalized Vortex

A nonisentropic generalized vortex is characterized by an arbitrary radial distribution of swirl factor $S_f = f_1(r)$ and total temperature $T_t = f_2(r)$, which results from both heat transfer and rotational work transfer. The functions $f_1(r)$ and $f_2(r)$ are piecewise polynomial functions between radii r_1 and r_2. The static pressure variation in this vortex is governed by Equation 2.176, which can be expressed in terms of rotational Mach number as

$$\frac{dP_s}{P_s} = \frac{V_\theta^2}{RT_s}\frac{dr}{r} = \kappa \left(\frac{M_\theta^2}{r}\right) dr \tag{2.197}$$

where $M_\theta = V_\theta / \sqrt{\kappa RT_s}$ in which $V_\theta = S_f \Omega r = f_1 \Omega r$ and $T_s = T_t - V_\theta^2/(2c_p) = f_2 - f_1^2 \Omega^2 r^2/(2c_p)$.

Integrating Equation 2.197 between radii r_1 and r_2, we can write

$$\int_{r_1}^{r_2} \frac{dP_s}{P_s} = \kappa \int_{r_1}^{r_2} \left(\frac{M_\theta^2}{r}\right) dr$$

$$\ln\left(\frac{P_{s_2}}{P_{s_1}}\right) = \kappa G \tag{2.198}$$

where G is computed by numerically integrating $\int_{r_1}^{r_2} \left(\frac{M_\theta^2}{r}\right) dr$, for example, using the Simpson's one-third rule. Thus, we finally obtain the following equation to compute static pressure distribution in a nonisentropic generalized vortex:

$$P_{s_2} = P_{s_1} e^{\kappa G} \tag{2.199}$$

2.3.2   Windage versus Vortex Temperature Change

In internal flow systems of a gas turbine, the fluid total temperature will change as a result of heat transfer from both rotor and stator surfaces and work transfer only from the rotor surfaces. As a result of the fact that the stator wall is not moving relative to the fluid (no-slip boundary condition as a result of nonzero fluid viscosity), the wall shear stress cannot participate in any work transfer between the fluid and the wall. For example, the total temperature of the fluid flowing through an adiabatic stationary duct with friction will remain constant in the duct. Windage is, therefore, defined as rotor power input into the surrounding fluid. It should not be confused with viscous dissipation, which results from viscosity-driven transfer of external flow energy into the internal energy with no change in fluid total temperature under adiabatic conditions.

Figure 2.20 shows a typical rotor-stator cavity where the windage is generated by the rotor surface. For the windage analysis between sections 1 and 2, let us consider the control volume ABCD. The angular momentum balance over this control volume yields

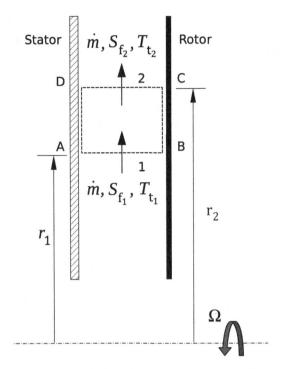

Figure 2.20 A typical windage situation in a gas turbine rotor-stator cavity.

$$\Gamma_{BC,\,rotor} - \Gamma_{AD,\,stator} = \dot{m}V_{\theta_2}r_2 - \dot{m}V_{\theta_1}r_1 = \dot{m}\left(S_{f_2}r_2^2 - S_{f_1}r_1^2\right)\Omega$$

$$\Gamma_{BC,\,rotor} = \Gamma_{AD,\,stator} + \dot{m}\left(S_{f_2}r_2^2 - S_{f_1}r_1^2\right)\Omega \qquad (2.200)$$

where $\Gamma_{BC,rotor}$ and $\Gamma_{AD,stator}$ are, respectively, the rotor torque and the stator torque within the control volume, determined using empirical equations. Windage power is then computed as

$$\dot{W}_{windage} = \Gamma_{BC,\,rotor}\Omega \qquad (2.201)$$

Combining Equations 2.200 and 2.201 yields

$$\dot{W}_{windage} = \Gamma_{AD,\,stator}\Omega + \dot{m}\left(S_{f_2}r_2^2 - S_{f_1}r_1^2\right)\Omega^2 \qquad (2.202)$$

Equation 2.202 is often misinterpreted to imply that the first term on its right-hand indicates the work done by the stator, which cannot do any work. To avoid such a nonphysical interpretation, one should use Equations 2.200 and 2.201 separately, the first originating from the angular momentum equation and the second from the energy equation.

The change in total temperature as a result of windage power can be obtained from Equation 2.201 as

$$\Delta T_{t_{windage}} = T_{t_2} - T_{t_1} = \frac{\dot{W}_{windage}}{\dot{m}c_p} \qquad (2.203)$$

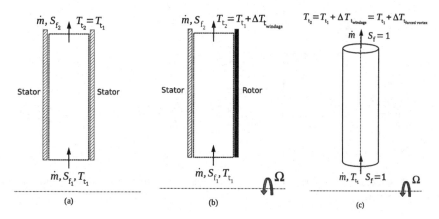

Figure 2.21 Total temperature change in: (a) all-stator cavity, (b) rotor-stator cavity, and (c) rotating pipe.

Note that the process of windage generation is nonisentropic. Any design practice to calculate forced vortex temperature change using Equation 2.193 in the control volume and adding to the temperature change computed from Equation 2.203 will be in violation of the energy equation and must be avoided. To further illustrate this flawed design practice, if it exists, let us consider an all-stator cavity shown in Figure 2.21a. Because there is no rotor surface, although the swirl factor will decrease ($S_{f_2} < S_{f_1}$) from inlet to outlet as a result of the torque from the stator walls, the total temperature will remain constant under adiabatic conditions. In the case of rotor-stator cavity shown in Figure 2.21b, one must use Equation 2.203 to compute the change in total temperature from inlet to outlet.

For the flow through a rotating pipe, shown in Figure 2.21c, where the constant swirl factor corresponds to solid-body rotation ($S_f = 1$), the change in total temperature from inlet to outlet can be established as isentropic forced vortex temperature rise (Equation 2.193), which also equals the windage temperature rise based on the work transfer computed using the Euler's turbomachinery equation. Readers are encouraged to verify on their own that $\Delta T_{t_{windage}} = \Delta T_{t_{forced\ vortex}}$ in this case.

2.3.3 Loss Coefficient versus Discharge Coefficient

In incompressible internal flows, it is customary to simulate minor losses using the incompressible loss coefficient K, which is defined as

$$K = \frac{P_{t_{inlet}} - P_{t_{outlet}}}{0.5\rho V^2} \qquad (2.204)$$

In Equation 2.204, the dynamic pressure used as the denominator of the right-hand side term corresponds to either inlet or outlet, whichever is higher. Values of K for various flow elements are reported for example in Idelchik (2005). No such comprehensive empirical data for minor losses for compressible flows yet exist. A common engineering practice, which may not be physics-based, is to extend the use of

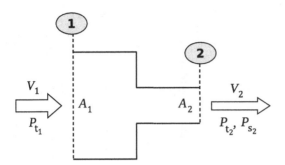

Figure 2.22 Relation between loss coefficient and discharge coefficient of a sudden pipe contraction.

Equation 2.204 for compressible flows by replacing the denominator with $\left(P_{t_{inlet}} - P_{s_{inlet}}\right)$ or $\left(P_{t_{outlett}} - P_{s_{outlet}}\right)$, whichever is higher. As the Mach number increases, so does the dynamic pressure over the corresponding incompressible value.

Another approach to simulate minor losses in both incompressible and compressible flows is through the discharge coefficient C_d, which is simply defined by the equation

$$C_d = \frac{\dot{m}_{actual}}{\dot{m}_{ideal}} \tag{2.205}$$

where the ideal mass flow rate $\dot{m}_{ideal}$ through the element is computed under no loss in total pressure. Note that the discharge coefficient C_d is usually determined empirically.

Relation between K and C_d for an incompressible flow. Let us derive a relationship between K and C_d for an incompressible flow through a sudden pipe contraction shown in Figure 2.22. According to Equation 2.204, we can write in this case

$$P_{t_1} - P_{t_2} = K\frac{\rho V_2^2}{2}$$

$$P_{t_1} - P_{s_2} - \frac{\rho V_2^2}{2} = K\left(\frac{\rho V_2^2}{2}\right)$$

$$P_{t_1} - P_{s_2} = (K+1)\left(\frac{\rho V_2^2}{2}\right)$$

Using this, we can write the mass flow rate through a sudden contraction pipe flow element as

$$\dot{m}_K = \rho A_2 V_2 = A_2\sqrt{\frac{2\rho\left(P_{t_1} - P_{s_2}\right)}{K+1}} \tag{2.206}$$

To calculate the ideal mass flow rate through the flow element, we assume that the inlet total pressure remains constant. Accordingly, we can write

$$P_{t_1} - P_{s_2} = \frac{\rho V_{2_{ideal}}^2}{2}$$

which yields

$$\dot{m}_{\text{ideal}} = \rho A_2 V_{2_{\text{ideal}}} = A_2 \sqrt{2\rho \left(P_{t_1} - P_{s_2} \right)} \tag{2.207}$$

Thus, using Equation 2.205, we can express the actual mass flow rate as

$$\dot{m}_{\text{actual}} = C_d \dot{m}_{\text{ideal}} = A_2 C_d \sqrt{2\rho \left(P_{t_1} - P_{s_2} \right)} \tag{2.208}$$

Equating mass flow rates computed by Equations 206 and 208, we finally obtain

$$C_d = \frac{1}{\sqrt{K+1}} \tag{2.209}$$

which is identical to the equation obtained by Sultanian (2015) for an orifice with incompressible flow. It is, however, important to note that the relation between K and C_d given by Equation 2.209 is not a universal one. For example, using the foregoing procedure for a sudden expansion pipe flow will yield a relationship between K and C_d with dependence on the expansion area ratio (see Problem 2.14).

Relation between K and C_d for a compressible flow. Let us now derive a relationship between K and C_d for a compressible flow through the sudden pipe contraction shown in Figure 2.22 and compare it with Equation 2.209. In this case, we define the compressible loss coefficient by the equation

$$K = \frac{P_{t_1} - P_{t_2}}{P_{t_2} - P_{s_2}} \tag{2.210}$$

which yields

$$P_{t_2} = \frac{P_{t_1} + K P_{s_2}}{K+1} \tag{2.211}$$

and

$$\frac{P_{t_2}}{P_{s_2}} = \frac{\dfrac{P_{t_1}}{P_{s_2}} + K}{K+1} \tag{2.212}$$

Using Equation 2.103, we can compute mass flow rate through the sudden contraction flow element as

$$\dot{m}_K = \frac{A_2 \hat{F}_{f_t} P_{t_2}}{\sqrt{R T_{t_2}}} \tag{2.213}$$

where $\hat{F}_{f_t}$ is the dimensionless total-pressure mass flow function given by

$$\hat{F}_{f_t} = M_2 \sqrt{\frac{\kappa}{\left(1 + \dfrac{\kappa - 1}{2} M_2^2 \right)^{\frac{\kappa+1}{\kappa-1}}}} \tag{2.214}$$

Where the Mach number M_2 is to be determined by the equation

$$M_2 = \sqrt{\frac{2}{\kappa - 1}\left[\left(\frac{P_{t_2}}{P_{s_2}}\right)^{\frac{\kappa-1}{\kappa}} - 1\right]} \qquad (2.215)$$

which uses outlet pressure ratio P_{t_2}/P_{s_2} from Equation 2.212.

Again for computing ideal mass flow rate through the flow element, we assume no loss in total pressure under adiabatic conditions, implying $P_{t_2} = P_{t_1}$ and $T_{t_2} = T_{t_1}$, and yielding

$$\dot{m}_{ideal} = \frac{A_2 \tilde{\hat{F}}_{f_t} P_{t_1}}{\sqrt{RT_{t_2}}} \qquad (2.216)$$

where

$$\tilde{\hat{F}}_{f_t} = \tilde{M}_2 \sqrt{\frac{\kappa}{\left(1 + \frac{\kappa - 1}{2}\tilde{M}_2^2\right)^{\frac{\kappa+1}{\kappa-1}}}} \qquad (2.217)$$

where

$$\tilde{M}_2 = \sqrt{\frac{2}{\kappa - 1}\left[\left(\frac{P_{t_1}}{P_{s_2}}\right)^{\frac{\kappa-1}{\kappa}} - 1\right]} \qquad (2.218)$$

Expressing the actual mass flow rate in terms of the discharge coefficient as

$$\dot{m}_{actual} = C_d \dot{m}_{ideal} = \frac{C_d A_2 \tilde{\hat{F}}_{f_t} P_{t_1}}{\sqrt{RT_{t_1}}} \qquad (2.219)$$

and equating it with the mass flow rate computed from Equations 2.213, we obtain

$$C_d = \left(\frac{\hat{F}_{f_t}}{\tilde{\hat{F}}_{f_t}}\right)\frac{P_{t_2}}{P_{t_1}}$$

which with the substitution of P_{t_2} from Equation 2.211 becomes

$$C_d = \left(\frac{\hat{F}_{f_t}}{\tilde{\hat{F}}_{f_t}}\right)\frac{P_{t_1} + KP_{s_2}}{P_{t_1}(K + 1)}$$

$$C_d = \left(\frac{\hat{F}_{f_t}}{\tilde{\hat{F}}_{f_t}}\right)\left(\frac{\frac{P_{t_1}}{P_{s_2}} + K}{\frac{P_{t_1}}{P_{s_2}}(K + 1)}\right) \qquad (2.220)$$

Thus, for a compressible flow through a sudden pipe contraction, Equation 2.220 relates K and C_d. Unlike for the case of incompressible flow (Equation 209), Equation

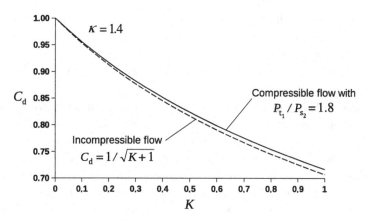

Figure 2.23 Variation of discharge coefficient with loss coefficient in a sudden pipe contraction for both incompressible and compressible flows.

2.220 for the compressible flow shows that the relationship between K and C_d also depends on the overall pressure ratio P_{t_1}/P_{s_2}.

Figure 2.23 compares the variation of discharge coefficient with loss coefficient for both incompressible flow (Equation 209) and compressible flow (Equation 2.220) with $\kappa = 1.4$ and $P_{t_1}/P_{s_2} = 1.8$. For $P_{t_1}/P_{s_2} < 1.1$, when a compressible flow can be treated as an incompressible flow with constant density, two curves are virtually identical for all values of K. A mathematical reduction of Equation 2.220 to Equation 2.209 for low values of P_{t_1}/P_{s_2} is not evident.

2.4   Heat Transfer

In gas turbine internal flow systems, both heat transfer and work transfer, separately or simultaneously, are responsible for any change the fluid temperature. We define heat transfer between the fluid and solid structure, or within the solid itself, as energy transfer by virtue of a temperature difference. Work transfer into the fluid will only occur if the solid wall in contact is rotating. There can be no work transfer when the fluid flows over any static structure. It is important to note that heat transfer always occurs from high to low temperature, and never the other way around. A fluid flow, however, can occur against an increasing static pressure, as in a diffuser, as a result of the associated inertia. There is no such thing as thermal inertia associated with heat transfer, and the direction of decreasing temperature uniquely determines the direction of heat transfer.

Now, consider a turbine airfoil with internal cooling. Often a thermal barrier coating (TBC) with very low thermal conductivity is used on the airfoil surface exposed to hot gases to reduce heat transfer to the internal cooling flow. To improve turbine efficiency, one may be tempted to a heat transfer design solution that eliminates the internal cooling flow by increasing TBC thickness or by using a TBC with much lower thermal conductivity. Because the temperature difference exists between the hot gases and the airfoil, some heat transfer will definitely occur and will eventually (in a steady state) heat the entire airfoil to the hot-gas temperature (thermal equilibrium).

There are three modes of heat transfer: conduction, convection, and radiation. Conduction, also called thermal diffusion, occurs at the molecular level as a result of temperature gradients within a material, be it solid, liquid, or gas. In most engineering applications, including in gas turbine design, conduction (diffusion) within the fluid is neglected. Convection refers to heat transfer between a solid and a fluid flowing over it. Fluid flow is the key differentiator between conduction and convection. In forced convection, the flow occurs as a result of an imposed pressure gradient. A free convection on the other hand is driven by density stratification as a result of temperature variation in a conservative force field such as gravity or the centrifugal force as a result of rotation (centrifugally-driven buoyant convection). In an enclosure, for example, if the bottom surface is hotter than the top surface, hot air will rise toward the top surface and the heavy colder air will take its place at the bottom surface. This will set a free convection current in the enclosure. If, on the other hand, the top surface is hotter than the bottom surface, the fluid in the enclosure will be stably stratified, and no free convection will take place. Note that the heat transfer between the solid and the fluid in contact at the wall is always by conduction.

While both conduction and convection require a material medium, the radiation heat transfer between two objects occurs via electromagnetic wave through randomly moving photons without the need for a medium. Our earth is mostly heated by thermal radiation from the sun.

Here is a situation that provides a simple layman's depiction for each mode of heat transfer. A teacher needed to distribute pencils to each student in a class room. She could do it in one of three ways. In one way, like conduction heat transfer, she hands over the pencil to the nearest student and asks him to pass it to his neighbor who passes it to his neighbor, and so on. This process continues until everyone in the class room has a pencil. In the second way, which is analogous to convection heat transfer, she asks each student to come to her one by one and get the pencil. In the third way, which is analogous to radiation heat transfer, she throws one pencil to each student in the class room until all of them have a pencil.

In the bibliography at the end of this chapter, we have included leading textbooks devoted to the three modes of heat transfer. In the following sections, we provide a brief review of these modes of heat transfer.

2.4.1   Conduction

Conduction heat flux (heat transfer rate per unit area) vector in a material region as a result of temperature gradient is governed by the Fourier's law, given in the equation form as

$$\vec{q} = -k\nabla T \qquad (2.221)$$

where k is the material thermal conductivity, which is usually a function a temperature. The negative sign on the right-hand side of Equation 2.221 ensures that heat transfer

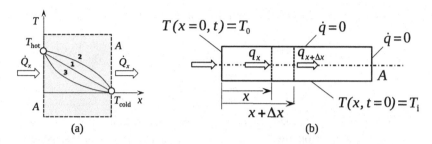

Figure 2.24 (a) 1-D steady heat conduction in a solid slab and (b) 1-D transient heat condition in a long insulated rod, initially at temperature T_i, whose un-insulated end is suddenly exposed to constant temperature T_0.

occurs from higher to lower temperature, in conformity with the second law of thermodynamics. When we multiply the heat flux vector with the area normal to it, we obtain the net heat transfer rate through the area, that is

$$\dot{Q}_n = \vec{q} \cdot \hat{n} A \qquad (2.222)$$

In order to develop a better intuitive understanding of conduction heat transfer, we present here some simple steady and unsteady one-dimensional (1-D) heat conduction situations shown in Figure 2.24. It is important to note that the key design-relevant information can often be gleaned from simple 1-D heat transfer analyses, which play an important role in conceptual and preliminary design phases.

2.4.1.1 One-Dimensional Steady Heat Conduction

Figure 2.24a shows 1-D steady heat conduction through a solid slab having temperature gradients only along the x direction. Over the slab area A, the magnitude of the steady heat transfer rate $\dot{Q}_x = Ak|dT/dx|$ for each of the three temperature profiles must remain constant through the slab. If the material thermal conductivity k is constant, the temperature gradient across the slab must be constant as depicted by the temperature profile 1. For the temperature profiles 2 and 3, k must be a function of T. For profile 2 $|dT/dx|$ increases with x. To keep $\dot{Q}_x$ constant, k must decrease with x. Because T also decreases with x for this temperature profile, k must be an increasing function of T, that is k increases as T increases and vice versa. For profile 3, however, both $|dT/dx|$ and T decease with x. Following the argument of constant $\dot{Q}_x$ through the slab, we conclude that for this temperature profile k must be a decreasing function of T.

2.4.1.2 One-Dimensional Unsteady Heat Conduction

Figure 2.24b shows 1-D unsteady heat conduction through a long solid rod of constant cross-section area A and thermal conductivity k. Except at one end, the rod is insulated over its entire surface and has an initial uniform temperature T_i. The un-insulated end of the rod is suddenly exposed and maintained to a constant surface temperature T_0 corresponding to $x = 0$, as shown in the figure.

From energy balance on the control volume between x and $x + \Delta x$, we can write

$$(\rho c A \Delta x)\frac{\partial T}{\partial t} = A(\dot{q}_x - \dot{q}_{x+\Delta}) = A\left(-\frac{\partial \dot{q}_x}{\partial x}\right)\Delta x$$

$$\frac{\partial T}{\partial t} = \alpha \frac{\partial^2 T}{\partial x^2} \tag{2.223}$$

where $\alpha = k/(\rho c)$ is the thermal diffusivity of the rod material, c its specific heat. All thermophysical properties are assumed constant in this analysis. Note that a partial differential equation similar to Equation 2.223 arises in Stokes' first problem involving a suddenly accelerated flat plate in a quiescent fluid.

The applicable initial and boundary conditions in this case are:

$$\text{IC: } T(x,t) = T_i$$
$$\text{BC1: } T(0,t) = T_0$$
$$\text{BC2: } T(\infty,t) = T_i$$

An effective way to solve Equation 2.223 with the specified initial and boundary conditions is to seek a similarity solution using the following dimensionless similarity variables:

$$\xi = \frac{x}{2\sqrt{\alpha t}} \qquad \text{and} \qquad \theta(\xi) = \frac{T(x,t) - T_0}{T_i - T_0}$$

Various derivatives in Equation 2.223 are evaluated as follows:

$$\frac{\partial T}{\partial t}: \qquad \frac{\partial T}{\partial t} = (T_i - T_0)\frac{d\theta}{d\eta}\frac{\partial\xi}{\partial t} = (T_i - T_0)\theta'\frac{\partial\xi}{\partial t}$$

which with $\partial\xi/\partial t = -\xi/2t$ becomes

$$\frac{\partial T}{\partial t} = -\frac{(T_i - T_0)\xi}{2t}\theta'$$

$$\frac{\partial T}{\partial x}: \qquad \frac{\partial T}{\partial x} = (T_i - T_0)\theta'\frac{\partial\xi}{\partial x}$$

which with $\partial\eta/\partial x = 1/(2\sqrt{\alpha t})$ becomes

$$\frac{\partial T}{\partial x} = \frac{(T_i - T_0)}{2\sqrt{\alpha t}}\theta'$$

$$\frac{\partial^2 T}{\partial x^2}: \qquad \frac{\partial^2 T}{\partial x^2} = \frac{(T_i - T_0)}{2\sqrt{\alpha t}}\theta''\frac{\partial\xi}{\partial x} = \frac{(T_i - T_0)}{4\alpha t}\theta''$$

Substituting these derivatives in Equation 2.223 finally yields

$$\theta'' + 2\xi\theta' = 0 \tag{2.224}$$

which is a second-order ordinary differential equation (ODE). This confirms that the mathematical problem governed by Equation 2.223 has a similarity solution when transformed using the nondimensional variables ξ and θ. In fact, the solutions are expected to be self-similar in that the temperature profiles $T(x, t)$ that evolve with time can all be collapsed into a single profile given by $T(x, t) = T_0 + (T_i - T_0) \theta(\xi)$. The transformed boundary conditions are:

$$\theta(0) = 0 \quad \text{and} \quad \theta(\infty) = 1$$

Integrating Equation 2.224 once yields

$$\theta' = C_1 e^{-\xi^2}$$

where C_1 is the integration constant. Integrating the aforementioned equation one more time, we obtain

$$\theta = C_1 \int_0^\xi e^{-\xi^2} d\xi + C_2 \tag{2.225}$$

where C_2 is the second integration constant. The integral on the right-hand side of Equation 2.225 may be replaced by the Gaussian error function, which is defined as

$$\text{erf}(\xi) = \frac{2}{\sqrt{\pi}} \int_0^\xi e^{-\xi^2} d\xi$$

giving

$$\theta = \frac{\sqrt{\pi}}{2} C_1 \text{erf}(\xi) + C_2$$

Applying the two boundary conditions to this equation yields the integration constants as

$$C_1 = 2/\sqrt{\pi} \quad \text{and} \quad C_2 = 0$$

where we have used the limiting values of the error function: $\text{erf}(0) = 0$ and $\text{erf}(\infty) = 1$. Thus, we finally obtain the following simple solution:

$$\theta = \text{erf}(\xi) \quad \text{or} \quad \frac{T(x, t) - T_0}{T_i - T_0} = \text{erf}\left(\frac{x}{2\sqrt{at}}\right) \tag{2.226}$$

in which the error function is either obtained from mathematical tables or evaluated numerically.

2.4.2   Convection

Convective heat transfer between a solid surface and the fluid flowing over it is governed by Newton's law of cooling expressed by the equation

$$\dot{Q} = hA(T_w - T_f) \tag{2.227}$$

where

$\dot{Q} \equiv$ Rate of convective heat transfer
$h \equiv$ Heat transfer coefficient

$A \equiv$ Heat transfer area

$T_w \equiv$ Wall temperature

$T_f \equiv$ Reference fluid temperature

In Equation 2.227, heat transfer coefficient h clearly depends on the choice of the reference fluid temperature T_f. In experiments involving measurements of heat transfer rate and wall temperature, perhaps as a matter of convenience, the flow inlet temperature is often used as a reference temperature, which at times leads to negative heat transfer coefficient. Any application of these heat transfer coefficients in heat transfer design is certainly problematic. Because the convective heat transfer is locally driven by conduction between the fluid and solid in the wall boundary layer, a consistent way to define the heat transfer coefficient in Equation 2.227 is by using the local adiabatic wall temperature as the reference fluid temperature. We discuss this temperature in the next section.

In forced convection correlations, heat transfer coefficient h typically appears in two dimensionless numbers: Nusselt number ($Nu = hL/k$ or $Nu = hD/k$) and Stanton number ($St = h/(\rho V c_p)$). These are usually correlated with the Reynolds number ($Re = VL\rho/\mu$ or $Re = VD\rho/\mu$) and Prandtl number ($Pr = \mu c_p/\kappa$). In these dimensionless numbers L and D are characteristic lengths. In free or natural convection, the Nusselt number is usually related to the Prandtl number and Grashof number, which is defined by

$$Gr = \frac{g\beta(\Delta T)L^3\rho^2}{\mu^2} \tag{2.228}$$

where β is the isobaric compressibility ($\beta = (\partial\rho/\partial T)/\rho$) of the fluid. Note that for a perfect gas with $P_s = \rho R T_s$ as its equation of state, we obtain $\beta = 1/T_s$.

When $\sqrt{Gr}$ and Re are of comparable magnitude, a mixed convection situation prevails. In this case, the resultant Nusselt number is obtained by (see Becker (1986))

$$Nu = \left(Nu_{fc}^3 + Nu_{nc}^3\right)^{1/3} \tag{2.229}$$

where Nu_{fc} and Nu_{nc} are Nusselt numbers for forced convection and natural convection, respectively. Each one is calculated under the assumption that the other one is negligible.

Thermophysical properties used in various dimensionless numbers used in empirical Nusselt number correlations are usually temperature dependent and are recommended to be evaluated at Eckert's (1961) reference temperature given by

$$T_{ref} = T_\infty + 0.5(T_w - T_\infty) + 0.22(T_{aw} - T_\infty) \tag{2.230}$$

Note that Equation 2.230 simplifies to

$$T_{ref} = 0.28T_\infty + 0.5T_w + 0.22T_{aw}$$

which for $T_{aw} \approx T_\infty$ leads to $T_{ref} = (T_\infty + T_w)/2$, the average of the free stream and wall temperatures.

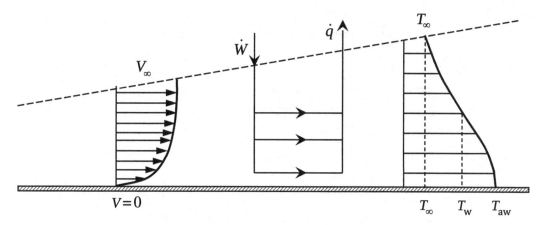

Figure 2.25 Hydrodynamic and thermal boundary layers on a flat plate.

2.4.2.1 Adiabatic Wall Temperature

When the dynamic temperature associated with a flow is significant, the concept of adiabatic wall temperature plays an important role in determining both the direction and magnitude of heat transfer between the fluid and solid wall in contact. As a result of the no-slip condition at the wall, the velocity of a viscous fluid equals that of the wall, and the fluid and wall in contact will have the same temperature under thermal equilibrium.

The behavior of both hydrodynamic and thermal boundary layers on a stationary wall is depicted in Figure 2.25. As the flow slows down toward the wall in the boundary layer, the work transfer as a result of viscous dissipation increases its temperature, which results in heat transfer toward the edge of the boundary layer. If we assume an adiabatic wall (zero heat transfer), it will attain a temperature higher than the free stream temperature under the equilibrium effects of viscous dissipation and reverse conduction in the boundary layer. This temperature is called the adiabatic wall temperature T_{aw} and is computed by

$$T_{aw} = T_\infty + \frac{r^* V_\infty^2}{2c_p} \tag{2.231}$$

where r^* is the recovery factor, which depends on the fluid Prandtl number. For a laminar boundary layer we have $r^* = Pr^{1/2}$ and for a turbulent boundary layer $r^* = Pr^{1/3}$. For $r^* = 1$, we obtain $T_{aw} = T_t$.

As shown in Figure 2.25, for a wall temperature T_w, which is higher than the free stream temperature T_∞ and lower than the adiabatic wall temperature T_{aw}, both the direction and magnitude of heat transfer will depend upon the choice of the reference fluid temperature whether $T_f = T_{aw}$ or $T_f = T_\infty$. For a physically consistent heat transfer calculation, it is therefore imperative that we always use $T_f = T_{aw}$ in Equation 2.227.

2.4.2.2 Forced Convection over Isothermal Walls

Let us first consider a 1-D steady flow through a pipe with uniform wall temperature and constant heat transfer coefficient, as shown in Figure 2.26a. Assuming a recovery factor

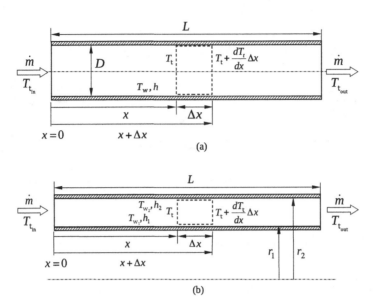

Figure 2.26 (a) Total temperature change in a steady flow through a pipe with uniform wall temperature and constant heat transfer coefficient and (b) Total temperature change in a steady flow through an annulus with each wall at uniform wall temperature and constant heat transfer coefficient.

of unity ($T_{aw} = T_t$), the steady flow energy balance on the control volume between x and $x + \Delta x$ yields

$$\dot{m}c_p \left(T_t + \frac{dT_t}{dx}\Delta x \right) - \dot{m}c_p T_t = (\pi D \Delta x)h\left(T_w - T_t \right)$$

$$\frac{dT_t}{dx} = \frac{\pi D h}{\dot{m}c_p}\left(T_w - T_t \right)$$

$$\frac{d\left(T_w - T_t \right)}{\left(T_w - T_t \right)} = -\left(\frac{DhL}{\dot{m}c_p} \right) d\left(\frac{x}{L} \right) \qquad (2.232)$$

Integrating Equation 2.232 from pipe inlet to a distance x along the pipe yields the following expression for the fluid total temperature at x:

$$T_t = T_w - \left(T_w - T_{t_{in}} \right)e^{-\eta x/L} \qquad (2.233)$$

where $\eta = DhL/(\dot{m}c_p)$. This solution shows that the variation of fluid total temperature in the flow through an isothermal pipe is not linear. Instead, from pipe inlet to outlet, the difference between the wall temperature and fluid total temperature decays exponentially with η, also known as the number of transfer units (NTU) in the literature on heat exchangers. In agreement with the second law of thermodynamics, this solution also ensures that the fluid total temperature never exceeds the pipe wall temperature, regardless of how long the pipe is. How can an isothermal pipe wall ever heat the fluid to a temperature higher than its own temperature?

Using Equation 2.233, the fluid total temperature at the pipe outlet $(x = L)$ is obtained as

$$T_{t_{out}} = T_w - \left(T_w - T_{t_{in}}\right) e^{-\eta} \tag{2.234}$$

which yields $T_{t_{out}} = T_w$ in the limit of zero mass flow rate $(\eta \to \infty)$.

Figure 2.26b shows a steady 1-D flow through an annulus whose inner and outer walls are held at constant wall temperatures T_{w_1} and T_{w_2} with heat transfer coefficients h_1 and h_2, respectively. Again with $T_{aw} = T_t$, the steady flow energy balance on the control volume between x and $x + \Delta x$ yields

$$\dot{m}c_p \left(T_t + \frac{dT_t}{dx}\Delta x\right) - \dot{m}c_p T_t = \left(2\pi r_1 \Delta x\right)h_1\left(T_{w_1} - T_t\right) + \left(2\pi r_2 \Delta x\right)h_2\left(T_{w_2} - T_t\right)$$

$$\dot{m}c_p \frac{dT_t}{dx} = \frac{A_{w_1}h_1}{L}\left(T_{w_1} - T_t\right) + \frac{A_{w_2}h_2}{L}\left(T_{w_2} - T_t\right)$$

$$\frac{d\left(\overline{T}_w - T_t\right)}{\left(\overline{T}_w - T_t\right)} = -\overline{\eta}\,d\left(\frac{x}{L}\right) \tag{2.235}$$

where

$$\overline{T}_w = \frac{h_{w_1}A_{w_1}T_{w_1} + h_{w_2}A_{w_2}T_{w_2}}{h_{w_1}A_{w_1} + h_{w_2}A_{w_2}} \quad \text{and} \quad \overline{\eta} = \frac{h_{w_1}A_{w_1} + h_{w_2}A_{w_2}}{\dot{m}c_p}$$

Integrating Equation 2.235 from the inlet to a distance x along the annulus yields

$$T_t = \overline{T}_w - \left(\overline{T}_w - T_{t_{in}}\right) e^{-\overline{\eta}x/L} \tag{2.236}$$

which results in the fluid total temperature at the outlet as

$$T_{t_{out}} = \overline{T}_w - \left(\overline{T}_w - T_{t_{in}}\right) e^{-\overline{\eta}} \tag{2.237}$$

Equations 2.236 and 2.237 can be easily extended for the case where the steady internal flow is exposed to multiple isothermal surfaces, each with its own constant heat transfer coefficient and wall temperature.

2.4.3 Radiation

According to Stefan-Boltzmann's law, the radiation heat transfer from a grey surface is governed by the equation

$$\dot{Q} = \varepsilon\sigma A T_S^4 \tag{2.238}$$

where

$\dot{Q} \equiv$ Radiation heat transfer rate, W
$\sigma \equiv$ Stefan-Boltzmann constant, $5.6697 \times 10^{-8}\ \text{W}/\left(\text{m}^2\,\text{K}^4\right)$
$\varepsilon \equiv$ Surface emissivity, $0 \le \varepsilon \le 1$

$A \equiv$ Heat transfer area, m^2

$T_S \equiv$ Surface temperature, K

For black-body radiation we have $\varepsilon = 1$. Often in design applications, we need to calculate the net thermal radiation between two surfaces, one at temperature T_{S_1} and the other at T_{S_2}, where $T_{S_1} > T_{S_2}$. As detailed in Howell, Menguc, and Siegel (2015), the net radiation heat transfer between surfaces 1 and 2 is given by

$$\dot{Q}_{1-2} = \frac{\sigma\left(T_{S_1}^4 - T_{S_2}^4\right)}{\left[(1-\varepsilon_1)/\varepsilon_1 A_1\right] + \left(1/A_1 F_{1-2}\right) + \left[(1-\varepsilon_2)/\varepsilon_2 A_2\right]} \tag{2.239}$$

where

$\dot{Q}_{1-2} \equiv$ Net radiation heat transfer rate from surface 1 to surface 2

$\varepsilon_1 \equiv$ Surface 1 emissivity

$\varepsilon_2 \equiv$ Surface 2 emissivity

$A_1 \equiv$ Heat transfer area of surface 1

$A_2 \equiv$ Heat transfer area of surface 2

$F_{1-2} \equiv$ View factor between surfaces 1 and 2

$T_{S_1} \equiv$ Surface 1 temperature

$T_{S_2} \equiv$ Surface 2 temperature

2.4.4   Multimode Heat Transfer

Heat transfer in design applications seldom occurs in just one of its three modes. In gas turbine internal flow systems, conduction within the structure is often accompanied with convection at the boundary surfaces, forming the needed boundary conditions for the conduction solution. The air in simple-cycle gas turbines and steam in some designs of combined-cycle power plants are used for cooling and are assumed as nonparticipating medium for the radiation heat transfer. In most heat transfer analysis involving gas turbine internal flows, radiation heat transfer is often found negligible compared to convection occurring in parallel.

For the convective boundary conditions at any part of the gas turbine structure, we need both fluid local temperature and heat transfer coefficient. While the heat transfer coefficient is calculated using an empirical correlation, the fluid temperature depends on both temporal and spatial history of the flow from its properties known at the inlet, for example, the compressor bleed point in an air-cooled gas turbine. This essentially implies that these convective boundary conditions are themselves dependent on the conduction solution within the structure.

2.4.4.1   Electrical Analogy and Thermal Resistances

1-D heat transfer analyses play an important role in most conceptual and preliminary designs. Electrical analogy of heat transfer in these analyses simplifies the formulation and resulting solution. As shown in Figure 2.27a, just as the current through a resistor is related to the voltage across it ($I_{12} = (V_1 - V_2)/R_{12}$), analogously, we define heat transfer rate by

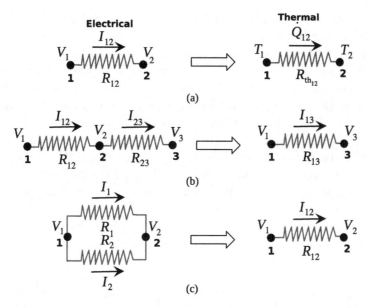

Figure 2.27 (a) Electrical analogue of heat transfer, (b) resistances in series, and (c) resistances in parallel.

$$\dot{Q}_{12} = \frac{T_1 - T_2}{R_{\text{th}_{12}}} \tag{2.240}$$

Equation 2.240 leads to the concept of thermal resistance, which for each mode of heat transfer can be expressed as follows:

Conduction: $\quad \dot{Q}_{\text{cond}} = \dfrac{kA}{t}\left(T_1 - T_2\right) = \dfrac{T_1 - T_2}{R_{\text{cond}}}; R_{\text{cond}} = \dfrac{t}{kA}$ \hfill (2.241)

Convection: $\quad \dot{Q}_{\text{conv}} = hA\left(T_{\text{w}} - T_{\text{aw}}\right) = \dfrac{T_{\text{w}} - T_{\text{aw}}}{R_{\text{conv}}}; R_{\text{conv}} = \dfrac{1}{hA}$ \hfill (2.242)

Radiation: $\quad \dot{Q}_{\text{rad}} = \dfrac{T_1 - T_2}{R_{\text{rad}}}; R_{\text{rad}} \dfrac{G_{1-2}}{\sigma\left(T_1 + T_2\right)\left(T_1^2 + T_2^2\right)}$ \hfill (2.243)

where

$$G_{1-2} = \left[\left(1 - \varepsilon_1\right)/\varepsilon_1 A_1\right] + \left(1/A_1 F_{1-2}\right) + \left[\left(1 - \varepsilon_2\right)/\varepsilon_2 A_2\right]$$

As shown in Figure 2.27b, when two resistances are in series, we have equal current passing through each ($I_{12} = I_{23}$), which leads to the equivalent total resistance $R_{13} = R_{12} + R_{23}$. Similarly, as shown in Figure 2.27c, when two resistances are in parallel, we have equal voltage drop ($V_1 - V_2$) through each, leading to the equivalent total resistance $R_{12} = R_1 R_2 / \left(R_1 + R_2\right)$. These series and parallel arrangements help us simplify complex multimode 1-D heat transfer situations into simple ones for ease of solution and intuitive understanding.

2.4.4.2 Coupling between Heat Transfer and Work Transfer in a Rotating Duct Flow

In gas turbines, internal cooling and sealing flows are often carried through rotating ducts. In these ducts, we have simultaneous heat transfer and rotational work transfer. The duct flow remains in solid-body rotation with the duct. While the rotational work transfer does not depend upon the simultaneous heat transfer, the heat transfer is influenced by the change in fluid total temperature as a result of simultaneous work transfer. In most current design practice, this effect of coupling between the heat transfer and rotational work transfer is neglected; the change in fluid total temperature is obtained by simply by adding the temperature changes computed separately as a result of heat transfer and rotational work transfer. Sultanian (2015) derives a closed-form analytical solution to compute the change in fluid total temperature under the coupled effects of heat transfer and rotational work transfer in a steady duct flow. The final equation obtained in this derivation is given by

$$
T_{t_{R_2}} = \left\{ T_{\mathrm{w}} - \left(T_{\mathrm{w}} - T_{t_{R_1}} \right) \mathrm{e}^{-\eta} \right\} + \frac{\Omega^2 (r_2 - r_1)}{c_p} \left[\frac{r_2 - r_1 \mathrm{e}^{-\eta}}{\eta} - \frac{(r_2 - r_1)(1 - \mathrm{e}^{-\eta})}{\eta^2} \right]
$$

$$(2.244)$$

where $\eta = hA_{\mathrm{w}}/(\dot{m}c_p)$. As in Equation 2.233, the surface area A_{w} equals πDL for a circular pipe

At times, some heat transfer engineers tend to simulate adiabatic case with zero heat transfer by assuming zero heat transfer coefficient, leading to $\eta = 0$. This practice is physically unrealistic, also reinforced by the second term, which becomes singular for $\eta = 0$, on the right-hand side of Equation 2.244. A physics-based approach to simulating zero convective heat transfer would be to set the adiabatic wall temperature equal to the wall temperature.

On the right-hand side of Equation 2.244, the terms within the curly brackets represent change in fluid total temperature as a result of heat transfer alone. Using Taylor series expansion of $\mathrm{e}^{-\eta}$, replacing it by the first three terms $(1 - \eta + \eta^2/2)$, and simplifying the resulting expression, we obtain

$$
T_{t_{R_2}} = \left\{ T_{\mathrm{w}} - \left(T_{\mathrm{w}} - T_{t_{R_1}} \right) \mathrm{e}^{-\eta} \right\} + \left\{ \frac{\Omega^2 (r_2^2 - r_1^2)}{2c_p} \right\} + \left(\Delta T_{t_R} \right)_{\mathrm{CCT}} \qquad (2.245)
$$

where the coupling correction term $\left(\Delta T_{t_R} \right)_{\mathrm{CCT}}$ is given by

$$
\left(\Delta T_{t_R} \right)_{\mathrm{CCT}} = -\frac{\Omega^2 (r_2 - r_1) r_1 \eta}{2c_p}
$$

Note that the terms within the second set of curly brackets on the right-hand side of Equation 2.245 represent the change in fluid total temperature (in the rotor reference frame) by rotation alone.

2.4.5   Numerical Heat Transfer

Closed-form analytical solutions for conduction heat transfer presented by Carslaw and Jaeger (1959) and for both conduction and convection presented by Han (2012) pertain only to simple geometries, constant material properties, and simple boundary conditions, which are seldom found in practical heat transfer design problems. For these problems, where analytical solutions are unobtainable or unavailable, we resort to semi-numerical and numerical solutions, taking full advantage of the computing power that has been constantly growing over the last several decades. For example, Sultanian (1978) and Sultanian and Sastri (1979, 1980) used the boundary collocation method involving harmonic functions for the solution of the Laplace equation governing 2-D conduction heat transfer in an arbitrary multiply-connected domain of coolant channels used in regeneratively-cooled liquid rocket engines. Because the energy equation governing a general conduction in a solid medium is not as nonlinear as the full Navier-Stokes equations in a fluid domain, especially involving turbulence, the former is routinely solved numerically, often using a commercial code, where convective boundary conditions are based on available empirical correlations. For example, Sultanian and Kotliarevsky (1981) present results of numerical heat transfer modeling of a steel ingot from end-of-teeming to start-of-rolling, including solidification (two-phase steel), cooling in the mould an in air, heating in the soaking pit, and subsequent air-cooling before rolling, all as a single continuous process. Minkowycz et al. (1988) and Minkowycz, Sparrow, and Murthy (2006) present a rich landscape of numerical heat transfer.

For steady two-dimensional heat conduction with mixed boundary conditions, consider a long slab of rectangular cross-section with no variations along its length, as shown in Figure 2.28a. For the purpose of our discussion of boundary conditions, let us assume that the thermal conductivity of the slab material is constant. The temperature distribution in the cross-section of the slab is governed by the Laplace equation

$$\frac{\partial^2 T}{\partial x^2} + \frac{\partial^2 T}{\partial y^2} = 0 \tag{2.246}$$

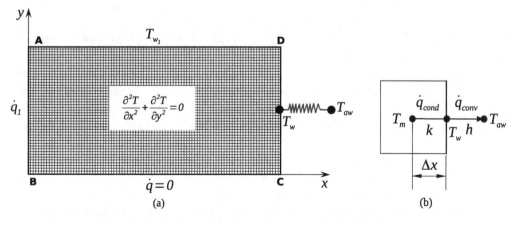

Figure 2.28 (a) Two-dimensional heat conduction with mixed boundary conditions and (b) numerical simulation of convective boundary condition.

with the boundary conditions:

Side AD: Fixed wall temperature T_{W_1} (Dirichlet type)
Side AB: Fixed heat flux $\dot{q}_1$ (Neumann type)
Side BC: Zero heat flux (Adiabatic)
Side CD: Convective heat transfer (Robin type)

A heat transfer solution at any point of the domain consists of both temperature and the heat flux vector. For a Dirichlet type boundary condition, the solution yields the boundary heat flux, and for the Neumann type boundary condition, the solution yields the boundary temperature. For the Robin type boundary condition, both the boundary temperature and heat flux are obtained from the numerical solution. For a boundary cell with its cell-center temperature T_m, as shown in Figure 2.28b, we enforce continuity of heat flux across the solid-fluid interface and write

$$\dot{q}_{\text{cond}} = \frac{k(T_m - T_w)}{\Delta x} = \dot{q}_{\text{conv}} = h(T_w - T_{aw})$$

$$T_w = \frac{kT_m + h\Delta x T_{aw}}{h\Delta x + k} \qquad (2.247)$$

where both T_{aw} and h are assumed constant, independent of the thermal solution in the slab. Once we know T_w from Equation 2.247, $\dot{q}_{\text{cond}} = \dot{q}_{\text{conv}} = h(T_w - T_{aw})$ is easily calculated.

2.4.5.1   Conjugate Heat Transfer

Durability, defined in terms of dependable operational life, of various components in gas turbines depends strongly on temperatures and their gradients in these components. Cooling these components essentially involves heat transfer from the hot gases of the primary flow path as the source to the coolant of internal flow systems as the sink. While heat transfer within the structural components is by conduction, that from the hot gases and into the coolant is by convection. In the conventional convective boundary condition (Robin type), the specified heat transfer coefficient and fluid temperature (adiabatic wall temperature) are assumed independent of the conduction solution within the structure. In gas turbines, while the hot gas-side adiabatic wall temperature, which is often assumed to equal the gas total temperature, does not depend upon the conduction solution within the structure, the heat transfer coefficient does depend on the wall temperature distribution, weakly coupling the two domains. On the coolant side heat transfer, on the other hand, both heat transfer coefficient and coolant temperature depend on the conduction solution within the structure, strongly coupling the two domains. When the solution of heat transfer (conduction) in a solid and that in the fluid in contact depend on each other, the situation is generally referred to as conjugate heat transfer, regardless of the solution scheme used.

The conjugate heat transfer situations are commonplace in internal flow systems of gas turbines. The traditional solution method consists of numerical conduction solution in the structure with heat transfer coefficients specified by empirical correlations along with the coolant side total temperature. At the solid-fluid interface, a common temperature and the continuity of heat flux is ensured by iterative solutions in the two domains. Note that the local coolant total temperature also depends on its parabolic journey from the coolant source with known properties as inlet conditions. This iteratively coupled method of solution for conjugate heat transfer analysis constitutes a dominant part of most current heat transfer design practices in gas turbines, being continuously improved for accuracy, speed, and robustness over the last several decades. Two primary factors favor the continued use of iteratively-coupled conjugate heat transfer in gas turbines. First, the numerical conduction solution using one of the commercially available finite element analysis (FEA) codes is both robust and fast, and so is the one-dimensional flow network solution for the coolant. Second, the method allows the use of heat transfer coefficients that are computed from empirical correlations, which may be adjusted based on the measurements from engine thermal survey for design validation.

The second solution method for a conjugate heat transfer problem requires using CFD on an integrated solid-fluid domain where all fluxes are conserved over the boundaries of adjacent computational cells. For a locally 1-D heat conduction and convection situation, Biot number, which is the ratio of thermal resistance of conduction to that of convection, emerges as a characteristic parameter. Ramachandran and Shih (2015) present a CFD-based conjugate heat transfer study to show how, through Biot number matching, one can deduce dimensionless metal temperature distributions at engine operating conditions by conducting experiments in a laboratory at less sever operating conditions. Their paper also lists a number of key references on this important topic.

Patankar (1980) presents control-volume-based CFD methods for both conduction and convection with unique physical insight and discussion. Ponnuraj et al. (2003) present a CFD-based conjugate heat transfer analysis of an industrial gas turbine compartment ventilation system to meet the requirements of ATEX certification. In this approach, both solid and fluid domains are directly coupled without the need to specify heat transfer coefficients at the solid-fluid boundaries. At first, this may appear as the strength of a CFD-based, directly-coupled method. But in reality it may not be so if the convection heat flux is not accurately computed as a result of inadequate turbulence modeling of heat transfer in the wall boundary layers, which in this case occur within the computational domain. In addition, this approach is computationally more intensive than the iteratively-coupled approach and may not meet the constraints of the ever-shrinking gas turbine design cycle time, especially when a robust design developed through probabilistic analyses is needed.

2.5 Concluding Remarks

In this chapter, we have briefly reviewed the essentials of thermodynamics, fluid mechanics, and heat transfer. Every reader should master this chapter to gain the required core competency for performing physics-based modeling of internal flow systems of gas turbines. In addition to some of the general concepts in these areas, we have introduced a number of novel concepts such as stream thrust, impulse pressure, rothalpy, nonisentropic generalized vortex, and windage in most simple terms.

The concept of rothalpy, which is easily derived from Euler's turbomachinery equation, plays an important role in a rotating system. Just as the total enthalpy remains constant in an adiabatic internal flow in a stationary duct, the rothalpy remains constant in an adiabatic flow in a rotating duct, automatically accounting for the flow energy change as a result of rotational work transfer. The chapter also presents a more accurate way to model both heat transfer and rotational work transfer in a duct, accounting for the nonlinear coupling between the two. A new method to evaluate pressure and temperature changes in a nonisentropic generalized vortex is also presented in this chapter. Windage is ubiquitous in gas turbine rotor-rotor and rotor-stator cavities. Its physics-based modeling, in contrast with the forced-vortex temperature rise, is clearly explained here. In addition, the chapter includes an easy-to-understand presentation of the Taylor-Proudman theorem and Ekman boundary layer, which features Ekman spiral.

In view of the inherent nonlinearity and geometric complexity, modern design practices in gas turbines favor numerical modeling of both flow and heat transfer, fully harnessing the available and growing computing power. To meet the needs of ever-shrinking design cycle time, these design practices are still dominated by locally one-dimensional analyses reinforced by product-validated empirical correlations. The classical closed-form analytical solutions; some of which are presented in this chapter, nevertheless, play an important role in validating various numerical methods and design tools. The chapter ends with a brief discussion of conjugate heat transfer, which is generally founded in the CFD technology.

Worked Examples

Example 2.1 Figure 2.29 shows an incompressible flow through a rotating duct of constant cross-section area. The duct is rotating at a constant angular velocity Ω. For a fluid of density ρ, determine the increase in static pressure in the duct from r_1 to r_2.

Solution

Because the duct cross-section and fluid density are constant, the radial flow velocity remains constant. As a result, the pressure force on a small control volume, shown in the figure, balances the centrifugal force in the radial direction. We can, therefore, write

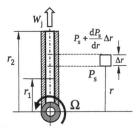

Figure 2.29 Incompressible flow through a radial, rotating duct of constant cross-section area (Example 2.1).

$$\left(P_s + \frac{dP_s}{dr}\Delta r\right)A - P_s A = A\Delta r \rho r \Omega^2$$

$$\frac{dP_s}{dr} = \rho r \Omega^2$$

Integrating from $r = r_1$ to $r = r_2$ yields

$$\int_1^2 dP_s = \int_{r_1}^{r_2} \rho r \Omega^2 dr$$

$$P_{s_2} - P_{s_1} = \frac{\rho\left(r_2^2 - r_1^2\right)\Omega^2}{2}$$

The aforementioned result shows that the change in static pressure equals the change in dynamic pressure associated with the fluid rotational velocity.

Example 2.2 Figure 2.30 shows a sprinkler with two unequal arms, each fitted with a nozzle. For the geometric and flow quantities shown in the figure, find an expression to calculate the maximum rotational velocity for the sprinkler with a constant frictional torque Γ_f. Assume that the jet area at outlet 1 is A_1 and that at outlet 2 is A_2.

Solution
In this sprinkler problem there is one inlet at the axis of rotation and two outlets at different radii. The flow at the inlet will have zero angular momentum. The mass velocity at each outlet corresponds to the jet velocity. Let us use the convention that the angular momentum is positive in the counterclockwise direction and negative in the clockwise direction. We can write various quantities at each outlet as follows:

Outlet 1

Mass flow rate: $\dot{m}_1 = A_1 \rho W_{j_1}$
Specific angular momentum: $R_1 V_{\theta_1} = R_1\left(R_1\Omega + W_{j_1}\right)$
Outflow rate of angular momentum: $\dot{H}_1 = \dot{m}_1 R_1\left(R_1\Omega + W_{j_1}\right)$

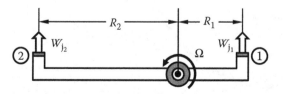

Figure 2.30 A sprinkler with two unequal arms (Example 2.2).

Outlet 2

Mass flow rate: $\dot{m}_2 = A_2 \rho W_{j_2}$

Specific angular momentum: $R_2 V_{\theta_2} = R_2\left(R_2\Omega - W_{j_2}\right)$

Outflow rate of angular momentum: $\dot{H}_2 = \dot{m}_2 R_2\left(R_2\Omega - W_{j_2}\right)$

Because the friction torque will act in the clockwise direction (opposite to the sprinkler arm rotation), the angular momentum equation yields

$$\dot{H}_1 + \dot{H}_2 = -\Gamma_f$$

$$\dot{m}_1 R_1\left(R_1\Omega + W_{j_1}\right) + \dot{m}_2 R_2\left(R_2\Omega - W_{j_2}\right) = -\Gamma_f$$

$$\Omega = \frac{\dot{m}_2 R_2 W_{j_2} - \dot{m}_1 R_1 W_{j_1} - \Gamma_f}{\dot{m}_1 R_1^2 + \dot{m}_2 R_2^2}$$

Example 2.3 Figure 2.31 shows a high-pressure rotary arm with three nozzles for air impingement cooling of a cylindrical surface. Air at total pressure of 3 bar and total temperature of 507.5 K exits each nozzle. The ambient pressure outside the rotary arm is 1 bar. At the maximum RPM, the rotary arm needs to overcome a frictional torque of 12.5 Nm. For the given geometric data, calculate the maximum RPM of the rotary arm. Note that each air nozzle operates under choked flow condition with identical jet velocity relative to the rotary arm. Assume $\kappa = 1.4$ and $R = 287\ \text{J/(kg K)}$ for air.

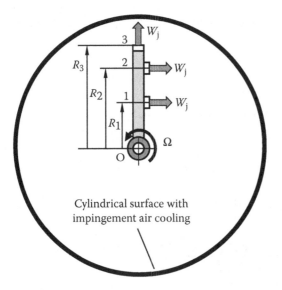

Figure 2.31 Impingement cooling of a cylindrical surface with three air jets (Example 2.3).

The geometric data are: jet diameter $(d_j) = 7$ mm, $R_1 = 0.50$ m, $R_2 = 1$ m, and $R_3 = 1.40$ m.

Solution

Mass flow rate through each air nozzle:

Jet area: $A_j = \frac{\pi d_j^2}{4} = \frac{\pi (0.007)^2}{4} = 3.8485 \times 10^{-5}$ m^2

Total pressure flow function at $M = 1$: $F_{f_t}^* = 0.0404$

$$\dot{m} = \frac{A_j F_{f_t}^* P_t}{\sqrt{T_t}} = \frac{3.8485 \times 10^{-5} \times 0.0404 \times 300000}{\sqrt{(134.5 + 273)}} = 0.0231 \text{ kg/s}$$

Air jet velocity relative to the rotary arm:

Static temperature at the nozzle throat: $T_s^* = \frac{2T_t}{\kappa + 1} = \frac{2 \times 407.5}{(1 + 1.4)} = 339.6$ K

For the choked flow through each nozzle, $M = 1$, giving

$$W_j = \sqrt{\kappa R T_s^*} = \sqrt{1.4 \times 287 \times 339.6} = 369.384 \text{ m/s}$$

Torque and angular momentum balance on the rotary arm control volume:

Absolute jet velocity contributing to the angular momentum flux at any radius is computed by

$$V_j = R\Omega - W_j \text{ (in the direction of tangential velocity)}$$

Net efflux of angular momentum in the counterclockwise direction

$$= \dot{m}R_1(R_1\Omega - W_j) + \dot{m}R_2(R_2\Omega - W_j) + \dot{m}R_3(R_3\Omega - 0)$$

Torque as a result of pressure force acting on the fluid control volume in the counterclockwise direction:

$$= A_j R_1 (P_s^* - P_{amb}) + A_j R_2 (P_s^* - P_{amb})$$

where

$$P_s^* = \frac{P_t}{\left(\dfrac{\kappa + 1}{2}\right)^{\frac{\kappa}{\kappa - 1}}} = \frac{300000}{1.893} = 158485 \text{ Pa}$$

If $Torque_{\text{arm-to-fluid}}$, the torque from the rotary arm, is acting in counterclockwise direction on the fluid control volume, then the torque-angular-momentum balance yields the following equation

$$Torque_{\text{arm-to-fluid}} + A_j R_1 (P_s^* - P_{amb}) + A_j R_2 (P_s^* - P_{amb})$$

$$= \dot{m}R_1(R_1\Omega - W_j) + \dot{m}R_2(R_2\Omega - W_j) + \dot{m}R_3(R_3\Omega - 0)$$

At the maximum RPM, the net torque acting on the rotary arm must be zero, i.e.,

$$Torque_{\text{fluid-to-arm}} - Torque_{\text{friction}} = 0$$

$$Torque_{\text{fluid-to-arm}} = Torque_{\text{friction}} = -Torque_{\text{arm-to-fluid}}$$

Thus,

$$-Torque_{\text{friction}} + A_j R_1 \left(P_s^* - P_{\text{amb}}\right) + A_j R_2 \left(P_s^* - P_{\text{amb}}\right)$$

$$= \dot{m} R_1 \left(R_1 \Omega_{\max} - W_j\right) + \dot{m} R_2 \left(R_2 \Omega_{\max} - W_j\right) + \dot{m} R_3 \left(R_3 \Omega_{\max} - 0\right)$$

Giving

$$\Omega_{\max} = \frac{\dot{m} W_j \left(R_1 + R_2\right) - Torque_{\text{friction}} + A_j \left(R_1 + R_2\right)\left(P_s^* - P_{\text{amb}}\right)}{\dot{m}\left(R_1^2 + R_2^2 + R_3^2\right)}$$

$$= \frac{\left\{\dot{m} W_j + A_j \left(P_s^* - P_{\text{amb}}\right)\right\}\left(R_1 + R_2\right) - Torque_{\text{friction}}}{\dot{m}\left(R_1^2 + R_2^2 + R_3^2\right)}$$

$$= \frac{(0.0231 \times 369.384 + 2.251) \times (0.5 + 1.0) - 12.5}{0.0231 \times \left((0.5)^2 + (1.0)^2 + (1.4)^2\right)} = 49.597 \text{ rad/s}$$

Rotary arm maximum $RPM = 473.6$

Example 2.4 Figure 2.32 shows an ejector pump used in many engineering applications. Describe the working principle of this pump. Making any simplifying assumptions, determine the ratio Q_2/Q_1.

Solution

The operating principle of an ejector pump schematically shown in Figure 2.32 is rather simple. The volumetric flow rate Q_1 in the convergent nozzle A enters the suction chamber B at a very high velocity. As a result, the static pressure of the stream as well the whole suction chamber drops considerably. This drop in static pressure in the

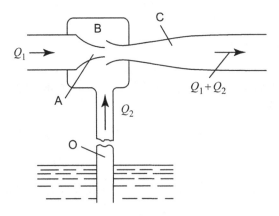

Figure 2.32 An ejector pump (Example 2.4).

chamber creates a suction pressure inducing the volumetric flow rate Q_2 in pipe O. The role of diffuser C is to allow the total volumetric flow rate $Q_1 + Q_2$ to exhaust at the ambient pressure with a subambient static pressure in chamber B.

For the present analysis, let us assume that the fluid of the nozzle flow is identical to that of the pumped (induced) flow with constant density ρ. For a total supply pressure of P_{t_1}, we can express the extended Bernoulli equation governing the nozzle flow from its inlet to outlet as

$$P_{t_1} = P_{s_B} + \frac{1}{2}\rho V_1^2 + K_1 \frac{1}{2}\rho V_1^2 = P_{s_B} + (1 + K_1)\frac{1}{2}\rho V_1^2$$

$$P_{t_1} = P_{s_B} + (1 + K_1)\frac{1}{2}\rho V_1^2$$

where K_1 is the minor loss coefficient in the nozzle based on its exit velocity V_1, and P_{s_B} is the subambient static pressure in chamber B.

Similarly, the extended Bernoulli equation for the induced flow from the ambient conditions yields

$$P_{amb} = P_{s_B} + \frac{1}{2}\rho V_2^2 + K_2 \frac{1}{2}\rho V_1^2 = P_{s_B} + (1 + K_2)\frac{1}{2}\rho V_2^2$$

Combining the aforementioned equation with the one previously obtained for the nozzle flow yields

$$P_{t_1} - P_{amb} = (1 + K_1)\frac{1}{2}\rho V_1^2 - (1 + K_2)\frac{1}{2}\rho V_2^2$$

$$P_{t_1} - P_{amb} = \frac{\rho(1 + K_1)}{2}\left(\frac{Q_1^2}{A_1^2}\right) - \frac{\rho(1 + K_1)}{2}\left(\frac{Q_2^2}{A_2^2}\right)$$

where A_1 and A_2 are the nozzle throat area and pipe area, respectively. By simplifying the aforementioned equation, we obtain

$$P_{t_1} - P_{amb} = C_1 Q_1^2 - C_2 Q_2^2$$

$$\frac{Q_2}{Q_1} = \sqrt{\frac{C_1}{C_2} - \frac{\left(P_{t_1} - P_{amb}\right)}{C_2 Q_1^2}}$$

where

$$C_1 = \frac{\rho(1 + K_1)}{2A_1^2} \text{ and } C_2 = \frac{\rho(1 + K_2)}{2A_2^2}$$

Example 2.5 Figure 2.32 shows a diverging duct whose diameter varies linearly from 0.1 m to 0.2 m over a length of 1.0 m. If the Darcy friction factor for the entire duct is constant at 0.005, find the major head loss as a result of friction within the duct for a volumetric flow rate of 1.0 m³/s.

Solution

Using the Darcy friction factor, the pressure loss as a result of friction in the fully developed flow in a circular duct of diameter D and length L can be written as

$$P_{f_{loss}} = f \frac{L}{D} \frac{1}{2} \rho V^2$$

which in terms of the volumetric flow rate Q becomes

$$P_{f_{loss}} = \frac{8fL\rho Q^2}{\pi^2 D^5}$$

In terms of major head loss, we can write the aforementioned equation as

$$h_{f_{loss}} = \frac{8fLQ^2}{\pi^2 g D^5}$$

Note that, for a given volumetric flow rate, the major head loss in the duct flow is independent of the fluid density.

For the diverging duct shown in Figure 2.33, the diameter increases in the flow direction, say the x direction. Assuming $x = 0$ at duct inlet, we can obtain its diameter at any x by the following equation:

$$D = 0.1(1 + x)$$

Noting that the component of the wall shear force along the flow direction is responsible for the frictional pressure loss in the diverging duct, we can express the differential major head loss over the differential duct length dx as

$$dh_{f_{loss}} = \frac{8fQ^2}{\pi^2 g D^5} dx$$

Substituting $D = 0.1(1 + x)$ in the aforementioned equation and integrating over the entire duct length, we obtain

$$h_{f_{loss}} = \frac{8fQ^2}{\pi^2 g (0.1)^5} \int_0^1 \frac{1}{(1 + x)^5} dx$$

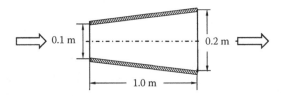

0.1 m   0.2 m

1.0 m

Figure 2.33 Head loss in a diverging duct flow (Example 2.5).

$$h_{f_{\text{loss}}} = \frac{8 \times 10^5 fQ^2}{\pi^2 g} \left[-\frac{1}{4} \frac{1}{(1+x)^4} \right]_0^1$$

$$h_{f_{\text{loss}}} = \frac{2 \times 10^5 fQ^2}{\pi^2 g} \left[1 - \frac{1}{(2)^4} \right]_0^1$$

$$h_{f_{\text{loss}}} = \frac{30 \times 10^5 fQ^2}{16\pi^2 g}$$

$$h_{f_{\text{loss}}} = \frac{30 \times 10^5 fQ^2}{16\pi^2 g}$$

$$h_{f_{\text{loss}}} = 9.683 \text{ m}$$

Example 2.6 Figure 2.34 shows an isentropic flow of air in a rubber pipe of constant-diameter 0.100 m with the inlet total pressure and total temperature of 1.2 bar and 300 K, respectively. The ambient pressure is 1.0 bar. The pipe is slowly deformed into a convergent-divergent nozzle, shown by the dotted line, until the flow just chokes ($M = 1.0$) at the throat. For the given boundary conditions, calculate the throat diameter. Assume air as a perfect gas with $\kappa = 1.4$ and $R = 287$ J/(kg K).

Solution

Area at pipe inlet and exit: $A_{\text{inlet}} = A_{\text{exit}} = \frac{\pi D_{\text{exit}}^2}{4} = \frac{\pi (0.1)^2}{4} = 7.854 \times 10^{-3} \text{ m}^2$

Pressure ratio at pipe exit: $\frac{P_{t_{\text{exit}}}}{P_{s_{\text{exit}}}} = \frac{P_{t_{\text{exit}}}}{P_{\text{amb}}} = \frac{1.2}{1.0} = 1.2$

Exit Mach number M_{exit}: $\frac{T_{t_{\text{exit}}}}{T_{s_{\text{exit}}}} = \left(\frac{P_{t_{\text{exit}}}}{P_{s_{\text{exit}}}} \right)^{\frac{\kappa-1}{\kappa}} = (1.2)^{0.286} = 1.0535$

$$\frac{T_{t_{\text{exit}}}}{T_{s_{\text{exit}}}} = 1 + \frac{\kappa - 1}{2} M_{\text{exit}}^2$$

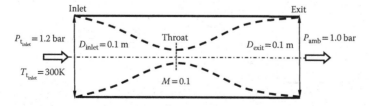

Figure 2.34 Isentropic air flow through a deformable rubber pipe (Example 2.6).

which yields

$$M_{\text{exit}} = \sqrt{\frac{2}{\kappa - 1}\left(\frac{T_{t_{\text{exit}}}}{T_{s_{\text{exit}}}} - 1\right)} = 0.517$$

Mass flow rate through the pipe:

$$\hat{F}_{f_{t_{\text{exit}}}} = M_{\text{exit}}\sqrt{\frac{\kappa}{R\left(1 + \frac{\kappa-1}{2}M_{\text{exit}}^2\right)^{\frac{\kappa+1}{\kappa-1}}}} = 0.517\sqrt{\frac{1.4}{287 \times \left(1 + \frac{1.4-1}{2}M_{\text{exit}}^2\right)^{\frac{1.4+1}{1.4-1}}}} = 0.5233$$

$$\dot{m} = \frac{\hat{F}_{f_{t_{\text{exit}}}} A_{\text{exit}} P_{t_{\text{exit}}}}{\sqrt{RT_{t_{\text{exit}}}}} = \frac{0.5233 \times 7.854 \times 10^{-3} \times 1.2 \times 10^5}{\sqrt{287 \times 300}} = 1.681 \text{ kg/s}$$

Throat diameter: For the choked flow at the throat, we have $M_{\text{throat}} = 1.0$ and $\hat{F}_{f_t}^* = 0.6847$, giving

$$A_{\text{throat}} = \frac{\dot{m}\sqrt{RT_{t_{\text{throat}}}}}{\hat{F}_{f_t}^* P_{t_{\text{throat}}}} = \frac{1.681 \times \sqrt{287 \times 300}}{0.6847 \times 1.2 \times 10^5} = 6.003 \times 10^{-3} \text{ m}^2$$

$$D_{\text{throat}} = 8.742 \times 10^{-2} \text{ m}$$

Example 2.7 Figure 2.35 shows air flow in a convergent-divergent (C-D) nozzle. A normal shock stands in the divergent section. The exit-to-throat area ratio $A_{\text{exit}}/A_{\text{throat}}$ of the C-D nozzle is known. For the given exit Mach number M_{exit}, write a step-by-step nongraphical and noniterative procedure to determine the ratio of nozzle area A_{NS} to the throat area A^* at the normal shock location.

Solution

Step 1: For the given value of M_{exit}, use isentropic flow Table A.2 of Sultanian (2015) to determine A_{exit}/A_2^*.

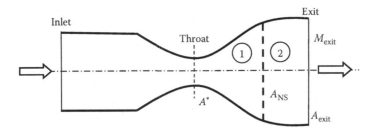

Figure 2.35 Normal shock located in a convergent-divergent nozzle (Example 2.7).

Step 2: Because $A_1^* = A^*$, calculate the ratio A_1^*/A_2^* as

$$\frac{A_1^*}{A_2^*} = \frac{\dfrac{A_{\text{exit}}}{A_2^*}}{\dfrac{A_{\text{exit}}}{A_1^*}}$$

Step 3: Because the mass flow rate and total temperature remain constant across a normal shock, we can write $A_1^*/A_2^* = P_{t_2}/P_{t_1}$; the flow on the either side of the normal shock is isentropic.

Step 4: Knowing P_{t_2}/P_{t_1}, find M_1 and M_2 from the normal shock Table A.6 of Sultanian (2015)

Step 5: Use M_1 to find A_{NS}/A_1^* from the isentropic flow Table A.2 of Sultanian (2015)

Example 2.8 In an attempt to reduce the heat loss from a metal hot-water tube of outer radius 6.5 mm, a plumber decides to insulate the water line with 12 mm thick insulation having thermal conductivity of 0.156 W/(m K). As a result of the high heat transfer coefficient between water flow and metal pipe and the high thermal conductivity of the metal, the entire metal pipe can be considered to remain at a uniform temperature of 350 K. The tube is surrounded by air at 300 K, for which $h = 9.0$ W/(m^2 K). In terms of percentage, how successful is the plumber in reducing the heat loss?

Solution

Let us compute the heat transfer rate per unit length of the tube, when it is bare (without insulation) and with insulation in place.

$$\dot{Q}_{\text{bare}} = h(2\pi r_1)(T_1 - T_\infty) = 9(2\pi \times 0.065)(350 - 300) = 18.378 \text{ W}$$

$$\dot{Q}_{\text{ins}} = \frac{2\pi(T_1 - T_\infty)}{\dfrac{\ln(r_2/r_1)}{k} + \dfrac{1}{hr_2}} = \frac{2\pi \times (350 - 300)}{\dfrac{\ln(0.0185/0.065)}{0.156} + \dfrac{1}{(9 \times 0.0185)}} = 24.716 \text{ W}$$

$$\text{Percentage reduction in heat loss} = \left(\frac{\dot{Q}_{\text{bare}} - \dot{Q}_{\text{ins}}}{\dot{Q}_{\text{bare}}}\right) \times 100 = -34.5\%$$

It is interesting to note that by insulating the tube, heat transfer increased by 34.5 percent; one would expect it to decrease. This situation is commonly known as the plumber paradox. The critical outer radius in this case is given by $r_c = k/h = 0.156/9 = 0.01733$, which corresponds to the maximum heat transfer from the insulated tube. It can be easily verified that, for the heat transfer from the insulated tube to be lower than that from a bare tube, the insulation thickness must be more than 67.5 mm.

Example 2.9 Figure 2.36 shows the rectangular cross-section of a plate, which is very long in the normal direction along which no property varies. As shown in the figure, one end of the plate is held at a constant temperature T_0 and the opposite end remains adiabatic ($\dot{q} = 0$). The bottom of the plate is heated with a uniform heat flux $\dot{q}_1$ and the top side is subject to convective boundary condition with constant heat transfer coefficient h and fluid temperature T_f. The Bio number ($Bi = ht/k$) is small enough that any temperature variation along the plat thickness may be neglected. The uniform volumetric heat generation rate within the plate equals $\dot{s}$. Find an analytical solution to compute plate temperature variation in the x direction, that is, find $T = T(x)$.

Solution
Assuming unit plate length in the normal direction, the energy balance over the control volume between x and $x + \Delta x$ yields

$$t\left(\dot{q}_x + \frac{\mathrm{d}\dot{q}_x}{\mathrm{d}x}\Delta x\right) - t\dot{q}_x = \dot{q}_1\Delta x + \dot{s}t\Delta x - h\Delta x\left(T - T_f\right)$$

$$t\frac{\mathrm{d}\dot{q}_x}{\mathrm{d}x} = \dot{q}_1 + \dot{s}t - h\left(T - T_f\right)$$

Using Fourier's law of heat conduction and simplifying the resulting expression, we finally obtain

$$\frac{\mathrm{d}^2\hat{\theta}}{\mathrm{d}\xi^2} - \hat{m}^2\hat{\theta} = 0$$

where

$$\hat{\theta} = \frac{T - \beta - T_f}{T_0 - \beta - T_f}; \beta = \frac{\dot{q}_1 + \dot{s}t}{h}; \xi = \frac{x}{L}; \text{and}\,\hat{m}^2 = \frac{hL^2}{kt}$$

With the boundary conditions $\theta = 1 @ \xi = 0$ and $\mathrm{d}\theta/\mathrm{d}\xi = 0 @ \xi = 1$, the solution of the aforementioned second-order homogeneous ordinary differential equation can be easily obtained as

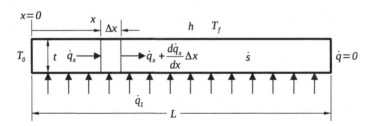

Figure 2.36 One-dimensional temperature variation in a rectangular plate with mixed boundary conditions and uniform volumetric heat generation (Example 2.9).

$$\hat{\theta} = \frac{e^{\hat{m}\zeta}}{1 + e^{2\hat{m}}} + \frac{e^{-\hat{m}\zeta}}{1 + e^{-2\hat{m}}} = \frac{\cos h\,\hat{m}(1 - \zeta)}{\cos h\,\hat{m}}$$

Problems

2.1 Consider an incompressible flow through a constant-area circular pipe with negligible wall friction. At the inlet, the pipe is connected to a plenum maintained constant pressure. At the outlet, the flow is discharged at the fixed ambient pressure. If a short ideal conical diffuser is appended to the pipe at its outlet, will the mass flow rate increase, decrease, or remain constant? How will your answer change for a compressible flow with subsonic and supersonic conditions at the pipe exit?

2.2 Based on how the static and total pressures vary in an adiabatic diffuser, shown in Figure 2.37, will the flow in the bypass duct occur from A to B or from B to A? Explain your reasoning.

2.3 For an incompressible flow through a variable-area duct, which is frictionless and rotates at a constant angular velocity, show that the change in total pressure relative to the duct equals the change in static pressure as a result of rotation alone, with no change in flow area.

2.4 Figure 2.38 shows steady water flow through a vertical pipe of radius R. A fully developed laminar flow with a parabolic velocity profile prevails at section 1. The flow transitions to a fully-developed turbulent flow with the 1/7th power-law profile at section 2. Over the pipe length L, the total shear force and gravitational body force correspond to F_f and W_g, respectively. For the mass flow rate $\dot{m}$

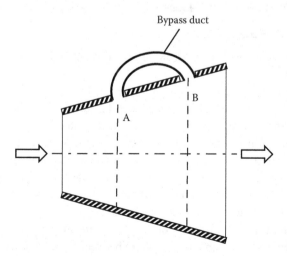

Figure 2.37 Flow direction in the bypass duct of an adiabatic diffuser (Problem 2.2).

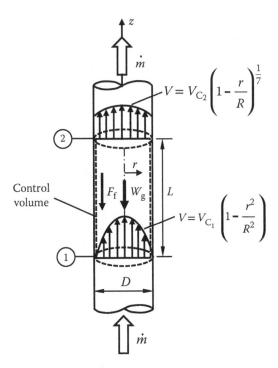

Figure 2.38 Steady flow of water through a vertical pipe (Problem 2.4).

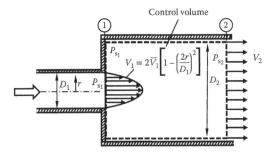

Figure 2.39 Sudden expansion pipe flow with a parabolic velocity profile at inlet and uniform velocity at outlet (Problem 2.5).

through the pipe, the static pressure at sections 1 and 2 are P_{s_1} and P_{s_2}, respectively. Express the total shear force F_f in terms of other known quantities.

2.5 Figure 2.39 shows an incompressible flow in a sudden pipe expansion with the upstream pipe diameter D_1 and downstream pipe diameter D_2. The laminar flow in the upstream pipe is fully developed with a parabolic velocity profile. The flow exiting the larger pipe is turbulent, which is assumed to have a uniform velocity profile. The static pressure at both sections 1 and 2 is assumed to be uniform. Neglecting any shear stress on the downstream pipe wall, find the change in both static pressure and total pressure between sections 1 and 2 of this flow system.

2.6 Figure 2.40 shows impingement air cooling of a cylindrical surface using a high-pressure rotary arm fitted with identical nozzles. Each nozzle is choked with equal mass flow rate $\dot{m}$ and exit static pressure P_s^*, which is higher than the ambient pressure P_{amb}. At the maximum rotational velocity Ω, the rotary arm needs to overcome a frictional torque Γ_f. Derive an expression to calculate Ω.

2.7 As shown in Figure 2.41, two air streams of different properties enter a long duct with the cross-sectional area of 0.002 m^2. At section 1 (inlet), the hotter stream forms the central core and is surrounded by the colder stream. Each stream occupies equal area (neglect the wall thickness of the inner duct). The uniform static pressure at section 1 equals 895614 Pa and that at section 2 (outlet) equals 10 bar. The total pressure and total temperature of each stream at section 1(inlet) are given later. Both streams are assumed to have been fully mixed before exiting the duct at section 2 with uniform properties.

	Stream 1	Stream 2
Total Temperature (K)	1173	1473
Total Pressure (bar)	10	15

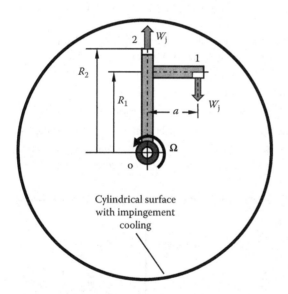

Figure 2.40 Impingement air cooling of a cylindrical surface (Problem 2.6).

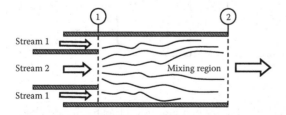

Figure 2.41 Mixing of two compressible flows in a duct (Problem 2.7).

Assuming $\kappa = 1.4$ and $R = 287 \, \text{J}/(\text{kg K})$, calculate:

(a)   Mass flow rate of each stream
(b)   Section-average total temperature and static temperature at inlet
(c)   Section-average total pressure at inlet
(d)   Percentage drop in total pressure from inlet to outlet
(e)   Total wall shear force acting on the flow between inlet and outlet

2.8   Figure 2.42 shows a high-pressure inlet bleed heat system (IBH) of a land-based gas turbine used for power generation. When the ambient air is cold, the inlet bleed heat system is used to raise its temperature to prevent ice formation on the compressor IGV's (inlet guide vanes). In this system, the hot air is bled from an intermediate compressor stage and mixed uniformly with air flow at the engine inlet. Let's consider such a system. The ambient pressure and temperature are 1 bar and 293 K, respectively. The Mach number of the air flow entering the IGV is 0.6. To prevent ice formation, the static temperature at this section is required to be 275 K. The air mass flow rate entering the engine inlet system (before the high-pressure bleed heat section) at a low Mach number ($M < 2$) is 275 kg/s. The total temperature and pressure of the air bled from the compressor are 542 K and 8 bar, respectively. Neglect any changes in the pressure and temperature in the bleed air supply system. Assuming air as a perfect gas with $\kappa = 1.4$ and $R = 287 \, \text{J}/(\text{kg K})$:

(a)   Calculate the mass flow rate of the compressor bleed air to achieve the design objective at compressor IGV inlet
(b)   To promote uniform mixing of the hot compressor bleed air with the cold inlet ambient air, 200 nozzles for the bleed air injection are uniformly placed in the inlet duct cross-section. Find the effective flow area of each nozzle.

2.9   In a land-based gas turbine used for power generation, the turbine exhaust enters an annular diffuser, shown in Figure 2.43, at the total-velocity Mach number of 0.60 and the total temperature of 723 K. The swirl velocity (tangential velocity) at

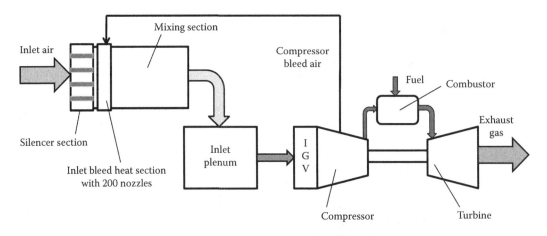

Figure 2.42 High-pressure inlet bleed heat (IBH) system of a land-based gas turbine power plant (Problem 2.8).

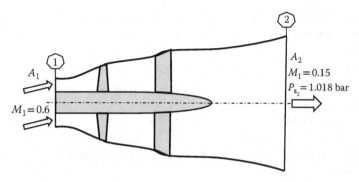

Figure 2.43 Exhaust diffuser of a land-based gas turbine for power generation (Problem 2.9).

the diffuser inlet equals15 percent of the total velocity. The flow exits the diffuser fully axially at a Mach number of 0.15 with no change in total temperature. The design calls for the static pressure at the diffuser exit to be 1.018 bar to allow the exhaust gases to discharge into the ambient air via a downstream duct system (not shown in the figure). The exit-to-inlet flow area ratio for the annular diffuser is 3.45. Find the pressure rise coefficient (C_p) for the annular diffuser. If the last stage turbine is redesigned for zero swirl velocity at its exit, how will the pressure rise coefficient of the annular diffuser change? Assume exhaust gases with $\kappa = 1.4$ and $R = 287$ J/(kg K).

2.10 An isentropic convergent-divergent nozzle having an area ratio (exit area/throat area) of 2 discharges air into an insulated pipe of length L and diameter D. The nozzle inlet air has the stagnation pressure of 7×10^5Pa and the stagnation temperature of 300 K, and the pipe discharges into a space where the static pressure is 2.8×10^5Pa. Calculate fL/D of the pipe and its mass flow rate per unit area for the cases when a normal shock stands: (a) at the nozzle throat, (b) at the nozzle exit, and (c) at the pipe exit.

2.11 A normal shock is detected in a convergent-divergent nozzle with the given exit-to- throat area ratio, inlet total pressure, and inlet total temperature. The air flow is isentropic on either side of the normal shock. For the given exit static temperature, write a noniterative, nongraphical step-by-step procedure to determine the ratio of the nozzle flow area, where the normal shock is located, to the nozzle throat area. Assume $\kappa = 1.4$ and $R = 287$ J/(kg K) for air.

2.12 A supersonic air-flow with known total pressure, total temperature, and static pressure enters a pipe of given constant diameter and constant friction factor. With its adiabatic and stationary wall, the pipe is 15 percent longer than the maximum needed to achieve the choked flow conditions (frictional choking). While the flow remains choked, a normal shock occurs within the pipe. Write a step-by-step procedure to determine the location (the distance from the pipe inlet) of the normal shock. Assume air to be a perfect gas with $\kappa = 1.4$ and $R = 287$ J/(kg K).

2.13 A normal shock is detected in a convergent-divergent nozzle with known exit-to-throat area ratio, inlet total pressure, and inlet total temperature. The air flow is isentropic on either side of the normal shock. For a given exit static temperature, write a noniterative, nongraphical step-by-step procedure for finding the nozzle area ratio (ratio of nozzle area to throat area) that corresponds to the normal shock location. For air, assume $\kappa = 1.4$ and $R = 287\,\mathrm{J/(kg\,K)}$.

2.14 Derive a relationship between the loss coefficient K and the discharge coefficient C_d for an incompressible flow in a sudden pipe expansion with the larger-to-smaller area ratio A_2/A_1.

2.15 For both radially inward and outward flows through two counterrotating disks, derive relevant equations to compute windage temperature rise between two radial locations. Assume both disks to be adiabatic (zero heat transfer). List all your assumptions.

References

Becker, M. 1986. *Heat Transfer: A Modern Approach*. New York: Plenum Press.

Carslaw, H. S., and J. C. Jaeger. 1959. *Conduction of Heat in Solids*, 2nd edn. Oxford: Oxford University Press.

Eckert, E. R. G. 1961. *Survey of heat transfer at high speeds*. A.R.L. Report 189.

Greitzer, E. M., C. S. Tan, and M. B. Graf. 2004. *Internal Flow Concepts and Applications*. Cambridge: Cambridge University Press.

Han, J. C. 2012. *Analytical Heat Transfer*. Boca Raton, FL: Taylor & Francis.

Howell, J. R., M. P. Menguc, and R. Siegel. 2015. *Thermal Radiation Heat Transfer*, 6th edn. Boca Raton, FL: Taylor & Francis.

Idelchik, I. E. 2005. *Handbook of Hydraulic Resistance*, 3rd edn. Delhi, India: Jaico Publishing House.

Lugt, H. J. 1995. *Vortex Flow in Nature and Technology*. Malabar: Krieger Publishing Company.

Minkowycz, W. J., E. M. Sparrow, G. E. Schneider, and R. H. Pletcher (eds.). 1988. *Handbook of Numerical Heat Transfer*. New York: Wiley.

Minkowycz, W. J., E. M. Sparrow, and J. Y. Murthy (eds.). 2006. *Handbook of Numerical Heat Transfer*, 2nd edn. Hoboken, NJ: John Wiley & Sons.

Patankar, S. V. 1980. *Numerical Heat Transfer and Fluid Flow*. Boca Raton, FL: Taylor & Francis.

Ponnuraj, B., B. K. Sultanian, A. Novori, and P. Pecchi. 2003. *3D CFD analysis of an industrial gas turbine compartment ventilation system*. Proc. ASME IMECE. Washington, DC.

Ramachandran, S. G., T. I-P. Shih. 2015. Biot number analogy for design of experiments in turbine cooling. *ASME J. Turbomachinery*. 137(6): 061002.1–061002.14.

Sultanian, B. K. 1978. *Influence of Geometry and Peripheral Heat Transfer Coefficients on Heat Conduction in Coolant Ducts*. MS thesis, Mechanical Engineering Department, Indian Institute of Technology, Madras.

Sultanian, B. K., and V. M. K. Sastri. 1979. Steady heat conduction in a circular duct with circumferentially varying heat transfer coefficients. *Regional J. Energy Heat Mass Transfer*. 1(2): 101–110.

Sultanian, B. K., and V. M. K. Sastri. 1980. Effect of geometry on heat conduction in coolant channels of a liquid rocket engine. *Wärme- und Stoffübertragung*. 14: 245–251.

Sultanian, B. K., and E. M. Kotliarevsky. 1981. On the mathematical heat transfer modeling of steel ingot with solidification and heating in the soaking pit. *Regional J. Energy Heat Mass Transfer*. 3(1): 11–27.

Sultanian, B. K. 1984. *Numerical modeling of turbulent swirling flow downstream of an abrupt pipe expansion*. PhD diss., Arizona State University.

Sultanian, B. K., G. P. Neitzel, and D. E. Metzger. 1986. A study of sudden expansion pipe flow using an algebraic stress transport model. Paper presented at the AIAA/ASME 4th Fluid Mechanics, Plasma Dynamics and Lasers Conference, Atlanta.

Sultanian, B. K. 2015. *Fluid Mechanics: An Intermediate Approach*. Boca Raton, FL: Taylor & Francis.

Bibliography

Bergman, T. L., A. S. Lavine, F. P. Incropera, and D. P. DeWitt. 2011. *Fundamentals of Heat and Mass Transfer*, 7th edn. New York: Wiley.

Blevin, P. R. N. 2003. *Applied Fluid Dynamics Handbook*. Malabar: Krieger Publishing Company.

Childs, P. D. 2011. *Rotating Flow*. Burlington, MA: Elsevier.

Colebrook, C. 1938–1939. Turbulent flow in pipes, with particular reference to the transition region between the smooth and rough pipe laws. J. Inst. Civ. Eng., *London*. 11: 133–156.

Daneshyar, H. 1976. *One-Dimensional Compressible Flow*. New York: Pergamon Press.

Dixon, S. L., and C. A. Hall. 2013. *Fluid Mechanics and Thermodynamics of Turbomachinery*, 7th edn. Oxford: Butterworth-Heinemann.

Fox, W., P. Prichard, and A. McDonald. 2010. *Introduction to Fluid Mechanics*, 7th edn. New York: John Wiley & Sons.

Greenspan, H .P. 1968. *The theory of rotating fluids*. Cambridge: Cambridge University Press.

Haaland, S. 1983. Simple and explicit formulas for friction factor in turbulent flow. Trans. ASME, J. *Fluids Eng*. 103: 89–90.

Holman, J. P. 2009. *Heat Transfer*, 10th edn. New York: McGraw-Hill.

Korpela, S. A. 2011. *Principles of Turbomachinery*. New York: John Wiley.

Kreith, F., R. M. Manglik, and M. S. Bohn. 2010. *Principles of Heat Transfer*, 7th edn. Stanford, CA: Cengage Learning

Kundu, P. K., I. M. Cohen, and D. R. Dowling. 2012. *Fluid Mechanics*, 5th edn. Waltham, MA: Elsevier.

Miller, R. W. 1996. *Flow Measurement Engineering Handbook*, 3rd edn. New York: McGraw-Hill.

Moody, L. 1944. Friction factors for pipe flow. *Trans. ASME*. 66(8): 671–684.

Moran, M. J., H. N. Shapiro, F. P. Incropera, D. D. Boettner., and M. B. Bailey 2011. *Fundamentals of Engineering Thermodynamics*, 7th edn. New York: Wiley.

Mott, R. L. 2006. *Applied Fluid Mechanics*, 6th edn. Upper Saddle River, NJ: Pearson Prentice Hall.

Saravanamutto, H. I. H., G. F. C. Rogers, and H. Cohen. 2001. *Gas Turbine Theory*, 5th edn. New York: Prentice Hall.

Sultanian, B. K., and D. A. Nealy. 1987. Numerical modeling of heat transfer in the flow through a rotor cavity. In D. E. Metzger, ed., *Heat Transfer in Gas Turbines*, HTD-Vol. 87, 11–24, New York: ASME.

Shapiro, A. H. 1953. *The Dynamics and Thermodynamics of Compressible Fluid Flow*, Vols. 1 and 2. New York: Ronald Press.

Swamee, P., and A. Jain. 1976. Explicit equations for pipe-flow problems. *J. Hydraul. Di. (ASCE).* 102 (5): 657–664.

Tritton, D. J. 1988. *Physical Fluid Dynamics.* Oxford: Clarendon Press.

Van Dyke, M. 1982. *An Album of Fluid Motion.* Stanford, CA: The Parabolic Press.

Vanyo, J. P. 2012. *Rotating Fluids in Engineering and Science.* Mineola, NY: Dover Publishing, Inc.

White, F. 2010. *Fluid Mechanics with Student DVD,* 7th edn. New York: McGraw-Hill.

Wilson, D. G., and T. Korakianitis. 1998. *The Design of High-Efficiency Turbomachinery and Gas Turbines,* 2nd edn. Upper Saddle River, NJ: Prentice-Hall.

Zucrow, M. J., and J. D. Hoffman. 1976. *Fundamentals of Gas Dynamics.* New York: John Wiley.

Nomenclature

A	Flow area
ATEX	Atmosphères Explosibles
A_c	Area of Vena contracta
$\vec{A}$	Flow area vector
c	Specific heat of solid
c_p	Specific heat of gas at constant pressure
c_v	Specific heat of gas at constant volume
C	Speed of sound
C_d	Discharge coefficient
C_f	Shear coefficient or Fanning friction factor
C_p	Diffuser pressure rise coefficient
D	Pipe diameter
f	Moody or Darcy friction factor
e	Specific total energy of a system
$\hat{e}_r$	Unit vector in the radial direction
E	Total energy of a system
Ek	Ekman number ($Ek = \mu/\rho\Omega L_{ref}^2$)
F	Total force
$\vec{F}$	Force vector
F_{f_s}	Static-pressure mass flow function with dimensions of $\sqrt{R}$
$\hat{F}_{f_s}$	Dimensionless static-pressure mass flow function
F_{f_t}	Total-pressure mass flow function with dimensions of $\sqrt{R}$
$\hat{F}_{f_t}$	Dimensionless total-pressure mass flow function
g	Acceleration as a result of gravity
Gr	Grashof number ($Gr = g\beta(\Delta T)L^3\rho^2/\mu^2$)
h	Specific enthalpy; heat transfer coefficient; height from a datum
H	Angular momentum
$\vec{H}$	Angular momentum vector
$\dot{H}$	Angular momentum flow rate
I	Rothalpy; current

I_{f_s}	Static-pressure impulse function
I_{f_t}	Total-pressure impulse function
k	Thermal conductivity
K	Loss coefficient
KE	Kinetic energy
L_{ref}	Reference length scale
m	Mass
$\dot{m}$	Mass flow rate
M	Mach number; linear momentum
$\vec{M}$	Linear momentum vector
$\dot{\vec{M}}$	Linear momentum flow rate
M_θ	Rotational Mach number $(M_\theta = r\Omega/\sqrt{\kappa R T_s})$
N	Normal shock function
N_∞	Asymptotic value of N as $M \to \infty$
N_{in}	Number of inlets
Nu	Nusselt number
N_{out}	Number of outlets
P	Pressure
P_i	Impulse pressure
$\tilde{P}$	Reduced pressure
$P_{d_{com}}$	Compressible dynamic pressure
$P_{d_{inc}}$	Incompressible dynamic pressure
$P_{d_{ratio}}$	Ratio of compressible dynamic pressure to incompressible dynamic pressure
$\dot{P}_{entropy}$	Rate of entropy production
PE	Potential energy
q	Specific heat transfer
$\dot{q}$	Heat flux
$\vec{q}$	Heat flux vector
Q	Heat transfer; volumetric flow rate
$\dot{Q}$	Heat transfer rate; cycle heat rate
r	Cylinder polar coordinate r
$\vec{r}$	Position vector from the origin (rotating coordinate system)
r^*	Recovery factor
R	Gas constant; radial distance; resistance
Re	Reynolds number
Ro	Rossby number $(Ro = W_{ref}/\Omega L_{ref})$
s	Specific entropy
$\dot{s}$	Volumetric heat generation rate
S	Entropy
S_f	Swirl factor
S_T	Stream thrust
t	Time; thickness
T	Temperature

T_f	Reference fluid temperature
u	Specific internal energy
U	Total internal energy of a system; rotor tangential velocity
U_i	Velocity in tensor notation
v	Specific volume
V	Total absolute velocity; voltage
$\vec{V}$	Absolute velocity vector
V	Volume
w	Specific work transfer
W	Total relative velocity; work transfer; weight
$\vec{W}$	Relative velocity vector
$\dot{W}$	Rate of work transfer (power)
W_{ref}	Reference relative velocity scale
x	Cartesian coordinate x; axial direction
x_i	Cartesian coordinates in tensor notation
y	Cartesian coordinate y
z	Cartesian coordinate z
Z	Complex velocity

Subscripts and Superscripts

aw	Adiabatic wall
B	Body
bare	Without insulation
com	Compressible
cond	Conduction
conv	Convection
cycle	Thermodynamic cycle
C	Compressor
CS	Control system
CV	Control volume
CCT	Coupling correction term
CVS	Control volume surface
D	Duct
f	Fluid; friction
fc	Forced convection
F_{1-2}	Radiation view factor between surfaces 1 and 2
g	Gravity
i	Isentropic; impulse
in	Inlet
inc	Incompressible
ins	With insulation
irrev	Irreversible

m	Metal; material
max	Maximum
nc	Natural convection
net	Net (turbine work output minus compressor work input)
out	Outlet
p	Polytropic
r	Component in r coordinate direction
rad	Radiation
ref	Reference
rev	Reversible
R	Rotor reference frame
s	Static
S	Stator reference frame; surface
sh	Shear
t	Total
T	Turbine
th	Thermal
w	Wall
x	Component in x coordinate direction
y	Component in y coordinate direction
z	Component in z coordinate direction
∞	Free stream; asymptotic value for $M \to \infty$
θ	Component in θ direction
*	Value at sonic condition ($M = 1$)

Greek Symbols

α	Thermal diffusivity ($\alpha = k/(\rho c)$)
β	Isobaric compressibility of fluid
ε	Surface emissivity
Γ	Torque
$\vec{\Gamma}$	Torque vector
$\vec{\zeta}$	Vorticity vector
η	Efficiency; number of transfer units (NTU): $\eta = \pi DhL/(\dot{m}c_p)$
θ	Cylindrical polar coordinate θ; dimensionless temperature
$\tilde{\theta}$	Angle between resultant velocity and tangential velocity within Ekman boundary layer
κ	Ratio of specific heats ($\kappa = c_p/c_v$)
λ	Second coefficient of viscosity
μ	Dynamic viscosity
ν	Kinematic viscosity ($\nu = \mu/\rho$)
ξ	Dimensionless similarity variable
π	Pressure ratio

Π	Change agent
ρ	Density
σ	Stefan-Boltzmann constant
τ	Shear stress
Φ	Extensive general property
ϕ	Intensive general property
$\vec{\omega}$	Local rotation vector
Ω	Rotational velocity around axial direction
$\vec{\Omega}$	Rotational velocity vector

3 1-D Flow and Network Modeling

3.0   Introduction

The main objective of an engineering analysis is to have the prediction results as close to the physical reality as possible. These results are said to be numerically accurate if the modeling equations are solved correctly. They are considered physically accurate if they also correctly predict the physical reality, which is independent of the method of analysis. In general, for prediction results to be physically accurate, they ought to be numerically accurate. In the context of computer code (design tool) development, the numerical accuracy of the computed results is ensured through code verification, and their physical accuracy is determined by code validation. In our discussion, we will tacitly assume that the computational method used for solving a flow network ensures numerical accuracy. The accuracy in the present context essentially means physical accuracy.

The physics-based one-dimensional (1-D) thermofluids modeling of various components of internal flow systems of a gas turbine is a design necessity, offering the best compromise of prediction accuracy, speed, and cost. Predictions are generally made using complex flow networks of these components in all three design phases: conceptual, preliminary, and detailed. The reliability of design predictions depends in large part on the company-proprietary empirical correlations. Each robust design is performed with compressible flow networks in a short available design cycle time. The CFD technology is leveraged in two ways; first, to develop a better understanding of the component flow physics, reinforcing its 1-D modeling, and second, to delineate areas of design improvement in the detailed design phase using an entropy map. In this chapter, we limit our discussion to steady compressible flow networks with the possibility of internal choking and normal shocks.

The accuracy of results from a flow network primarily depends upon two factors: (1) the core formulation that ensures that the conservation laws of mass, momentum, energy, and entropy are duly satisfied for each flow element and junction in the network and (2) the empirical correlations, which are based on the benchmark quality data, are representative of the physical reality. While there is always a need for improved empirical correlation for an existing or newly designed element, the core mathematical formulation based on the established laws of flow and heat transfer physics ought to remain invariant. Accordingly, the physics-based modeling in the context of our discussion in this chapter will imply that all the conservation laws are fully satisfied in the flow network. Thus, any lack of accuracy of a network solution can be entirely attributed to the deficiencies in one or more empirical correlations used in the network.

3.1   1-D Flow Modeling of Components

The physics-based 1-D flow modeling of each component of gas turbine internal flow systems is achieved through large control volume analysis of the conservation equations of mass, momentum, energy, and entropy. For a compressible flow, the momentum equation remains coupled with the energy equation through density, which is computed using the equation of state of a perfect gas. In this chapter, we present the modeling of a duct and an orifice, which are two basic components of an internal flow system. Modeling of other special components, such as vortex, rotor-rotor and rotor-stator cavities, and seals are presented in the following chapters.

3.1.1   Duct with Area Change, Friction, Heat Transfer, and Rotation

In this section, we present the most general modeling of one-dimensional compressible flow in a variable-area duct with wall friction, heat transfer, and rotation (constant angular velocity) about an axis different from the flow axis. The modeling methodology also allows for the presence of internal choking ($M = 1$) and normal shock, which features abrupt changes in flow properties. In the present approach, the long duct is divided into multiple control volumes, which are serially coupled such that the outlet flow properties of one control volume become the inlet flow properties for the downstream one. In each control volume, wall boundary conditions are assumed uniform, although they may vary over different control volumes. Thus, in order to model the entire duct, we need to develop modeling equations only for a representative control volume, as has been done in the following sections.

3.1.1.1   Mass Conservation (Continuity Equation)

For the duct control volume shown in Figure 3.1, the velocity, density, and area for the x-direction flow are assumed to vary linearly from inlet (section 1) to outlet (section 2). For a steady flow through the control volume, the mass conservation yields

$$\dot{m} = \rho_1 V_1 A_1 = \overline{\rho}\,\overline{V}\,\overline{A} = \rho_2 V_2 A_2 \qquad (3.1)$$

where

$$\overline{\rho} = \frac{\rho_1 + \rho_2}{2}$$

$$\overline{V} = \frac{V_1 + V_2}{2}$$

$$\overline{A} = \frac{A_1 + A_2}{2}$$

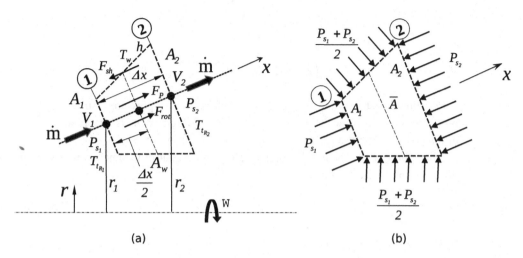

Figure 3.1 (a) Duct control volume with area change, friction, heat transfer, and rotation and (b) pressure distribution on the control volume.

In terms of the total pressure, total temperature, Mach number, and total-pressure mass flow function (presented in Chapter 2), we can compute mass flow rate at the duct control volume inlet and outlet as

$$\dot{m} = \frac{F_{f_{t_1}} A_1 P_{t_1}}{\sqrt{RT_{t_1}}} = \frac{F_{f_{t_2}} A_2 P_{t_2}}{\sqrt{RT_{t_2}}}$$

3.1.1.2   Linear Momentum Equation

For the duct control volume shown in Figure 3.1a, we can write the steady flow momentum equation as

$$F_P - F_{sh} + F_{rot} = \dot{m}V_2 - \dot{m}V_1 = \dot{m}\left(V_2 - V_1\right) \tag{3.2}$$

where

$F_P \equiv$ Pressure force acting on the control volume in the momentum direction

$F_{sh} \equiv$ Shear force acting on the control volume opposite to the momentum direction

$F_{rot} \equiv$ Rotational body force acting on the control volume in the momentum direction

Let us now evaluate the surface forces due to static pressure and wall shear and the body force due to rotation.

Pressure force *(F_P).* Figure 3.1b shows the static pressure distribution on the duct control volume, where we have assumed that the average pressure on the lateral surfaces is the average of inlet and outlet pressures. The total pressure force in the x direction, resulting from the surface pressure distribution, can be expressed as

$$F_P = P_{s_1}A_1 + \frac{1}{2}\left(P_{s_1} + P_{s_2}\right)\left(A_2 - A_1\right) - P_{s_2}A_2$$

$$F_P = P_{s_1}\left(\frac{A_1 + A_2}{2}\right) - P_{s_2}\left(\frac{A_1 + A_2}{2}\right)$$

$$F_P = \left(P_{s_1} - P_{s_2}\right)\left(\frac{A_1 + A_2}{2}\right) = \left(P_{s_1} - P_{s_2}\right)\bar{A} \tag{3.3}$$

Equation 3.3 shows that the net force resulting from a pressure distribution on the duct control volume with area change in the flow direction is the product of the difference in the inlet and outlet pressures and the mean flow area.

Shear force (F_{sh}). The net shear force acting on the control volume in the momentum direction can be expressed as

$$F_{sh} = \bar{A}\bar{f}\frac{\Delta x}{D_h}\frac{1}{2}\bar{\rho}\bar{V}^2 \tag{3.4}$$

where $\bar{f}$ is the average value of the Darcy friction factor over the lateral control volume surface.

Rotational body force (F_{rot}). The component of the rotational body force acting on the control volume in the momentum direction can be expressed as

$$F_{rot} = \bar{A}\bar{\rho}\Omega^2\left(\frac{r_2^2 - r_1^2}{2}\right) \tag{3.5}$$

Substituting the foregoing expressions for the pressure force, shear force, and rotational body force in Equation 3.2, we obtain

$$\left(P_{s_1} - P_{s_2}\right)\bar{A} - \bar{A}\bar{f}\frac{\Delta x}{D_h}\frac{1}{2}\bar{\rho}\bar{V}^2 + \bar{A}\bar{\rho}\Omega^2\left(\frac{r_2^2 - r_1^2}{2}\right) = \dot{m}\left(V_2 - V_1\right)$$

$$\left(P_{s_1} - P_{s_2}\right) - \bar{f}\frac{\Delta x}{D_h}\frac{1}{2}\bar{\rho}\bar{V}^2 + \bar{\rho}\Omega^2\left(\frac{r_2^2 - r_1^2}{2}\right) = \frac{\dot{m}\left(V_2 - V_1\right)}{\bar{A}} \tag{3.6}$$

Thus, knowing P_{s_1} at the inlet, we can use Equation 3.6 to compute the static pressure P_{s_2} at the outlet as

$$P_{s_2} = P_{s_1} - \bar{f}\frac{\Delta x}{D_h}\frac{1}{2}\bar{\rho}\bar{V}^2 + \bar{\rho}\Omega^2\left(\frac{r_2^2 - r_1^2}{2}\right) - \frac{\dot{m}\left(V_2 - V_1\right)}{\bar{A}}$$

$$P_{s_2} = P_{s_1} - \Delta P_{s_f} + \Delta P_{s_{rot}} - \Delta P_{s_{mom}} \tag{3.7}$$

In Equation 3.7, the terms representing changes in static pressure due to friction, rotation, and momentum-change are given as follows:

$$\Delta P_{s_f} = \bar{f}\frac{\Delta x}{D_h}\frac{1}{2}\bar{\rho}\bar{V}^2 \tag{3.8}$$

$$\Delta P_{s_{rot}} = \bar{\rho}\Omega^2\left(\frac{r_2^2 - r_1^2}{2}\right) \tag{3.9}$$

$$\Delta P_{s_{mom}} = \frac{\dot{m}\left(V_2 - V_1\right)}{\bar{A}} \tag{3.10}$$

3.1.1.3 Energy Equation

The steady flow energy equation for the duct control volume involves heat transfer at the wall and work transfer due to rotation. Both of these energy exchanges result in change in gas total temperature from inlet to outlet. It is important to note that, while the rotational work transfer does not depend upon the simultaneous heat transfer, the heat transfer does depend upon the simultaneous work transfer. This work transfer changes the gas total temperature and hence the difference between the wall temperature and the adiabatic wall temperature, affecting heat transfer. Sultanian (2015) provides a closed-form analytical solution for the coupled heat transfer and rotational work transfer in a compressible duct flow, including a correction term if one decides to simply add the separately-computed changes in gas total temperature due to heat transfer and rotation. Thus, the gas total temperature at the exit of the flow element with rotation can be computed as

$$T_{t_{R_2}} = T_{t_{R_1}} + \left(\Delta T_{t_R}\right)_{HT} + \left(\Delta T_{t_R}\right)_{rot} + \left(\Delta T_{t_R}\right)_{CCT} \tag{3.11}$$

where

$$\left(\Delta T_{t_R}\right)_{HT} \equiv \text{Change in gas total temperature due to heat transfer}$$

$$\left(\Delta T_{t_R}\right)_{rot} \equiv \text{Change in gas total temperature due to rotation}$$

$$\left(\Delta T_{t_R}\right)_{CCT} \equiv \text{Heat transfer and rotational work transfer coupling correction term}$$

Heat transfer: $\left(\Delta T_{t_R}\right)_{HT}$. For the convective heat transfer, we assume that the duct control volume walls are isothermal (constant wall temperature T_w) and the heat transfer coefficient h, which is either specified or computed from an empirical correlation, remains constant from inlet to outlet. We further assume that the adiabatic wall temperature at each section of the control volume equals the gas total temperature, implying a recovery factor of 1.0.

Based on the analysis presented in Chapter 2, the change in gas total temperature due to heat transfer from inlet to outlet can be expressed as

$$\left(\Delta T_{t_R}\right)_{HT} = \left(T_{t_{R_2}} - T_{t_{R_1}}\right)_{HT} = \left(T_w - T_{t_{R_1}}\right)\left(1 - e^{-\eta}\right) \tag{3.12}$$

where

$$\eta = \left(A_w h\right)/\left(\dot{m}c_p\right)$$

Rotational work transfer: $\left(\Delta T_{t_R}\right)_{rot}$. When the gas enters a rotating duct, it assumes the state of solid body rotation, that is, the gas starts rotating at the constant angular velocity of the duct. Under adiabatic conditions, the change in gas total temperature (relative to the control volume rotating at constant speed) between two radial locations can be obtained by equating gas rothalpy (see Chapter 2) at these locations. Thus, we can write

$$T_{t_{R_2}} - \frac{\Omega^2 r_2^2}{2c_p} = T_{t_{R_1}} - \frac{\Omega^2 r_1^2}{2c_p}$$

$$\left(\Delta T_{t_R}\right)_{rot} = \left(T_{t_{R_2}} - T_{t_{R_1}}\right)_{rot} = \frac{\Omega^2 \left(r_2^2 - r_1^2\right)}{2c_p} \tag{3.13}$$

Coupling Correction Term: $\left(\Delta T_{t_R} \right)_{\text{CCT}}$. Sultanian (2015) derived the heat transfer and rotational work transfer coupling correction term as

$$\left(\Delta T_{t_R} \right)_{\text{CCT}} = -\frac{\Omega^2 \left(r_2 - r_1 \right) r_1 \eta}{2 c_p} \tag{3.14}$$

3.1.1.4　Internal Choking and Normal Shock Formation

Compressible flow in a variable-area duct, for example in a convergent-divergent nozzle, can feature internal choking at a section where the flow velocity equals the local speed of sound ($M = 1$). If the flow area increases beyond this section, the flow becomes supersonic with the possibility of a normal shock if the duct exit conditions are subsonic. The flow properties vary continuously across a section where the flow is choked; however, they vary abruptly across a normal shock. In the modeling of a long variable-area duct, a good way to simulate the choked-flow section is to make it coincide with an interface between adjacent control volumes. For simulating a normal shock, however, it is better to imbed a negligibly thin control volume within which the normal shock occurs. Using the normal shock equations presented in Chapter 2, we can then compute the properties at the outlet of this imbedded control volume.

3.1.1.5　Flexibility of Duct Flow Modeling

The comprehensive 1-D modeling of a duct flow presented in the foregoing is much more versatile that it first appears. An orifice can also be simulated using a short duct with no area change and with specified discharge coefficient C_d such that the mass flow rate through the orifice (duct) can be computed using the equation

$$\dot{m} = \frac{C_d A \hat{F}_{f_t} P_t}{\sqrt{RT_t}} = \frac{A_{\text{eff}} \hat{F}_{f_t} P_t}{\sqrt{RT_t}} \tag{3.14}$$

In Equation 3.14, the flow properties correspond to the mechanical area A that yields the effective area $A_{\text{eff}} = C_d A$. The total-pressure mass flow function $\hat{F}_{f_t}$ is a function of Mach number, which is uniquely computed from the isentropic pressure ratio at A. In terms of the static-to-total pressure ratio P_s/P_t, we can write Equation 3.14 as

$$\dot{m} = \frac{C_d A P_t}{\sqrt{RT_t}} \sqrt{\frac{2\kappa}{\kappa - 1}} \left(\frac{P_s}{P_t} \right)^{\frac{1}{\kappa}} \sqrt{1 - \left(\frac{P_s}{P_t} \right)^{\frac{\kappa - 1}{\kappa}}} \tag{3.15}$$

which holds good for $0.5283 \le P_s/P_t < 1$ where $P_s/P_t = 0.5283$ corresponds to the choked flow condition ($M = 1$) for air with $\kappa = 1.4$.

The discrete duct flow modeling presented in this section can also be used to simulate and be validated against the duct flow with each separate effect presented in Chapter 2, namely, isentropic flow with area change (without the friction, heat transfer, and rotation), Fanno flow (without the area change, heat transfer, and rotation), and Rayleigh flow (without the area change, friction, and rotation). We can certainly simulate any combination of various effects on the duct flow. It is interesting to note that a rotating duct flow can also be used to simulate a forced vortex.

3.1.2   Orifice

Orifices are the most ubiquitous element of a gas turbine internal flow system. They are used in both stationary and rotating components either to restrict the flow or to meter it. A choked-flow ($M = 1$) orifice designed with negligible vena contra can be used as a device with constant mass flow rate, not affected by downstream flow conditions. For constant source and discharge pressures in a passive bleed flow line, orifices in the form of short nozzles are generally used to obtain the desired flow rate. Valves may be modeled as a variable-area orifice.

3.1.2.1   Sharp-Edged Orifice

Figure 3.2a shows a sharp-edged orifice in which the flow through the orifice area at section 2 is driven by the upstream total pressure at section 1 and the downstream static pressure at section 3. Due to flow contraction at the orifice, the flow initially converges to a smaller area, called vena contracta, before expanding to the larger downstream area. The area (A_{vc}) of the vena contracta is a strong function of A_3/A_2, but it does not depend on the pressure ratio across it for an incompressible flow. The overall loss in the total pressure between sections 1 and 3 mainly results from the sudden-expansion loss downstream of the vena contracta. Using the control volume analysis of an incompressible flow, this loss is calculated to be the dynamic pressure of the difference in velocities at the vena contracta and section 3.

For a compressible flow, there are essentially two approaches to compute mass flow rate through a sharp-edged orifice. The first approach (a classical one) is based on an extension of the incompressible flow method and is given by (see Benedict (1980))

$$\dot{m} = \frac{YC_{d_{inc}}AP_t}{\sqrt{RT_t}}\sqrt{2\left(1 - \left(\frac{P_s}{P_t}\right)_{inc}\right)}$$ (3.16)

where Y is called the adiabatic expansion factor to account for the decrease in density as the flow expands (static pressure decreases) in a compressible flow. Buckingham (1932) proposed the following empirical relation for computing Y:

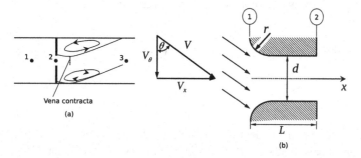

Figure 3.2 (a) Sharp-edged orifice and (b) parameters influencing orifice discharge coefficient.

$$Y = 1 - \frac{0.41 + 0.35\beta^4}{\kappa} \left(1 - \left(\frac{P_s}{P_t}\right)_{inc}\right) \tag{3.17}$$

where β is the ratio of orifice diameter to housing pipe diameter. Parker and Kercher (1991) recommended the use of Equation 3.17 for $0.63 \leq (P_s/P_t)_{inc} \leq 1$ and $0 \leq \beta \leq 0.5$, and for $0 \leq (P_s/P_t)_{inc} \leq 0.63$, they proposed the following semi-empirical equation:

$$Y = Y_{0.63} - \left(0.3475 + 0.1207\beta^2 - 0.3177\beta^4\right)\left(0.63 - \left(\frac{P_s}{P_t}\right)_{inc}\right) \tag{3.18}$$

where $Y_{0.63}$ is computed from Equation 3.17 with $(P_s/P_t)_{inc} = 0.63$.

To compute the incompressible flow discharge coefficient $C_{d_{inc}}$ used in Equation 3.15, Stolz (1975) proposed the following "universal" equation:

$$C_{d_{inc}} = 0.5959 + 0.0312\beta^{2.1} - 0.1840\beta^8$$

$$+ 0.0900L_1 \frac{\beta^4}{1 - \beta^4} - 0.0337L_2\beta^3 + 91.71 \frac{\beta^{2.5}}{\left(\beta Re_d\right)^{0.75}} \tag{3.19}$$

where

$L_1 \equiv$ Dimensionless upstream pressure-tap location with respect to the orifice upstream face

$L_2 \equiv$ Dimensionless downstream pressure-tap location with respect to the orifice downstream face

$Re_d \equiv$ Reynolds number based on the orifice diameter

The main motivation behind using the foregoing approach in calculating the mass flow rate of a compressible flow through a sharp-edged orifice is to leverage the vast amount of incompressible flow data available for $C_{d_{inc}}$. In addition, as discussed next, the method also captures the compressibility effect as the downstream static pressure decreases beyond the nominal choking at the vena contracta, increasing its area with pressure ratio. This results in a higher mass flow rate beyond the critical pressure ratio under the same upstream stagnation conditions.

The second approach, which will be used in this book, to calculating compressible mass flow rate through a sharp-edged orifice is to use Equation 3.15. From Equations 3.15 and 3.16, we can compute the discharge coefficient C_d as

$$C_d = \frac{YC_{d_{inc}}\sqrt{\left(1 - \left(\frac{P_s}{P_t}\right)_{inc}\right)}}{\sqrt{\frac{\kappa}{\kappa - 1}}\left(\frac{P_s}{P_t}\right)^{\frac{1}{\kappa}}\sqrt{1 - \left(\frac{P_s}{P_t}\right)^{\frac{\kappa-1}{\kappa}}}} \tag{3.20}$$

Note that in the numerator of Equation 3.20, $0 \leq (P_s/P_t)_{inc} \leq 1$ while in the denominator we have $0.5283 \leq P_s/P_t < 1$, which means the denominator of this equation has a maximum value of 0.6847 that corresponds $P_s/P_t = 0.5283$.

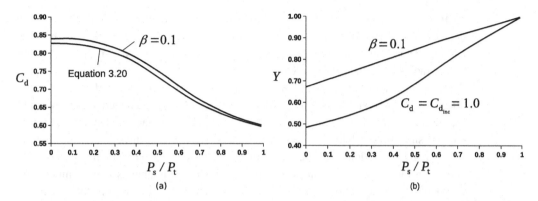

Figure 3.3 (a) Comparison of C_d prediction by Equation 3.20 with the measurements of Perry (1949) for $\beta = 0.1$ and (b) variation of adiabatic expansion factor Y with static-to-total pressure ratio.

Figure 3.3a compares the compressible flow C_d predictions by Equation 3.20 with the measurements of Perry (1949) for $\beta = 0.1$ over the entire range of the static-to-total pressure ratio P_s/P_t. The corresponding linear variation of the adiabatic expansion factor Y is shown in Figure 3.3b, which also shows how Y varies with P_s/P_t under isentropic conditions when $C_d = C_{d_{inc}} = 1.0$. Under these conditions, Equation 3.20 yields

$$Y_{\text{ideal}} = \frac{\sqrt{\dfrac{\kappa}{\kappa - 1}\left(\dfrac{P_s}{P_t}\right)^{\frac{1}{\kappa}}}\sqrt{1 - \left(\dfrac{P_s}{P_t}\right)^{\frac{\kappa-1}{\kappa}}}}{\sqrt{1 - \left(\dfrac{P_s}{P_t}\right)_{\text{inc}}}} \tag{3.21}$$

A serious drawback of the method of computing C_d using Equation 3.20 is that for $\beta > 0.5$ and low pressure ratios it yields $C_d > 1.0$, which is unacceptable. Parker and Kercher (1991) present a noniterative semi-empirical prediction method to generate intermediate values of C_d between 0.5959 and 1.0 for $0.5959 \leq C_{d_{inc}} \leq 1.0$ over the entire range of P_s/P_t. The method, however, has limited experimental validation. High-fidelity CFD, including LES and DNS, may be used to generate C_d for the entire range of orifice geometry and pressure ratio variations while validating these results against limited available experimental data.

3.1.2.2 Generalized Orifice

Figure 3.2b shows a generalized orifice featuring two geometric parameters, namely, the radius of curvature r/d at the inlet and the length-to-diameter ratio L/d and one inlet flow parameter representing the ratio of the tangential velocity to axial velocity V_θ/V_x. All three parameters, either individually or in combination, significantly affect the orifice discharge coefficient. The maximum benefit of the radius of curvature to minimize flow separation at the orifice inlet and the adverse effect of the vena contracta on discharge coefficient occurs at $r/d \approx 1.0$. For a thick orifice with $L/d \approx 2$, the sudden-expansion loss at the vena contracta is reduced, as a result, the discharge coefficient increases. For $L/d > 2$, the adverse affect of the orifice wall friction tends to decrease C_d. The effect of V_θ/V_x is always to increase flow separation at the inlet causing deterioration in the orifice discharge coefficient. This can occur even with corner radiusing at the orifice inlet.

McGreehan and Schotsch (1988) present a design-friendly method to predict the effects of r/d, L/d and V_θ/V_x on discharge coefficient for a generalized orifice shown in Figure 3.2b. Their approach uses the classical adiabatic expansion factor to generate compressible flow discharge coefficient for a sharp-edged orifice. While the method shows good validation with the available test data for each individual parameter, its validation for the case featuring interaction effects of parameters is rather limited, and the method may at times predict $C_d > 1$. Parker and Kercher (1991) further improved the method of McGreehan and Schotsch (1988) by using the total-pressure mass flow function and ensuring that $C_d \leq 1$ with a semi-empirical technique.

Existing methods of predicting orifice discharge coefficient for a rotating orifice are generally not reliable due a highly three-dimensional flow field caused by rotation within the orifice. These methods, therefore, tend to overpredict discharge coefficients for a rotating orifice. An obvious effect of rotation is to change V_θ/V_x at the orifice inlet. In addition, if the orifice inlet and outlet radii are appreciably different, the fluid total temperature relative to the orifice will change such that the rothalpy remains constant under adiabatic conditions. According to the isentropic relation, this change in fluid total temperature will results in a change in total pressure. Both these changes will affect the isentropic flow rate through the rotating orifice. Idris and Pullen (2005) present correlations to compute discharge coefficients for rotating orifices. A rotating long orifice with heat transfer and wall friction may alternatively be modeled as a duct, presented in Section 3.1.

From the foregoing discussion it is clear that the need for accurate and dependable 1-D modeling and prediction method for a general compressible flow orifice continues to exist. The lack of benchmark quality measurements in this area remains a serious impediment to further improvement – a situation that is not likely to change in the foreseeable future. Most of the available prediction methods are semi-empirical in nature with limited experimental validation. Another approach to mitigate the current situation is to leverage a CFD-based DOE using all key parameters that influence the discharge coefficient of a generalized orifice and correlating the results in the form a response surface. The response surface correlation can be easily implemented in a flow network code, which is used for modeling internal flow systems of modern gas turbines. The orifice response surface will predict both the direct and interaction effects of various parameters that influence the orifice discharge coefficient.

3.1.3   Vortex

The vortex is an important feature of a gas turbine internal flow system with rotor surfaces. In Chapter 2, we discussed isentropic free vortex, forced vortex, and non-isentropic generalized vortex and their modeling equations to compute pressure and temperature changes across them. The flow in a rotating duct creates a forced vortex that is coupled with other effects of friction and heat transfer. When an internal flow passes over rotor surfaces, like in rotor disk cavities, which we discuss in Chapter 4, it becomes a vortex with concurrent generation of windage.

It is important to note that, unlike a duct or an orifice, the vortex is not a physical component. It is simply a flow feature associated with its bulk circular motion (swirl) about the machine axis of rotation. Accordingly, unlike in a component, knowing the end conditions of a vortex, one may not be able to compute a unique mass flow rate associated with it. Therefore, a vortex will not qualify to be an element (discussed in the next section), which is directly connected to junctions in a flow network. We will henceforth call vortex a pseudo element. For its inclusion in a flow network, it must be sandwiched between two elements, either of which could be a duct or an orifice. The resulting super element, which becomes a mass flow metering component, can be connected to junctions in a flow network.

3.2   Description of a Flow Network: Elements and Junctions

Gas turbine internal flow systems such as blade cooling system, rotor-stator or rotor-rotor cooling system, rotor-stator sealing system, discourager or rim seal system to minimize or prevent hot gas ingestion, inlet bleed heat system, and others are essentially handled using a flow network in which various components (ducts, orifices, seals, vortices, etc.) are modeled as 1-D flow element. Two basic entities of a flow network, shown in Figure 3.4, are element (also called link or branch) and junction (also called chamber or node). An element in this flow network is depicted by a solid line along with an arrow to represent the positive flow direction. A flow network has two types of junctions: internal junctions, depicted by an open circle, and the boundary junctions, depicted by an open square.

3.2.1   Element

An element of a flow network typically represents a component of the gas turbine internal flow system. A general flow network may include a variety of elements: duct, orifice, seal, vortex (pseudo component), heat exchanger (super element), and others. As shown in Figure 3.4, each element in the flow network connects two junctions. It is

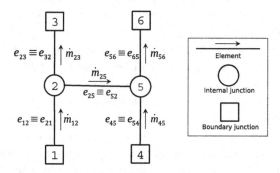

Figure 3.4 A flow network.

characterized by a mass flow rate, which in steady state remains constant from inlet to outlet. By its very function, an element in a network represents a flow metering device. The element connecting junctions i and j is represented by e_{ij}, which is the same as e_{ji}. Depending on the flow direction in the element, either of the junctions could be an inlet or an outlet. Note that the state variables at both junctions of an element uniquely determine the actual flow direction in the element. Later in this chapter, we will discuss a physics-based criterion to determine this flow direction. Even if one element in the network is assigned a wrong flow direction, the entire network solution is corrupted, which demonstrates the elliptic nature of the flow field modeled by a flow network, although a 1-D flow is assumed in each element.

Each element in a flow network is analogous to a thermodynamic path connecting two states of a system. All the path variables, evaluated in terms of the amount of work transfer and heat transfer, are associated with the flow through the element. In addition to the geometric parameters, empirical and semi-empirical correlations are specified for each element in the network to determine the friction factor for major loss, discharge coefficient for minor loss, and heat transfer coefficient to compute heat transfer in or out of the flow, etc. Various thermal and hydrodynamic boundary conditions applied to an element are assumed uniform; two- or three-dimensional variations of flow properties within the element are not discernible. These variations, if needed to develop a better understanding of the component, may be obtained through a CFD analysis, which may also be used readily to generate needed data to facilitate 1-D flow modeling of the component in the network to carry out a robust design.

At times, it becomes import to capture variations in geometry and boundary conditions over a flow element connecting two adjacent junctions. Examples include a long pipe line for bleed and coolant supply and the serpentine passage used for internal cooling of turbine airfoils. In such situations, the flow element may be modeled using serially-connected small control volumes without creating additional junctions in the overall network. This modeling practice offers significant economy in the network solution and preserves the dynamic pressure between adjacent control volumes in the flow direction for higher prediction accuracy.

There is often a debate among gas turbine engineers about which pressure, static or total, should be used at the junctions in a flow network to compute mass flow rate through the connecting elements. In steady state, the mass flow rate at any section in an internal flow requires that $P_t > P_s$ or $P_t/P_s > 1$, and this mass flow rate remains constant through the element or the gas turbine component that it represents, regardless of how fluid properties change from section to section. As discussed in Chapter 2, we can compute the element mass flow rate at any section using either the total-pressure mass flow function or the static-pressure mass flow function, both of which are functions of the section Mach number and the ratio of gas specific heats. This mass flow rate should be computed at the section, typically outlet, inlet, or where the flow is choked with $M = 1$. This approach, however, doesn't determine the flow direction across the section, which must be determined from the entropy change over the element.

It is important to note that the junction pressure at the inlet must be interpreted as the total pressure and that at the outlet as the static pressure. A junction with no associated dynamic pressure behaves like a plenum in which $P_t = P_s$ such that, for all flows leaving the junction, the plenum pressure becomes the total pressure, and for all flows entering the junction, the plenum pressure becomes the back pressure. For a subsonic flow through the element, the static pressure at the element exit section must equal the plenum back pressure (static). If, however, the flow is choked at the element exit, its static pressure becomes decoupled from the connected downstream junction, and the element mass flow rate is entirely determine by the upstream conditions. As an example, for the flow network shown in Figure 3.4, consider element e_{25} with the mass flow rate $\dot{m}_{25}$. Even in the presence of dynamic pressure at junction 2 due to the flow along e_{12} and e_{23}, the total pressure at the inlet of element e_{25} equals the static pressure at junction 2. Similarly, for a subsonic flow through the element e_{25}, its exit static pressure must be equal to the static pressure at junction 5.

3.2.2   Internal Junction

An internal junction connects two or more elements in a network. The flow network shown in Figure 3.4 has two internal junctions, namely, 2 and 5. The internal junction 2, for example, connects elements e_{12}, e_{23}, and e_{25}. In steady state, the continuity equation for this junction yields

$$\dot{m}_{23} + \dot{m}_{25} = \dot{m}_{12} \tag{3.22}$$

which, with $\dot{m}_{21} = -\dot{m}_{12}$, can be written as

$$\dot{m}_{23} + \dot{m}_{25} + \dot{m}_{21} = 0 \tag{3.23}$$

Thus, at an arbitrary internal junction i connected through elements to multiple junctions $j = 1$ to $j = k_i$, as shown in Figure 3.5, we can write the steady continuity equation as

$$\sum_{j=1}^{j=k_i} \dot{m}_{ij} = 0 \tag{3.24}$$

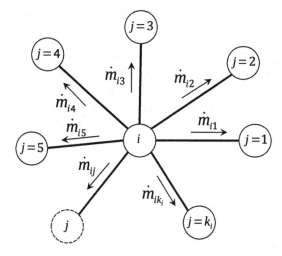

Figure 3.5 Mass conservation at an internal junction.

where $j \neq i$. In this equation, we have adopted the convention that $\dot{m}_{ij}$ is nominally positive if the flow direction in the element e_{ij} is from junction i to junction j. According to this convention, we obtain $\dot{m}_{ij} = -\dot{m}_{ji}$.

To ensure energy conservation at each internal junction, we can compute the mixed mean total temperature of all incoming flows. All outflows take place at this mixed mean total temperature. For example, we can write at junction i

$$T_{t_i} = \frac{\sum_{j=1}^{j=k_i} \delta_{ij}\dot{m}_{ij}T_{t_{ij}}}{\sum_{j=1}^{j=k_i} \delta_{ij}\dot{m}_{ij}} \tag{3.25}$$

where we have $\delta_{ij} = 1$ for $\dot{m}_{ij} < 0$ (inflows) and $\delta_{ij} = 0$ for $\dot{m}_{ij} > 0$ (outflows).

The foregoing discussion makes it clear that the net mass flow rate associated with an internal junction must be zero. All the state variables like pressure and temperature are associated with a junction. While the static pressure is uniform within a junction, the assumption of a uniform total pressure is a matter of modeling assumption depending upon the accuracy and rigor one needs in a design application. In this book, we limit our discussion to dump-type or plenum junctions, where we neglect the dynamic pressure associated with each incoming flow. This is akin to the "tank-and-tube" approach used in modeling a flow network. Sultanian (2015) presents more accurate and detailed modeling of junctions in a compressible flow network.

Thus, we model an internal junction with zero dynamic pressure in it, making the static and total pressures equal. In this case, the junction static temperature also equals the total temperature, which is determined by Equation 3.25 as the mixed mean total temperature of all flows entering the junction. All flows leaving the junction are assumed to be at this total temperature. By contrast, the static pressure of all subsonic flows entering the plenum junction equals the junction pressure, which becomes the total pressure for all flows leaving the junction.

3.2.3   Boundary Junction

A boundary junction in a flow network is either a source or a sink where the boundary conditions are specified. The network solution corresponds to these boundary conditions. Based on the flow directions shown in Figure 3.4, the boundary junctions 1 and 4 are the sources and the boundary junctions 3 and 6 are the sinks. Being sources and sinks, boundary junctions are allowed to violate the steady continuity equation. At a boundary junction, all state properties are generally assumed uniform.

3.3   Compressible Flow Network Solution

A flow network solution yields the mass flow rate through each element and the state properties, such as the total temperature and pressure, at each internal junction. Except for a very simple flow network, hand calculations to obtain a flow network solution could be very tedious and time-consuming. Therefore, the flow networks of gas turbine internal flow systems are typically solved using a computer with the help of a computer code based on a solution method similar to the one presented in this section. Because of nonlinear dependence of an element mass flow rate through on the difference in pressures at the connected junctions, a commonly used solution strategy is to iteratively perform the tasks of element solutions and junction solutions until the continuity equation at each internal junction is satisfied within an acceptable error.

In a flow network, for an assumed initial solution, all the mass flow rates generated in the connected elements will most likely not satisfy the continuity equation at each junction. The continuity Equation 3.24 becomes

$$\sum_{j=1}^{j=k_i} \dot{m}_{ij} = \sum_j \dot{m}_{ij} = \Delta \dot{m}_i \tag{3.26}$$

where $\Delta \dot{m}_i$ is the residual error in the continuity equation at junction i. For the junction solution, let us assume that the mass flow rate $\dot{m}_{ij}$ through each element e_{ij} depends only on P_{s_i} and P_{s_j}. At each iteration, our goal is to change the junction static pressures so as to annihilate $\Delta \dot{m}_i$ in Equation 3.26. Accordingly, we can write

$$\sum_j d\left(\dot{m}_{ij}\right) = d\left(\Delta \dot{m}_i\right) = 0 - \Delta \dot{m}_i$$

$$\sum_j \left(\frac{m_{ij}}{P_{s_i}} \Delta P_{s_i} + \frac{m_{ij}}{P_{s_j}} \Delta P_{s_j} \right) = -\Delta \dot{m}_i \tag{3.27}$$

Writing Equation 3.27 at each internal junction will result in a system of nonlinear algebraic equations for which no known direct solution method exists. One must therefore solve such a system using an iterative numerical method. At the current iteration, the system is treated as a linear system for which there are many numerical solution methods, for example, the direct method presented by Sultanian (1980); see also Appendix F.

As shown in the foregoing discussion, we can express each element solution in the form $\dot{m}_{ij} = a_{ij}\left(P_{s_i} - P_{s_j}\right)$, and assuming a_{ij} as constant at each iteration, we obtain $\dot{m}_{ij}/P_{s_i} = a_{ij}$ and $\dot{m}_{ij}/P_{s_j} = -a_{ij}$. Substituting them in Equation 3.27 yields

$$\sum_j \left(a_{ij}\Delta P_{s_i} - a_{ij}\Delta P_{s_j} \right) = -\Delta \dot{m}_i \tag{3.28}$$

Writing Equation 3.28 for all internal junctions n in the flow network results in the following system of algebraic equations:

$$\begin{bmatrix} \sum_j a_{1j} & \cdot & \cdot & -a_{1n} \\ \cdot & & \cdot & \cdot \\ \cdot & \cdot & \cdot & \cdot \\ -a_{n1} & \cdot & \cdot & \sum_j a_{nj} \end{bmatrix}_{old} \begin{bmatrix} \Delta P_{s_1} \\ \cdot \\ \cdot \\ \Delta P_{s_n} \end{bmatrix}_{new} = \begin{bmatrix} -\Delta \dot{m}_1 \\ \cdot \\ \cdot \\ -\Delta \dot{m}_n \end{bmatrix}_{old} \tag{3.29}$$

where the subscripts old and new refer to previous iteration and current iteration, respectively.

Note that the coefficient matrix a_{ij}, although assumed constant during the current iteration, varies from iteration to iteration until both the element and junction solutions are converged.

We can write Equation 3.29 in the following alternative form:

$$
\begin{bmatrix} J_{11} & \cdot & \cdot & J_{1n} \\ \cdot & \cdot & \cdot & \cdot \\ \cdot & \cdot & \cdot & \cdot \\ J_{n1} & \cdot & \cdot & J_{nn} \end{bmatrix}_{\text{old}}
\begin{bmatrix} \Delta P_{s_1} \\ \cdot \\ \cdot \\ \Delta P_{s_n} \end{bmatrix}_{\text{new}}
=
\begin{bmatrix} -\Delta \dot{m}_1 \\ \cdot \\ \cdot \\ -\Delta \dot{m}_n \end{bmatrix}_{\text{old}}
\tag{3.30}
$$

where the coefficient matrix is known as the Jacobian matrix expressed as

$$
\begin{bmatrix} J_{11} & \cdot & \cdot & J_{1n} \\ \cdot & \cdot & \cdot & \cdot \\ \cdot & \cdot & \cdot & \cdot \\ J_{n1} & \cdot & \cdot & J_{nn} \end{bmatrix}
=
\begin{bmatrix} \sum_j a_{1j} & \cdot & \cdot & -a_{1n} \\ \cdot & \cdot & \cdot & \cdot \\ \cdot & \cdot & \cdot & \cdot \\ -a_{n1} & \cdot & \cdot & \sum_j a_{nj} \end{bmatrix}
\tag{3.31}
$$

Because each junction in a typical flow network is connected to only a few neighboring junctions, the coefficient matrix in Equation 3.29 or the Jacobian in Equation 3.31 is generally very sparse with only a handful of nonzero entries.

3.3.1   Initial Solution Generation

An iterative solution method for a flow network requires an initial solution either as user input or generated automatically from the specified boundary conditions. For developing a robust design, which requires multiple probabilistic flow network solutions, it becomes imperative to eliminate any user intervention in obtaining each converged numerical solution.

For automatic generation of initial solution of the static pressure distribution at all internal junctions in a flow network, we can use one of the two methods. For the first method, we treat the flow network as a network of heat conduction rods each with the fixed thermal conductivity and geometry. The resulting system of equations will be linear for which we can directly solve for temperatures at all internal junctions. From the analogy between temperature and static pressure, we thus obtain the distribution of static pressure at all internal junctions of the flow network. For the second method, we use the fact that the relation between the pressure drop and mass flow rate in a fully developed laminar pipe flow is linear, given by the equation

$$
\dot{m}_{ij} = a_{ij} \left(P_{s_1} - P_{s_2} \right)
\tag{3.32}
$$

where

$$
a_{ij} = \frac{\pi}{128} \left(\frac{\rho D^4}{\mu L} \right)
$$

For the initial solution generation, we replace each element in the flow network by an equivalent laminar flow pipe. For the entire network, we will again obtain a system of linear equations that can be solved directly for static pressure at each internal junction.

3.3.2 Determination of Element Flow Direction

For a compressible flow network to yield accurate predictions, it is important that the flow direction in each element must be determined on physical grounds. We find that the entropy change over an element provides a sound physical basis to determine the element flow direction. For an adiabatic flow in an element when one or more effects of area change, friction, and rotation are present, the entropy must always increase in the flow direction. Accordingly, for the flow from section 1 to section 2 of such an element, we must have

$$\frac{s_2 - s_1}{R} = \left(\frac{c_p}{R} \ln \frac{T_{s_2}}{T_{s_1}} - \ln \frac{P_{s_2}}{P_{s_1}} \right) > 0 \qquad (3.33)$$

which involves static properties at these sections. In terms of total properties at these sections, we can write Equation 3.33 as

$$\frac{s_2 - s_1}{R} = \left(\frac{c_p}{R} \ln \frac{T_{t_2}}{T_{t_1}} - \ln \frac{P_{t_2}}{P_{t_1}} \right) > 0 \qquad (3.34)$$

Based on the second law of thermodynamics, heat transfer always accompanies entropy transfer. Thus, heating of the element flow will increase its entropy downstream, and cooling will decrease it. Although heat transfer by itself is unlikely to change the flow direction in an element, for applying the entropy-based criterion, it is important to remove the contribution of heat transfer to the total entropy change over the element.

Equating the change in gas total temperature due to heat transfer across an element to the convective heat transfer through the element walls, we can write

$$hA_w \left(T_w - \overline{T}_t \right) = \dot{m} c_p \left(T_{t_2} - T_{t_1} \right)_{HT}$$

$$\overline{T}_t = T_w - \frac{\dot{m} c_p}{hA_w} \left(T_{t_2} - T_{t_1} \right)_{HT} = T_w - \frac{\left(T_{t_2} - T_{t_1} \right)_{HT}}{\eta} \qquad (3.35)$$

According to Equation 3.35, the convective heat transfer in the element occurs at the average gas temperature $\overline{T}_t$. Thus, the entropy change due to heat transfer can be written as

$$\frac{(\Delta s)_{HT}}{R} = \frac{c_p}{R} \frac{\left(T_{t_2} - T_{t_1} \right)_{HT}}{\overline{T}_t} \qquad (3.36)$$

The positive value of the modified entropy change occurring in an element in the flow direction is given by

$$\frac{s_2 - s_1}{R} - \frac{(\Delta s)_{HT}}{R} = \left(\frac{c_p}{R} \left\{ \ln \frac{T_{t_2}}{T_{t_1}} - \frac{\left(T_{t_2} - T_{t_1} \right)_{HT}}{\overline{T}_t} \right\} - \ln \frac{P_{t_2}}{P_{t_1}} \right) > 0 \qquad (3.37)$$

which is the new criterion we can use to determine the flow direction in an element in the presence of heat transfer.

3.3.3   Newton-Raphson Method

Equation 3.29 represents the standard form used in the Newton-Raphson solution method, for example, presented by Carnahan, Luther, and Wilkes (1969) along with a computer program in FORTRAN.

The following key steps constitute the Newton-Raphson solution method:

1. Using one of the methods discussed in the foregoing, generate an initial distribution of static pressure at each internal junction of the flow network.
2. Compute mass flow rate through each element.
3. Compute Jacobian matrix from element solutions.
4. Compute mass flow rate error at each internal junction, yielding the right-hand side vector of Equation 3.29.
5. Use a direct solution method, for example Sultanian (1980), to obtain the vector of changes in static pressure at all internal junctions.
6. Obtain the new static pressure at each internal junction:

$$\left(P_{s_i}\right)_{new} = \left(P_{s_i}\right)_{old} + \alpha_i \left(\Delta P_{s_i}\right)_{new}$$

where α_i is an under-relaxation parameter specified for each junction to help the solution convergence.

7. Repeat steps from 2 to 6 until $\left|\Delta \dot{m}_i\right|_{max} \leq \delta_{tol}$, where δ_{tol} is the acceptable maximum error in the mass flow rate at an internal junction.
8. Solve the energy equation at each internal junction.
9. Repeat steps from 2 to 8 until $\left|\left(T_{t_i}\right)_{old} - \left(T_{t_i}\right)_{new}\right|_{max} \leq \hat{\delta}_{tol}$, where $\hat{\delta}_{tol}$ is the maximum acceptable difference between the total temperatures at any internal junction in successive iterations.

The standard Newton-Raphson method outlined earlier has two main shortcomings. First, the solution convergence is found sensitive to the initial solution assumed for the network. Second, when using a computer code to perform the computations, the user-specified under-relation parameters may need adjustments when the network solution starts diverging. In an integrated design environment, when a designer is required to generate multiple solutions of a flow network for a probabilistic (robust) design, these shortcomings of the solution method may become a serious handicap. The modified Newton-Raphson method presented in the next section provides a robust solution methodology, ideally suited for a complex compressible flow network used in gas turbine design.

3.3.4   Modified Newton-Raphson Method

In the modified Newton-Raphson method, a positive damping parameter λ is added to each diagonal element of the Jacobian matrix. Equation (3.30) now becomes

$$\begin{bmatrix} J_{11}+\lambda & . & . & J_{1n} \\ . & . & . & . \\ . & . & . & . \\ J_{n1} & . & . & J_{nn}+\lambda \end{bmatrix}_{old} \begin{bmatrix} \Delta P_{s_1} \\ . \\ . \\ \Delta P_{s_n} \end{bmatrix}_{new} = \begin{bmatrix} -\Delta \dot{m}_1 \\ . \\ . \\ -\Delta \dot{m}_n \end{bmatrix}_{old} \qquad (3.38)$$

where the damping parameter λ, which is auto-adjusted during the iterative solution process, ensures that the Jacobian matrix remains diagonally dominant, preventing it from becoming singular. In this form, the modified Newton-Raphson method is equivalent to the linear least-squares optimization based on Levenberg-Marquardt method presented in Nash and Sofer (1996). This modified Newton-Raphson solution method consists of the following key steps:

1. Using one of the methods discussed in the foregoing, generate an initial distribution of static pressure at each internal junction of the flow network.
2. Set $\lambda_{\text{old}} = 0.5$ and $\lambda_{\text{new}} = \lambda_{\text{old}}$.
3. Compute mass flow rate through each element.
4. Compute residual error of the continuity equation at each internal junction and the corresponding error norm

$$E_{\text{old}} = \sqrt{\frac{1}{n}\sum_{i=1}^{n}\left(\Delta \dot{m}_i\right)^2}$$

5. Compute Jacobian matrix from solution for each element in the flow network.
6. Modify the Jacobian matrix by adding λ_{new} to all its diagonal components.
7. Use a direct solution method, for example, Sultanian (1980), to obtain the vector of changes in static pressure at all internal junctions.
8. Obtain the new static pressure at each internal junction:

$$\left(P_{s_i}\right)_{\text{new}} = \left(P_{s_i}\right)_{\text{old}} + \left(\Delta P_{s_i}\right)_{\text{new}}$$

9. Compute mass flow rate through each element.
10. Compute error of the continuity equation at each internal junction and the error norm

$$E_{\text{new}} = \sqrt{\frac{1}{n}\sum_{i=1}^{n}\left(\Delta \dot{m}_i\right)^2}$$

11. If $E_{\text{new}} > E_{\text{old}}$, set $\lambda_{\text{new}} = 2\lambda_{\text{old}}$ and repeat steps from 6 to 11.
12. Set $\lambda_{\text{new}} = \lambda_{\text{old}}/2$.
13. Repeat steps from 5 to 12 until $E_{\text{new}} \le \tilde{\delta}_{\text{tol}}$, where $\tilde{\delta}_{\text{tol}}$ is the acceptable maximum value of E_{new}
14. Solve the energy equation at each internal junction.
15. Repeat steps from 5 to 14 until $\left|\left(T_{t_i}\right)_{\text{old}} - \left(T_{t_i}\right)_{\text{new}}\right|_{\text{max}} \le \hat{\delta}_{\text{tol}}$, where $\hat{\delta}_{\text{tol}}$ is the maximum acceptable difference between the total temperatures at any internal junction in successive iterations.

3.4   Concluding Remarks

Physics-based 1-D flow modeling of each component of a gas turbine internal flow system satisfies the conservation equations of mass, momentum, energy, and entropy, which provides a robust criterion to determine the flow direction in the flow. Any elliptic flow feature involving flow separation and reversal within the component is accounted for using empirical correlations to compute pressure loss and heat transfer coefficient. Understanding of these flow features in a component may be discerned from a three-dimensional CFD analysis, reinforcing an improved 1-D flow modeling of the component.

Because the internal flow system boundary conditions are generally known only at the source (gas turbine main flow path) and sink (ambient or downstream gas turbine flow path), intermediate flow properties are obtained from the solution of a flow network. In this network, each element, whose mass flow rate is a single-valued function of the properties at the connected end junctions, represents a system component. Because a unique mass flow rate cannot be obtained from the property changes across a vortex, it does not qualify to be an element of the flow network. It is therefore named here a pseudo element and must be sandwiched between two elements for its integration into the flow network.

The compressible flow network formulations presented in this chapter have no Mach number limitation. In addition to the effects of area-change, friction, and heat transfer, the formulation also includes the effects of rotation in each element. For the development of a computer code to handle a flow network, we have discussed here key details of the physics-based modeling methods for elements and junctions that form the network. We have also presented here a modified Newton-Raphson method for the robust iterative numerical solution of a system of nonlinear algebraic equations, which generally arise from the flow network modeling of a gas turbine internal flow system.

Worked Examples

Example 3.1 Consider an orifice with diameter $D = 30$ mm, upstream total pressure $P_{t_1} = 12$ bar, constant total temperature $T_{t_1} = 781$ K, and downstream static pressure $P_{s_2} = 10$ bar. Assume the orifice discharge coefficient $C_d = 0.8$ for the following:
(1)   Compute air mass flow rate through the orifice
(2)   Compute the total pressure (P_{t_2}) at the orifice exit
(3)   Defining the orifice loss coefficient $K = \left(P_{t_1} - P_{t_2}\right)/\left(P_{t_2} - P_{s_2}\right)$ and using the relation $C_d = 1/\sqrt{K+1}$ derived in Chapter 2, compute the orifice exit total pressure and find the percentage error from its value computed in (2)

Solution
Duct flow area: $A = \dfrac{\pi D^2}{4} = \dfrac{\pi \times (0.030)^2}{4} = 0.707 \times 10^{-3}\,\mathrm{m}^2$

Loss coefficient: From $C_d = 1/\sqrt{K+1}$, we can write

$$K = \frac{1}{C_d^2} - 1 = \frac{1}{(0.8)^2} - 1 = 0.563$$

(1)   Mass flow rate through the orifice

$$\frac{P_{t_1}}{P_{s_2}} = \frac{12 \times 10^5}{10 \times 10^5} = 1.2$$

$$\frac{T_{t_1}}{T_{s_2}} = \left(\frac{P_{t_1}}{P_{s_2}}\right)^{\frac{\kappa-1}{\kappa}} = (1.2)^{0.286} = 1.0535$$

$$M_1 = \sqrt{\left(\frac{2}{\kappa-1}\right) \times \left(\frac{T_{t_1}}{T_{s_2}} - 1\right)} = \sqrt{5 \times (1.0535 - 1)} = 0.517$$

$$\hat{F}_{f_{t_1}} = M_1 \sqrt{\frac{\kappa}{\left(1 + \frac{\kappa-1}{2}M_1^2\right)^{\frac{\kappa+1}{\kappa-1}}}} = 0.517 \times \sqrt{\frac{1.4}{(1.0535)^6}} = 0.5233$$

$$\dot{m} = \frac{C_d A \hat{F}_{f_{t_1}} P_{t_1}}{\sqrt{RT_{t_1}}} = \frac{0.8 \times 0.707 \times 10^{-3} \times 0.5233 \times 12 \times 10^5}{\sqrt{287 \times 781}}$$

$$\dot{m} = 0.750 \, \text{kg/s}$$

(2)   Total pressure at orifice exit (section 2)

We can write the mass flow rate in terms of static-pressure mass flow function at orifice exit (section 2) as

$$\dot{m} = \frac{A \hat{F}_{f_{s_2}} P_{s_2}}{\sqrt{RT_{t_1}}}$$

which yields

$$\hat{F}_{f_{s_2}} = \frac{\dot{m}\sqrt{RT_{t_1}}}{AP_{s_2}} = \frac{0.750 \times \sqrt{287 \times 781}}{0.707 \times 10^{-3} \times 10 \times 10^5} = 0.5024$$

Using Equation 2.106, we obtain the Mach number at orifice exit (section 2) as

$$M_2 = \sqrt{\frac{-\kappa + \sqrt{\kappa^2 + 2\kappa(\kappa-1)\hat{F}_{f_{s_2}}^2}}{\kappa(\kappa-1)}}$$

$$= \sqrt{\frac{-1.4 + \sqrt{(1.4)^2 + 2 \times 1.4 \times (1.4-1) \times (0.5024)^2}}{1.4 \times (1.4-1)}}$$

$$M_2 = 0.4174$$

Using isentropic relations, we obtain

$$\frac{P_{t_2}}{P_{s_2}} = \left(1 + \frac{\kappa-1}{2}M^2\right)^{\frac{\kappa}{\kappa-1}} = \left(1 + \frac{1.4-1}{2}(0.4174)^2\right)^{3.5} = 1.1273$$

which yields

$$P_{t_2} = 1.1273 \times P_{s_2} = 1.1273 \times 10 \times 10^5$$

$$P_{t_2} = 11.273 \times 10^5 \, \text{Pa}$$

(3) Total pressure at orifice exit using the loss coefficient $K = \left(P_{t_1} - P_{t_2}\right)/ \left(P_{t_2} - P_{s_2}\right)$

We rearrange the given definition for the loss coefficient $K = \left(P_{t_1} - P_{t_2}\right)/ \left(P_{t_2} - P_{s_2}\right)$ to compute the orifice exit total pressure as

$$P_{t_2} = \frac{P_{t_1} + K P_{s_2}}{1 + K} = \frac{12 \times 10^5 + 0.563 \times 10 \times 10^5}{1 + 0.563}$$

$$P_{t_2} = 11.280 \times 10^5 \, \text{Pa}$$

which is within an acceptable error of 0.0587 percent from the value computed in (2) using the static-pressure mass flow function computed at the orifice exit with the specified static pressure.

Example 3.2 Cooling air flows through a circular duct of diameter 0.025 m and length 1.0 m at inlet conditions of $P_{t_1} = 1.5$ bar and $T_{t_1} = 573 \, \text{K}$. The duct wall temperature remains uniform at 1023 K. Divide the duct in ten equal segments. For the air mass flow rate of 0.1 kg/s through the duct, compute the following for the first duct segment of length 0.1 m:

1. At section 1 (duct inlet): Mach number (M_1), static pressure (P_{s_1}), and static temperature (T_{s_1})
2. At section 2 (exit of the first duct segment): Mach number (M_2), static pressure (P_{s_2}), static temperature (T_{s_2}), total pressure (P_{t_2}), and total temperature (T_{t_2})
3. Change in entropy. Verify that the criterion for determining the flow direction in the duct is satisfied.

For the computations, assume the following air properties and empirical correlations:

Air properties: $\kappa = 1.4$; $R = 287 \, \text{J/(kg K)}$; $k = 0.0429 \, \text{W/(m K)}$; $Pr = 0.71$

Empirical equation to compute friction factor: $f = \dfrac{1}{\left(1.82 \log_{10} Re - 1.64\right)^2}$

Empirical equation to compute heat transfer coefficient (Dittus-Boelter equation):

$$Nu = 0.023 Re^{0.8} Pr^{0.334} \quad \text{or} \quad h = 0.023 \left(\frac{k}{D}\right) Re^{0.8} Pr^{0.334}$$

Solution

In this example, the cooling air flow through the duct is subjected to both friction and heat transfer. The solution method presented here, however, can be easily extended to include change in duct flow area and duct rotation about an axis other than its own axis; see Problem 3.5.

Some Preliminary Calculations:

$$\frac{\kappa}{\kappa - 1} = \frac{1.4}{1.4 - 1} = 3.5$$

$$c_p = \frac{\kappa R}{\kappa - 1} = 3.5 \times 287 = 1004.5 \, \mathrm{J/(kg\,K)}$$

$$\mu = \frac{kPr}{c_p} = \frac{0.0429 \times 0.71}{1004.5} = 3.0323 \times 10^{-5} \, \mathrm{kg/(ms)}$$

Duct flow area: $A = \dfrac{\pi D^2}{4} = \dfrac{\pi \times (0.025)^2}{4} = 0.491 \times 10^{-3} \, \mathrm{m}^2$

Duct surface area for heat transfer: $A_{\mathrm{w}} = \pi D L = \pi \times 0.025 \times 0.1 = 7.854 \times 10^{-3} \, \mathrm{m}^2$

1. Quantities at section 1 (duct inlet)

    Total-pressure mass flow function: $\hat{F}_{f_{t_1}} = \dfrac{\dot{m}\sqrt{RT_{t_1}}}{A P_{t_1}} = \dfrac{0.1\sqrt{287 \times 573}}{(0.491 \times 10^{-3}) \times (1.5 \times 10^5)} = 0.551$

    Using an iterative solution method, for example, "Goal Seek" in EXCEL, we obtain $M_1 = 0.553$.

    From isentropic relations we obtain

    $$T_{s_1} = \frac{T_{t_1}}{\left(1 + \dfrac{1 + \kappa}{2} M_1^2\right)} = \frac{573}{1 + 0.5(1 + 1.4)(0.553)^2} = 539.9 \, \mathrm{K}$$

    $$P_{s_1} = P_{t_1}\left(\frac{T_{s_1}}{T_{t_1}}\right)^{\frac{\kappa}{\kappa-1}} = 1.5 \times 10^5 \left(\frac{539.9}{573}\right)^{3.5} = 121831 \, \mathrm{Pa}$$

    Inlet velocity: $V_1 = M_1 \sqrt{\kappa R T_{s_1}} = 0.553 \times \sqrt{1.4 \times 287 \times 539.9} = 257.717 \, \mathrm{m/s}$

    Therefore, at section 1 (duct inlet), we obtain $M_1 = 0.553$, $T_{s_1} = 539.9 \, \mathrm{K}$, and $P_{s_1} = 121831 \, \mathrm{Pa}$.

2. Quantities at section 2

    Let us first calculate the total temperature T_{t_2} as follows:

    Reynolds number: $Re = \dfrac{\rho V D}{\mu} = \dfrac{(\rho V A)D}{\mu A} = \dfrac{\dot{m} D}{\mu A} = \dfrac{0.1 \times 0.025}{3.0323 \times 10^{-5} \times 0.491 \times 10^{-3}}$

    $$Re = 1.68 \times 10^5$$

    Nusselt number: $Nu = 0.023 Re^{0.8} Pr^{0.334} = 0.023 \times (1.68 \times 10^5)^{0.8} \times (0.71)^{0.334}$
    $$= 310.61$$

    Heat transfer coefficient: $h = \dfrac{k}{D} Nu = \left(\dfrac{0.0429}{0.025}\right) \times 310.61 = 533 \, \dfrac{\mathrm{W}}{\mathrm{m}^2 \, \mathrm{K}}$

Number of transfer units (NTU):

$$\eta = \frac{A_w h}{\dot{m} c_p} = \frac{7.854 \times 10^{-3} \times 533}{0.1 \times 1004.5} = 4.167 \times 10^{-2}$$

Using Equation 3.12, we compute the air total temperature at section 2 as

$$T_{t_2} = T_w - \left(T_w - T_{t_1}\right) e^{-\eta} = 1023 - (1023 - 573) e^{-0.04167}$$

$$T_{t_2} = 591.4 \text{ K}$$

To compute remaining quantities at the duct outlet, we present in the following an iterative solution method. The converged value of each quantity is given within parentheses.

Step 1: Assume M_2 $(M_2 = 0.593)$

Step 2: Compute $\hat{F}_{f_2}$: $\quad \hat{F}_{f_2} = M \sqrt{\dfrac{\kappa}{\left(1 + \frac{\kappa-1}{2} M_2^2\right)^{\frac{\kappa+1}{\kappa-1}}}}$ $\left(\hat{F}_{f_2} = 0.572\right)$

Step 3: Compute P_{t_2}: $\quad P_{t_2} = \dfrac{\dot{m}\sqrt{RT_{t_2}}}{A \hat{F}_{f_2}}$ $\left(P_{t_2} = 146706 \text{ Pa}\right)$

Step 4: Compute T_{s_2}: $\quad T_{s_2} = \dfrac{T_{t_2}}{\left(1 + \dfrac{1+\kappa}{2} M_2^2\right)}$ $\left(T_{s_2} = 552.5 \text{ K}\right)$

Step 5: Compute P_{s_2}: $\quad P_{s_2} = P_{t_2} \left(\dfrac{T_{s_2}}{T_{t_2}}\right)^{\frac{\kappa}{\kappa-1}}$ $\left(P_{s_2} = 115670 \text{ Pa}\right)$

Step 6: Compute V_2: $\quad V_2 = M_2 \sqrt{\kappa R T_{s_2}}$ $(V_2 = 279.29 \text{ m/s})$

Step 7: Compute ΔP_{s_f}: $\quad \Delta P_{s_f} = \dfrac{f}{2} \dfrac{L}{D} \dfrac{\dot{m}}{A} \left(\dfrac{V_1 + V_2}{2}\right)$ $\left(\Delta P_{s_f} = 1766 \text{ Pa}\right)$

Step 8: Compute $\hat{P}_{s_2}$ from the force-momentum balance on the duct control volume using

$$\hat{P}_{s_2} = P_{s_1} - \Delta P_{s_f} - \frac{\dot{m}}{A}\left(V_2 - V_1\right) \qquad (\hat{P}_{s_2} = 115670 \text{ Pa})$$

Step 9: Repeat Steps 1 to 8 until P_{s_2} computed in Step 5 is within an acceptable difference with $\hat{P}_{s_2}$ computed in Step 8.

The required quantities computed at section 2 are summarized as follows:

$$M_{\text{out}} = 0.593, \ P_{s_{\text{out}}} = 115670 \text{ Pa}, \ T_{s_{\text{out}}} = 552.5 \text{ K}, \ P_{t_{\text{out}}} = 146706 \text{ Pa}, \text{ and } T_{t_{\text{out}}} = 591.4 \text{ K}$$

3. Entropy change between sections 1 and 2:

$$\frac{s_2 - s_1}{R} = \frac{c_p}{R} \ln \frac{T_{t_2}}{T_{t_1}} - \ln \frac{P_{t_2}}{P_{t_1}} = \frac{1004.5}{287} \ln \left(\frac{591.4}{573}\right) - \ln \left(\frac{146706}{150000}\right) = 0.1326$$

For the entropy-based criterion for determining the flow direction in an element in the presence of heat transfer, we need to use Equation 3.37 for which the average gas total temperature for heat transfer is first computed as

$$\overline{T}_t = T_w - \frac{\left(T_{t_2} - T_{t_1}\right)_{HT}}{\eta} = 1023 - \frac{591.4 - 573}{4.167 \times 10^{-2}} = 581.44 \text{ K}$$

Then, Equation 3.36 yields

$$\frac{(\Delta s)_{HT}}{R} = \frac{1004.5}{287} \frac{(591.4 - 573)}{581.44} = 0.1107$$

Finally, using Equation 3.37 we obtain

$$\frac{s_2 - s_1}{R} - \frac{(\Delta s)_{HT}}{R} = 0.1326 - 0.1107 = 0.022$$

which is positive, satisfying the criterion for the element flow direction from section 1 to section 2.

Example 3.3 Consider a Fanno flow in a pipe of 0.0508 m diameter and average Darcy friction factor of 0.022. The total pressure and temperature at the pipe inlet are 11 bar and 459 K, respectively. Assuming $\kappa = 1.4$ and $R = 287 \text{ J/(kg K)}$ for air, calculate the maximum pipe length to deliver a mass flow rate of 1.75 kg/s for both subsonic and supersonic inlet conditions. In each case, determine the total pressure at the pipe exit.

Solution
In this example, for the given total pressure and total temperature at the pipe inlet, both subsonic and supersonic flow solutions are possible for the specified mass flow rate through the pipe. In both cases, the flow chokes at the pipe exit ($M_{exit} = 1$), albeit requiring different pipe lengths.

(a) Subsonic flow at inlet

Pipe flow area: $A = \frac{\pi D^2}{4} = \frac{\pi \times (0.0508)^2}{4} = 2.027 \times 10^{-3} \text{ m}^2$

Inlet Mach number: We can compute the mass flow rate at pipe inlet by the equation

$$\dot{m} = \frac{\hat{F}_{f_{t(a)}} A P_{t_{inlet}}}{\sqrt{RT_{t_{inlet}}}}$$

where $\hat{F}_{f_{t(a)}}$ is the total-pressure mass flow function for case (a), and it is given by the equation

$$\hat{F}_{f_{t(a)}} = M_{inlet(a)} \sqrt{\frac{\kappa}{\left(1 + \frac{\kappa-1}{2}M^2_{inlet(a)}\right)^{\frac{\kappa+1}{\kappa-1}}}}$$

which; upon using, for example, the "Goal Seek" function in Excel with a subsonic Mach number as the initial guess, yields $M_{inlet(a)} = 0.25$.

Maximum pipe length: Using the inlet Mach number, we compute $fL_{max\,(a)}/D$ by the Fanno flow equation from Table 2.4, we obtain

$$\frac{fL_{max\,(a)}}{D} = \left(\frac{1.4+1}{2\times 1.4}\right) \ln \left(\frac{(1.4+1)(0.25)^2}{\{2+(1.4-1)(0.25)^2\}}\right) + \frac{1}{1.4}\left(\frac{1}{(0.25)^2}-1\right) = 8.491$$

which gives

$$L_{max\,(a)} = \frac{8.491 \times 0.0508}{0.022} = 19.606 \text{ m}$$

Total pressure at pipe exit: Using the Fanno flow equation in Table 2.4, we compute the inlet-to-exit total pressure ratio as

$$\frac{P_{t_{inlet}}}{P^*_{t(a)}} = \frac{1}{M_{inlet(a)}}\left(\frac{2+(\kappa-1)M^2_{inlet(a)}}{\kappa+1}\right)^{\frac{\kappa+1}{2(\kappa-1)}} = \frac{1}{0.25}\left(\frac{2+(1.4-1)(0.25)^2}{1.4+1}\right)^{\frac{1.4+1}{2(1.4-1)}} = 2.404$$

which yields

$$P^*_{t(a)} = \frac{P_{t_{inlet}}}{2.404} = \frac{11 \times 10^5}{2.404} = 457670 \text{ Pa}$$

(b)   Supersonic flow at inlet

Inlet Mach number: We again use an iterative method (e.g., the "Goal Seek" function in Excel with a supersonic Mach number as the initial guess) to solve the following equations

$$\dot{m} = \frac{\hat{F}_{f_{t(b)}} A P_{t_{inlet}}}{\sqrt{RT_{t_{inlet}}}}$$

and

$$\hat{F}_{f_{t(a)}} = M_{inlet(b)} \sqrt{\frac{\kappa}{\left(1 + \frac{\kappa-1}{2}M^2_{inlet(b)}\right)^{\frac{\kappa+1}{\kappa-1}}}}$$

and obtain $M_{inlet(b)} = 2.4$.

Maximum pipe length: Using $M_{inlet(b)} = 2.4$, we compute $fL_{max\,(a)}/D$ by the Fanno flow equation from Table 2.4 as

$$\frac{fL_{max\,(b)}}{D} = \frac{1.4+1}{2\times 1.4} \ln\left(\frac{(1.4+1)(2.4)^2}{\{2+(1.4-1)(2.4)^2\}}\right) + \frac{1}{1.4}\left(\frac{1}{(2.4)^2}-1\right) = 0.410$$

which gives

$$L_{\text{max (b)}} = \frac{0.410 \times 0.0508}{0.022} = 0.947 \text{ m}$$

Total pressure at pipe exit: Using the Fanno flow equation in Table 2.4, we compute

$$\frac{P_{t_{\text{inlet}}}}{P_{t(b)}^*} = \frac{1}{M_{\text{inlet(b)}}} \left(\frac{2 + (\kappa - 1)M_{\text{inlet(b)}}^2}{\kappa + 1} \right)^{\frac{\kappa+1}{2(\kappa-1)}} = \frac{1}{2.4} \left(\frac{2 + (1.4 - 1)(2.4)^2}{1.4 + 1} \right)^{\frac{1.4+1}{2(1.4-1)}} = 2.404$$

which yields

$$P_{t(b)}^* = \frac{P_{t_{\text{inlet}}}}{2.404} = \frac{11 \times 10^5}{2.404} = 457670 \text{ Pa}$$

The results obtained in this example indicate that the maximum pipe length with subsonic inlet is 19.606 m and that with supersonic inlet 0.947 m, which provides a much shorter delivery system. For creating a subsonic inlet flow, we can use an isentropic convergent nozzle; for creating a supersonic inlet flow, we can use an isentropic convergent-divergent nozzle. In the latter case, however, the flow system will have two sonic sections, one at the physical throat of the nozzle and the other at the pipe exit.

The fact that in both cases we obtain the same total pressure at the pipe exit should not be surprising. For both subsonic-inlet and supersonic-inlet cases, we have at the pipe inlet identical total pressure, total temperature, and mass flow rate, which remains constant throughout the pipe. At the pipe exit, the flow is choked with $M_{\text{exit}} = 1$, yielding a constant value of total-pressure mass flow function ($F_{f_t}^* = 0.6847$). Because the total temperature remains constant in a Fanno flow, the mass flow rate equation should yield the same value for the exit total pressure for both cases. Also, from isentropic relations, we can conclude that the static pressure at the pipe exit should also be identical for both cases. Because the frictional pressure loss per unit length of the pipe for the supersonic flow will be much higher, due to its higher velocities, than for the subsonic flow, we obtain a much shorter length for the same overall loss in total pressure and change in static pressure over the pipe from its inlet to exit.

Example 3.4 Reconsider the Fanno flow of Example 3.3. Without changing the specified total pressure and temperature at the pipe inlet, calculate and describe the properties of the new Fanno flow of air if the maximum pipe lengths with sonic exit computed for the subsonic and supersonic inlet conditions are extended by 10 percent. In both cases, shown in Figure 3.6, the flow remains choked at the extended pipe exit.

Solution
Figure 3.6 schematically (not to scale) shows pipe extensions in initially choked Fanno flows with subsonic and supersonic inlet conditions. As shown in Figure 3.6(a), when we extend a choked Fanno flow with a subsonic inlet, the sonic section moves

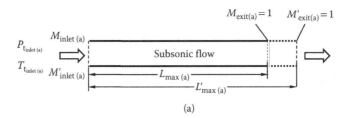

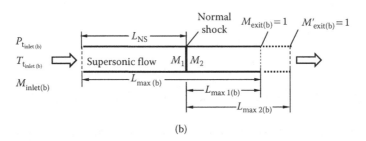

Figure 3.6 (a) Pipe extension in an initially choked Fanno flow with subsonic inlet and (b) pipe extension in an initially choked Fanno flow with supersonic inlet (Example 3.4).

downstream to the extended pipe exit with reduction in both mass flow rate and exit total pressure. A choked Fanno flow with a supersonic inlet, shown in Figure 3.6(b), features a different flow behavior when the pipe is extended beyond its maximum length. Because the downstream changes are not communicated upstream in a supersonic flow, the pipe extension results in a normal shock at an intermediate section. The supersonic flow upstream of the normal shock becomes subsonic downstream and reaches the sonic condition at the extended pipe exit. In this case, interestingly, the mass flow rate and total pressure at the new exit remain unchanged.

(a)   Subsonic flow at inlet

With reference to Figure 3.6(a), the inlet conditions and calculated properties from Example 3.3 are summarized here:

$$P_{t_{\text{inlet (a)}}} = 11 \times 10^5 \text{ Pa}, \ T_{t_{\text{inlet (a)}}} = 459 \text{ K}, \ M_{\text{inlet (a)}} = 0.25,$$

$$L_{\text{max (a)}} = 19.606 \text{ m, and } P_{t_{\text{exit (a)}}} = 457670 \text{ Pa}$$

With 10 percent extension of the pipe, the new maximum pipe length becomes

$$L'_{\text{max (a)}} = 1.1 \times 19.606 = 21.566 \text{ m}$$

which yields

$$\frac{f L'_{\text{max (a)}}}{D} = 0.022 \times 21.566/0.054 = 9.34$$

Using the "Goal Seek" iterative solution method in Excel, we obtain the new inlet Mach number from the equation in Table 2.4 as $M'_{\text{inlet (a)}} = 0.241$. With this value of inlet Mach number, we calculate the new mass flow rate as $\dot{m}'_{(a)} = 1.689 \text{ kg/s}$, which is

3.51% lower than the initial value of 1.75 kg/s without the pipe extension. Similarly, at $M'_{inlet\,(a)} = 0.241$, we compute the new exit total pressure as $P'^{*}_{t_{exit\,(a)}} = 441616\,\text{Pa}$, which is also 3.51% lower than its value calculated in Example 3.3.

(b) Supersonic flow at inlet

With reference to Figure 3.6(b), the inlet conditions and calculated properties from Example 3.3 are summarized here:

$$P_{t_{inlet\,(b)}} = 11 \times 10^5\,\text{Pa},\ \ T_{t_{inlet\,(b)}} = 459\,\text{K},\ \ M_{inlet\,(b)} = 2.4,$$

$$L_{max\,(b)} = 0.947\,\text{m, and}\ \ P_{t_{exit\,(b)}} = 457670\,\text{Pa}$$

With 10 percent extension of the pipe, the new maximum pipe length becomes

$$L'_{max\,(b)} = 1.1 \times 0.947 = 1.0412\,\text{m}$$

From the flow physics presented earlier for this case, it is clear that the conditions at the pipe inlet and new exit remain the same as before the pipe is extended. We only need to find the distance L_{NS} from the pipe inlet where the normal shock occurs. From Figure 3.6(b), we can write

$$L_{NS} = L_{max\,(b)} - L_{max\,1\,(b)}$$

and

$$L'_{max\,(b)} = L_{NS} + L_{max\,2\,(b)}$$

where

$L_{max\,(b)} \equiv$ Maximum pipe length that corresponds to the supersonic Mach number at the pipe inlet

$L_{max\,1(b)} \equiv$ Maximum pipe length computed at M_1

$L_{max\,2(b)} \equiv$ Maximum pipe length computed at M_2.

Note that M_1 is the supersonic Mach number upstream of the normal shock, and M_2 is the corresponding downstream subsonic Mach number.

We present next the steps of an iterative solution method to calculate L_{NS}.

1. Assume M_1.
2. Calculate M_2 with the following normal shock equation in Table 2.6:

$$M_2^2 = \frac{(\kappa - 1)M_1^2 + 2}{2\kappa M_1^2 - (\kappa - 1)}$$

3. Calculate $fL_{max\,1(b)}/D$ at M_1, hence $L_{max\,1(b)}$.
4. Calculate $L_{NS} = L_{max\,(b)} - L_{max\,1\,(b)}$.
5. Calculate $fL_{max\,2(b)}/D$ at M_2, hence $L_{max\,2(b)}$.
6. Repeat steps from 1 to 6 until $L'_{max\,(b)} = L_{NS} + L_{max\,2\,(b)}$.

Using "Goal Seek" in Excel, the aforementioned iterative method yields the following values for various quantities.

$M_1 = 1.3967,\ M_2 = 0.7411,\ L_{\max 1(b)} = 0.228,\ L_{\mathrm{NS}} = 0.719,$ and $L_{\max 2(b)} = 0.322$

When the initially choked pipe flow with the supersonic inlet is extended, the aforementioned results show that the normal shock within the pipe converts the upstream supersonic flow into a subsonic flow downstream, keeping the loss in total pressure from pipe inlet to exit constant. This feature of a choked supersonic Fanno flow makes it an ideal constant mass flow delivery system that is robust to variations in downstream conditions.

Problems

3.1 Consider the orifice of Example 3.1 under adiabatic compressible flow. Six of such orifices are connected in series. The inlet boundary conditions for this arrangement are the same as given for the single orifice in Example 3.1. The fixed static pressure at the exit of the sixth orifice is 10 bar. This flow network has five internal nodes and two boundary nodes, where the required boundary conditions are known. The flow being adiabatic, the total temperature remains constant throughout the network.

(1) Assuming total loss of the dynamic pressure at the exit of each orifice, sequentially carry out the solution for each orifice to determine the network mass flow rate, including the static and total pressure at each internal node. You may have to repeat this a few times with an initial guess for the mass flow rate to ensure that the computed exit static pressure equals the specified value.

(2) Repeat the calculations of (1) to find the minimum total pressure at inlet needed to achieve the choked flow condition in the orifice network. Where does the flow choke? What is the resulting network mass flow rate in this case?

(3) Find the network mass flow rate if the inlet total pressure is 10 percent higher than the minimum inlet total pressure for choking computed in (2).

Carry out solutions of (1), (2), and (3) using the following alternate approach. For the compressible flow through an orifice whose upstream end is connected to node i and the downstream end connected to node j, we can express the mass flow rate as

$$\dot{m}_{ij} = \frac{\hat{F}_{f_{t_i}} A C_d P_{t_i}}{\sqrt{RT_{t_i}}} = \left[\frac{\hat{F}_{f_{t_i}} A C_d P_{t_i}}{\left(P_{s_i} - P_{s_j}\right)\sqrt{RT_{t_i}}} \right] \left(P_{s_i} - P_{s_j}\right) = a_{ij}\left(P_{s_i} - P_{s_j}\right)$$

where

$$a_{ij} = \frac{\hat{F}_{f_{t_i}} A C_d P_{t_i}}{\left(P_{s_i} - P_{s_j}\right)\sqrt{RT_{t_i}}}$$

$$\hat{F}_{f_{t_{ij}}} = M_i \sqrt{\frac{\kappa}{\left(1 + \frac{\kappa-1}{2} M_i^2\right)^{\frac{\kappa+1}{\kappa-1}}}}$$

$$M_i = \sqrt{\left(\frac{2}{\kappa-1}\right)\left\{\left(P_{t_i}/P_{s_j}\right)^{\frac{\kappa-1}{\kappa}} - 1\right\}}$$

Note that the coefficient a_{ij} is not a constant, giving a nonlinear relation between the mass flow rate and difference in static pressures across the orifice. Although we have assumed a constant value for the discharge coefficient C_d in this problem, experimental measurements suggest that it is a function of the pressure ratio across the orifice.

Writing the aforementioned equation for the mass flow rate through each orifice in the series network and noting that, at each internal node, the mass flow rate of the upstream orifice equals that of the downstream orifice, leads a simultaneous system of equations for the unknown nodal pressures where the coefficient matrix in tridiagonal. Such a system of equations is easily solved using the tridiagonal matrix algorithm, a FORTRAN routine for which is listed in Appendix E. In this approach, however, we need an initial guess for the internal nodal pressures. In addition, because the coefficients are solution-dependent, we need to iterate them to reach convergence.

Compare the results obtained using both approaches and comment on their relative advantages and disadvantages.

3.2 Verify the solution presented for Example 3.2 for the first duct segment of length 0.1 m. Repeat the solution sequentially for the remaining duct segments to compute the duct outlet conditions. Do these results change significantly if you divide the duct into 20 and 40 segments?

3.3 Having developed the numerical solution method in Problem 3.2, it is important to compare its computed results against the known exact solutions for Fanno and Rayleigh flows. For the Fanno flow, set $T_w = T_t$ in each pipe segment to achieve zero heat transfer. Although physically unrealistic, you may also artificially set $h = 0$ to achieve an adiabatic pipe flow. Compare your results at the pipe outlet with the results obtained using the Fanno flow equations presented in Chapter 2. For the Rayleigh flow, set pipe friction factor $f = 0$ in your solution method and compare the results at the pipe outlet with the results obtained using the Rayleigh flow equations presented in Chapter 2. Note that achieving a good comparison with the exact Fanno and Rayleigh flow solutions is a necessary condition, but it does not constitute the sufficient validation of the numerical solution method, which computes the combined effects of friction and heat transfer on the duct flow.

3.4 Repeat Example 3.2 for the case of a variable-area duct. First, consider a linearly converging duct with inlet-to-outlet area ratio of 2. Second, consider a linearly diverging duct of outlet-to-inlet area ratio of 2.0.

3.5 Consider the design situation where the variable-area duct used in Problem 3.4 is rotating at an angular velocity $\Omega = 300$ rad/s. The angle between the duct axis and the axis of rotation is $60°$. Including the effects of duct rotation, repeat your calculations in Problem 3.4 for both converging and diverging ducts.

3.6 The numerical solution method developed Problem 3.5 allows one to simulate air flow in a variable-area duct under the coupled influence of friction, heat transfer, and rotation. Using this solution method, one can also obtain the solution with each effect separately or for various combinations of these effects. For the same inlet conditions of total pressure, total temperature and mass flow rate, obtain solutions for pressure and temperature variation throughout the duct for each individual effect of area change, friction, heat transfer, and rotation. Over each duct segment, add the changes in static pressure due to each effect of area change, friction, heat transfer, and rotation and compare the resulting variation in static pressure over the duct with that obtained in Problem 3.5. Again, over each duct segment, add the changes in total pressure due to each effect of area change, friction, heat transfer, and rotation and compare the resulting variation in total pressure over the duct with that obtained in Problem 3.5. Comment on your results of comparison.

3.7 In Problem 3.5, numerical results are obtained for the specified inlet boundary conditions of total pressure, total temperature, and mass flow rate. In most design applications, and in a flow network solution, we need to compute the mass flow rate through the duct under the boundary conditions of total pressure and temperature at its inlet and static pressure at its outlet. For an outlet pressure of 1 bar, modify the numerical method used in Problem 3.5 to compute the resulting mass flow rate through the duct under the specified inlet total pressure and total temperature. At what inlet total pressure will the flow choke in the duct? Find the corresponding mass flow rate and the location of choking ($M = 1$).

3.8 Divide the duct of Example 3.2 into 40 equal segments and carry out the numerical solution under adiabatic conditions with wall friction only, which corresponds to a Fanno flow. Representing each duct segment as an orifice with its mechanical area equal to the duct area and its pressure and temperature conditions from the numerical Fanno flow solution, find the variation in orifice discharge coefficient along the duct from inlet to outlet.

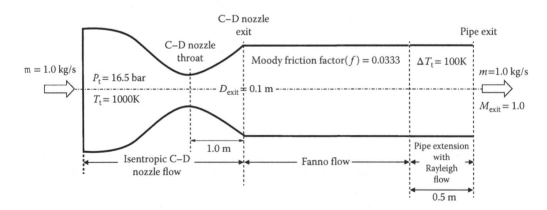

Figure 3.7 Serially-coupled isentropic and nonisentropic compressible flows (Problem 3.10).

3.9   In a flow network, when orifices are connected in series, it is generally assumed that the dynamic pressure from the upstream orifice is lost in the junction connecting two orifices. In Problem 3.8, we simulated a Fanno flow using orifices in series, one orifice for each duct segment, without losing the interorifice dynamic pressure. Repeat the calculations to determine the discharge coefficient of each orifice under the assumption that the static pressure downstream of each orifice (duct segment) becomes the total inlet pressure of the next orifice. Compare the variation in orifice discharge coefficients along the duct with that obtained in Problem 3.8.

3.10   The diagram here shows an air flow system consisting of an isentropic flow through a convergent-divergent (C-D) nozzle with exit diameter of 0.1 m, a Fanno flow, and 0.5 m long Rayleigh flow. The pipe diameter of both the Fanno flow and the Rayleigh flow equals the exit diameter of the C-D nozzle. Without the 0.5 m pipe extension (Rayleigh flow), the flow system remains free from any normal shock, and delivers the required mass flow rate of 1.0 kg / s with $M_{exit} = 1$ at the end of the Fanno flow when the inlet total pressure and total temperature are 16.5 bar and 1000 K, respectively. With the 0.5 m pipe extension (Rayleigh flow), which is needed to further increase the air total temperature by 100 K through uniform heating, the flow undergoes a normal shock somewhere between the C-D nozzle throat and the end of the Rayleigh flow where $M_{exit} = 1$ with the same inlet total pressure and total temperature of 16.5 bar and 1000 K, respectively.

   Assumptions and data:

   1.   The compressible air flow is one-dimensional and air is considered a calorically perfect gas having the equation of state: $P_s = \rho_s R T_s$, where $R = 287 \ \text{J}/(\text{kg K})$.
   2.   Ratio of air specific heats: $\kappa = c_p/c_v = 1.4$.
   3.   In the divergent section of the C-D nozzle, the diameter varies linearly from throat to nozzle exit.
   4.   The normal shock has zero thickness.

3.11   Figure 3.8 shows a pipe of diameter $D = 50\,\text{mm}$ with six equispaced converging nozzles, which are to be sized to yield equal mass flow rate of hot air through them to promote uniform mixing with a large cross flow. The hot air enters the mixing arm at the total pressure and temperature of 1.1 bar and 1000 K, respectively. All converging nozzles discharge into the cross flow at a static pressure of 1.0 bar. Assume that the flow is adiabatic in the entire mixing arm

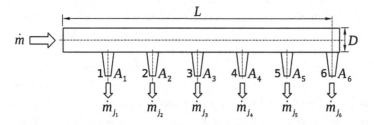

Figure 3.8 A low-pressure uniform mixing arm with converging nozzles (Problem 3.11).

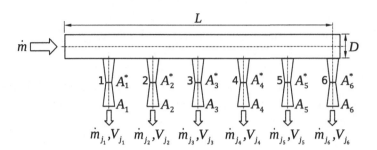

Figure 3.9 A high-pressure jet impingement cooling arm with converging-diverging nozzles (Problem 3.12).

with constant Darcy fiction factor $f = 0.02$ in the main pipe. Furthermore, assume that the flow is isentropic in each nozzle.

(a) Setting each nozzle exit area equal to one-sixth of the pipe flow area, find the mass flow rate through each nozzle.

(b) Adjust the nozzle jet area to ensure equal mass flow rate through each nozzle. What is the total mass flow rate through the mixing arm? What is the exit area of each nozzle?

3.12 Figure 3.9 shows a pipe of diameter $D = 50$ mm with six equispaced converging-diverging (C-D) nozzles, which are to be sized to yield equal mass flow rate and jet velocity of cold air for impingement cooling of a hot surface. The high-pressure cold air enters the impingement arm at the total pressure and temperature of 11 bar and 500 K, respectively. All converging-diverging nozzles discharge the coolant air on a surface maintained at a uniform static pressure of 1.5 bar. Assume that the flow is adiabatic in the entire impingement arm with constant Darcy fiction factor $f = 0.02$ in the main pipe. Furthermore, assume that the flow is isentropic in each nozzle.

(a) Determine the total mass flow rate at the pipe inlet and critical throat area of each C-D nozzle to deliver one-sixth of this flow rate.

(b) Determine each nozzle exit area so that the impingement jet velocities of all nozzles are equal.

3.13 Repeat the solution of Example 3.1 for four values of the discharge coefficient C_d, namely, 0.6, 0.7, 0.8, and 0.9, and, for each value, use four values of the orifice inlet total pressure P_{t_1}, namely, 11 bar, 13 bar, 15 bar, and 17 bar, keeping all other given data unchanged. Based on your calculation results, will you recommend in a design application the method of calculating the orifice exit total pressure using the loss coefficient K?

3.14 Repeat the solution of Example 3.1 with the orifice inlet total pressure $P_{t_1} = 20$ bar, keeping all other given data unchanged. Note that the pressure ratio across the orifice in this case will exceed the critical pressure ratio for a choked flow condition. Comment on the validity of calculating the orifice exit total pressure using the loss coefficient K?

3.15 Solve Example 3.3 by discretizing the duct into small segments and using the numerical method used in Example 3.2 for the solution over a duct segment.

3.16 Solve Example 3.4 by discretizing the duct into small segments and using the numerical method used in Example 3.2 for the solution over a duct segment.

Project

3.1 Develop a stand-alone, general purpose computer code in FORTRAN, or the language of your choice, to solve for the properties of 1-D compressible flow in a variable-area duct under the coupled effects of friction, heat transfer, and duct rotation about an axis inclined to the duct axis. The code should work at all Mach numbers corresponding to subsonic, sonic (internal choking), and supersonic flow regimes, including the possibility of a normal shock, which is the nonisentropic mechanism a supersonic flow uses to abruptly become subsonic. How will you verify your computer code, and how will you validate it?

References

Benedict, R. P. 1980. *Fundamentals of Pipe Flow*. New York: John Wiley & Sons.

Buckingham, E. 1932. Notes on the orifice meter; the expansion factor for gases. *J. Res. Natl. Bur. Stand.* 9. p. 61.

Carnahan, B., H. A. Luther, and J. O. Wilkes. 1969. *Applied Numerical Methods*. New York: John Wiley & Sons.

Idris, A., and K.R. Pullen. 2005. Correlations for the discharge coefficient of rotating orifices based on the incidence angle. *Proc. IMechE Part A: J Power Energy*. 219: 333–352.

McGreehan, W. F., and M. J. Schotsch. 1988. Flow characteristics of long orifices with rotation and corner radiusing. ASME *J. Turbomach*. 110: 213–217.

Nash, S. G., and A. Sofer. 1996. *Linear and Nonlinear Programming*. New York: McGraw-Hill.

Parker, D. M., and D. M. Kercher. 1991. Enhanced method to compute the compressible discharge coefficient of thin and long orifices with inlet corner radiusing. Heat Transfer in Gas Turbine Engines. HTD-Vol. 188, ed. E. Elovic. pp. 53–63. New York: ASME.

Perry, J. A. 1949. Critical flow through sharp-edged orifices. *Trans. ASME*. 71(10):757–764.

Stolz, J. 1975. An approach towards a general correlation of discharge coefficients of orifice plate meters. Paper K-1, Conference on Fluid Flow in the Mid 1970's, 5–10 April 1975, East Kilbride, Glasgow.

Sultanian, B.K. 1980. HOME – A program package on Householder reflection method for linear least-squares data fitting. *J. Institution of Engineers (I)*. 60(pt Et. 3):71–75.

Sultanian, B. K. 2015. *Fluid Mechanics: An Intermediate Approach*. Boca Raton, FL: Taylor & Francis.

Bibliography

Alexiou, A., and K. Mathioudakis. 2009. Secondary air system component modeling for engine performance simulations. *ASME J. Engrg. Gas Turbine Power*. 131(3). pp.31202.1–9.

Blevin, R. D. 2003. *Applied Fluid Dynamics Handbook*. Malabar: Krieger Publishing Company.

Bragg, S. L. 1960. Effect of compressibility on the discharge coefficient of orifices and convergent nozzles. *J. Mechanical Engineering Science*. 2(1): 35–44.

Cunningham, R. G. 1951. Orifice meters with supercritical compressible flow. *Trans. ASME*. 73: 625–638.

Debler, W. R. 1990. *Fluid Mechanics Fundamentals*, 1st edn. Englewood Cliffs, NJ: Prentice Hall.

Deckker, B. E. L., and Y. F. Chang. 1965–1966. An investigation of steady compressible flow through thick orifices. *Proc. IMechE*. 180 (Part 3F):132.

Dittmann, M., K. Dullenkopf, and S. Wittig. 2003. Discharge coefficients of rotating short orifices with radiused and chamfered inlets. ASME Paper No. GT2003–38314.

Gritsch, M., A. Schulz, and S. Wittig. 2001. Effect of crossflows on the discharge coefficient of film cooling holes with varying angles of inclination and orientation. ASME Paper No. 2001-GT-0134.

Hay, N., D. Lamperd, and S. Benmansour. 1983. Effect of crossflows on the discharge coefficient of film cooling holes. *Trans. ASME J. Eng Power*.105: 243–248.

Huening, M. 2008. Comparison of discharge coefficient measurements and correlations for several orifice designs with cross-flow and rotation around several axes. ASME Paper No. GT2008–50976.

Huening, M. 2010. Comparison of discharge coefficient measurements and correlations for orifices with cross-flow and rotation. *ASME J. Turbomach*. 132(3): 031017.1–031017.10.

Idelchik, I. E. 2005. *Handbook of Hydraulic Resistance*, 3rd edn. Delhi: Jaico Publishing House.

Idris, A., K. Pullen, D. Barnes. 2004. An investigation into the flow within inclined and rotating orifices and the influence of incidence angle on the discharge coefficient. *Proc IMechE Part A: J. Power Energy*. 218: 55–69.

Idris, A., K. R. Pullen, and R. Read. 2004. The influence of incidence angle on the discharge coefficient for rotating radial orifices. ASME Paper No. GT2004–53237.

Jeppson, R. W. 1976. *Analysis of Flow in Pipe Networks*. Ann Arbor: Ann Arbor Science Publishers Inc.

Miller, D. S. 1990. *Internal Flow Systems*, 2nd edn. Houston: Gulf Publishing Company

Miller, R. W. 1996. *Flow Measurement Engineering Handbook*, 3rd edn. New York: McGraw-Hill.

Mott, R. L. 2006. *Applied Fluid Mechanics*, 6th edn. New Jersey: Pearson Prentice Hall.

Hay, N., and D. Lampard. 1998. Discharge coefficient of turbine cooling holes: a review. *ASME J. Turbomach*. 120. pp. 314–319.

Rohde, J. E., H. T. Richard, and G. W. Metzger. 1969. Discharge coefficients of thick plate orifices with approach flow perpendicular and inclined to the orifice axis. NASA TN D-5467.

Shapiro, A. H. 1953. *The Dynamics and Thermodynamics of Compressible Fluid Flow*, Vol. 1. New York: John Wiley & Sons.

Sousek, J. 2010. Experimental study of discharge coefficients of radial orifices in high-speed rotating shafts. ASME Paper No. GT2010–22691.

Nomenclature

a_{ij}	Ratio of mass flow rate to static pressure drop in the pipe flow element e_{ij}
A	Area
A_{eff}	Effective area

A_{vc}	Vena contracta area
c_p	Specific heat at constant pressure
C_d	Discharge coefficient
CFD	Computational fluid dynamics
d	Orifice diameter
D	Pipe diameter
D_h	Hydraulic mean diameter
DNS	Direct numerical simulation
DOE	Design of experiments
e_{ij}	Element connected to junctions i and j ($e_{ij} = e_{ji}$)
E	Error norm: $E = \sqrt{\frac{1}{n}\sum_{i=1}^{n}\left(\Delta\dot{m}_i\right)^2}$
f	Darcy friction factor
F	Force
F_P	Net pressure force acting on the control volume in the momentum direction
F_{rot}	Rotational body force acting on the control volume in the momentum direction
F_{sh}	Net shear force acting on the control volume in the momentum direction
$\hat{F}_{f_t}$	Dimensionless total-pressure mass flow function
g	Acceleration due to gravity
h	Heat transfer coefficient
k	Thermal conductivity
k_i	Number of junctions connected through elements to junction i in a flow network
K	Minor-loss coefficient
K_f	Major-loss coefficient ($K_f = fL/D$)
L	Pipe length
LES	Large eddy simulation
$\dot{m}$	Mass flow rate
$\Delta\dot{m}_i$	Residual error of continuity equation at junction i
$\dot{m}_{ij}$	Mass flow rate from junction i to j ($\dot{m}_{ij} = -\dot{m}_{ji}$)
M	Mach number
n	Total number of internal junctions of a flow network
P	Pressure
Pr	Prandtl number
r	Cylindrical polar coordinate r; radius of curvature at orifice inlet
R	Gas constant
Re	Reynolds number
Re_d	Reynolds number based on the orifice diameter
s	Specific entropy
T	Temperature
$\left(\Delta T_{t_R}\right)_{HT}$	Change in gas total temperature in the flow element due to heat transfer

$\left(\Delta T_{t_R}\right)_{rot}$	Change in gas total temperature in the flow element due to rotation
$\left(\Delta T_{t_R}\right)_{CCT}$	Heat transfer and rotational work transfer coupling correction term
V	Velocity (magnitude)
x	Cartesian coordinate x
Δx	Extent of the control volume in x direction
Y	Expansion factor

Subscripts and Superscripts

1	Location 1; section 1
2	Location 2; section 2
3	Location 3; section 3
f	Friction
g	Due to gravity
HT	Heat transfer
i	Junction i
in	Inlet
inc	Incompressible
ideal	Isentropic case with no loss in total pressure
j	Junction j
max	Maximum
mom	Momentum
new	Values obtained at the current iteration
old	Values belonging to the previous iteration
out	Outlet
R	Rotor reference frame
rot	Rotation
s	Static
t	Total (stagnation)
w	Wall
x	Axial direction
θ	Tangential direction
$(^-)$	Average

Greek Symbols

α_i	Under-relaxation parameter used at junction i
β	Ratio of orifice diameter to housing pipe diameter
δ_{ij}	Binary multiplier: $\delta_{ij} = 1$ for $\dot{m}_{ij} < 0$ (inflows) and $\delta_{ij} = 0$ for $\dot{m}_{ij} > 0$ (outflows)
δ_{tol}	Maximum error allowed in mass conservation at an internal junction
$\tilde{\delta}_{tol}$	Maximum acceptable value of the error norm E

$\hat{\delta}_{tol}$	Maximum error allowed in computed total temperatures at an internal junction in successive iterations
Δ	Change in quantity over the control volume
η	Number of transfer units $\eta = (hA_w)/(\dot{m}c_p)$
κ	Ratio of specific heats
λ	Auto-adjusted damping parameter used in the modified Newton-Raphson method
μ	Dynamic viscosity
ρ	Density
Ω	Angular velocity

4 Internal Flow around Rotors and Stators

4.0   Introduction

The critical load-bearing structural components of the compressor and turbine in a gas turbine engine are essentially airfoils (vanes and blades) and disks, which rotate at high angular velocity with the blades mounted on them; the vanes are mounted on the static structure. Although the blades directly participate in energy conversion in the primary flow paths of compressors and turbines, the disks are exposed to internal cooling and sealing flows. The failure of a rotor disk with its extremely high rotational kinetic energy is considered catastrophic for the entire engine. In Chapter 3, we discussed an arbitrary duct and orifice being the two core flow elements of an internal flow system and presented their 1-D flow modeling as needed to assemble a flow network model, including its robust numerical solution method. In this chapter, we will expound on some novel concepts and flow features associated with gas turbine internal flows over rotor disks and in cavities formed between a rotor disk and either another rotor disk or a static structure. As in Chapter 3, we continue here our emphasis on the 1-D modeling of disk pumping flow, swirl and windage distributions, and centrifugally-driven buoyant convection in compressor rotor cavities with or without a bore cooling flow.

Most of the concepts presented in this chapter with good physical insight are by and large outside the mainstream of thermofluids education at the senior undergraduate and graduate levels in most universities around the world. Nevertheless, these concepts are critically important for the design and analysis of gas turbine internal flow systems, for example, to design rim seals to minimize, or to prevent, hot gas ingestion; to develop an optimum preswirl system for the turbine blade cooling air; and to accurately compute rotor axial thrust for sizing the thrust bearing.

Although the references list a number of leading references on various topics covered in this chapter, readers may refer to Owen and Rogers (1989, 1995) and Childs (2011) for a comprehensive bibliography, particularly related to free disk, rotor-stator, and rotor-rotor systems.

4.1   Rotor Disk

In a gas turbine, compressor and turbine blades are mounted on rotor disks. These disks must have acceptable temperature distributions to ensure their structural integrity during both steady and transient engine operations. Unless the coolant flow over the disk co-rotates at the same angular velocity as the disk, it gets pumped in the disk boundary layer; radially outward if the flow rotates slower than the disk and radially inward if it rotates faster than the disk, as physically explained later in this chapter. In the following sections, we discuss two disk pumping situations. In one, the free disk pumping, the free-stream air next to the rotating disk is stagnant. In the second, the disk pumping beneath a forced vortex, the flow is co-rotating at a constant angular velocity, which is a fraction of the disk angular velocity.

4.1.1 Free Disk Pumping

Figure 4.1 depicts the boundary layers of radial and tangential velocities for a disk rotating in a quiescent fluid far away from the disk. The growth of the radial velocity boundary layer from $r = 0$ to $r = R$ occurs through fluid entrainment via the axial velocity. Note that the free rotating disk flow features zero radial pressure gradient imposed from the adjacent stagnant fluid outside the boundary layers. In the same way as the flat pate, the boundary layers in this case become turbulent for $Re(r) = \rho r^2 \Omega / \mu > 3 \times 10^5$ – the local rotational Reynolds number.

Schlichting (1979) presents von Karman momentum integral boundary layer solutions for both laminar and turbulent boundary layers in a free rotating disk. Here we present the key results only from the turbulent boundary solutions. For the one-seventh power law profile in the boundary layer, assuming turbulent boundary layer right from $r = 0$, the fee disk pumping mass flow rate is given by

$$\dot{m}_{\text{free disk}} = 0.219 \mu r \left(\frac{\rho r^2 \Omega}{\mu} \right)^{0.8} \tag{4.1}$$

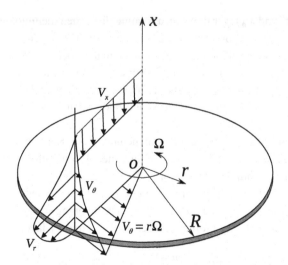

Figure 4.1 Free disk pumping.

which may be alternatively written as

$$\dot{m}_{\text{free disk}} = 0.219 \mu R \left(\frac{\rho R^2 \Omega}{\mu}\right)^{0.8} \left(\frac{r}{R}\right)^{2.6} \tag{4.2}$$

where $Re = \rho R^2 \Omega / \mu$ is the disk rotational Reynolds number based on the disk radius R. Although a free rotating disk is not found in a gas turbine design, Equations 4.1 or 4.2 estimates the upper limit for the disk pumping flow. Note that, for a laminar boundary layer, the disk pumping flow rate varies as r^2, see for example Schlichting (1979). Because the disk area also varies as r^2, the axial velocity of entrainment remains uniform over the disk. For the turbulent boundary layer, however, the disk pumping flow rate according to Equation 4.2 varies as $r^{2.6}$, as a result, the axial velocity of the entrained flow increases with radius for an incompressible flow.

Another design parameter of interest is the torque produced by the tangential wall shear stress distribution on the rotor disk. This torque for one side of the disk wetted by the fluid is given by

$$\Gamma = 2\pi \int_0^R \tau_{\text{w}\theta} r^2 \mathrm{d}r \tag{4.3}$$

which from the solution, expressed in terms of the moment coefficient, yields

$$C_{\text{M}} = \frac{\Gamma}{0.5 \rho \Omega^2 R^5} = 0.073/Re^{0.2} \tag{4.4}$$

Because the disk is rotating with angular velocity Ω, the total disk torque computed from Equation 4.3 will impart windage equal to $\Gamma\Omega$ to the boundary layer flow (pumping flow).

4.1.2   Disk Pumping Beneath a Forced Vortex

Figure 4.2a shows the boundary layer flows on a rotating disk when the fluid outside the boundary layer itself rotates as a forced vortex at a fraction of the disk angular velocity. The ratio of fluid core angular velocity and the disk angular velocity is represented by the swirl factor S_f. At $S_f = 0$, the flow field of Figure 4.2a reverts to that of the free rotating disk shown in Figure 4.1 and yields the maximum pumping flow rate in the boundary layer. When $S_f = 1$, the fluid gets into solid-body rotation with the disk with no pumping flow.

Newman (1983) extends the momentum integral method of von Karman, presented in Schlichting (1979) for a free rotating disk, to cases for which the outer flow is rotating at a constant angular velocity. Boundary layers for both radial and tangential velocities are assumed turbulent right from $r = 0$. From the solution results presented by Newman (1983), we obtain the following formula to compute disk pumping mass flow rate for the one-seventh power law velocity profile assumed in the boundary layer:

$$\dot{m}_{\text{disk pump}} = 0.219 \mu r \left(\frac{\rho r^2 \Omega}{\mu}\right)^{0.8} \zeta = \dot{m}_{\text{free disk}} \zeta \tag{4.5}$$

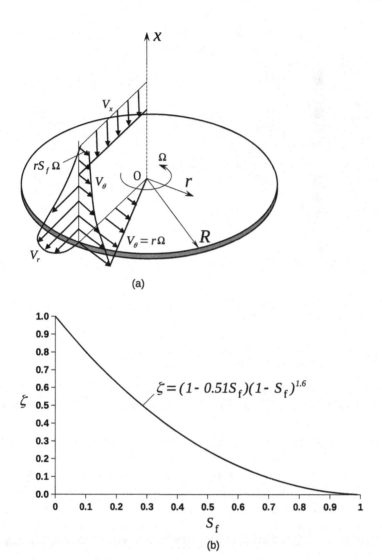

Figure 4.2 (a) Disk pumping beneath a forced vortex and (b) fraction of free disk pumping mass flow rate versus swirl factor.

where ζ, given by Equation 4.6, is the fraction of the free disk pumping mass flow rate computed by Equation 4.1.

$$\zeta = \left(1 - 0.51 S_{\mathrm{f}}\right)\left(1 - S_{\mathrm{f}}\right)^{1.6} \tag{4.6}$$

As shown in Figure 4.2b, ζ depends strongly on the swirl factor S_{f}, yielding the maximum value of the disk pumping mass flow rate for $S_{\mathrm{f}} = 0$ and no pumping for $S_{\mathrm{f}} = 1$. Equation 4.5 is a useful design equation to estimate pumping flow rate on a rotor surface between two radii.

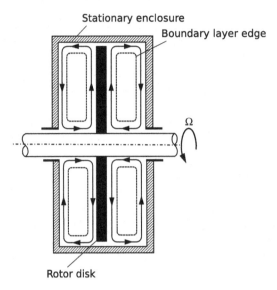

Figure 4.3 Rotor disk in an enclosed cavity.

4.1.3   Rotor Disk in an Enclosed Cavity

Figure 4.3 shows a disk rotating in a stationary housing with zero inflow and outflow. As a result of no-slip boundary conditions, the fluid assumes local velocity of the rotor and stator in contact. The rotor disk acts like a bladeless compressor or a pump (incompressible flow) and pumps the fluid radially outward. Because the disk pumping flow increases radially outward, it is continuously being fed (axial flow from the stator boundary layer to the rotor boundary layer) from the radially inward flow along the stator. The presence of the stationary shroud makes the stator torque somewhat higher than the rotor torque. As a result, the fluid core is expected to rotate at less than half of the disk angular velocity ($S_f < 0.5$). As the gap between the stator and the rotor decreases, the fluid core will tend to rotate at $S_f = 0.5$. The radial static pressure gradient in the enclosure will be established corresponding to the forced vortex with $S_f = 0.5$.

Although the torque on the stator surface does not do any work, there is continuous work transfer into the fluid from the rotor disk. As a result, the fluid temperature within a perfectly insulated enclosure will rise continuously.

4.2   Cavity

The rotor-rotor and rotor-stator cavities are the most dominant and ubiquitous features of internal flow systems of gas turbines. Assuming a turbulent cooling and sealing flow in these systems, the interplay of flow behavior on a rotor surface, a stator surface, and the mass flow rate associated with radially outward or inward flow is responsible for a variety of flow features found in theses cavities. A good understanding of these flow features is the key to their one-dimensional modeling for the flow network simulation of these internal flow systems.

A rotor surface tends to pump the flow radially outward and acts like a bladeless compressor if the adjacent fluid core rotates at a fraction of the disk angular velocity and like a bladeless turbine if the fluid core co-rotates faster than the disk. In the first case, the energy transfer occurs from the rotor disk to the fluid and in the second case from the fluid to the disk. On the disk itself, tangential velocity varies linearly with radius. Stator torque acts to reduce the angular momentum of the flow regardless of the flow direction and rotation. The stator does not partake directly in the energy transfer to or from the fluid. For a small flow influenced by rotor and stator torques, the core behaves like a forced vortex rotating at a fraction (around 0.5) of the rotor disk angular velocity. For a large flow, which is not influenced by rotor and stator torques, the flow behaves more like a free vortex, keeping a nearly constant angular momentum. In this case, the angular velocity of a radially outward flow decreases downstream and for a radially inward flow increases in the flow direction, at times exceeding the rotor angular velocity.

In general, a fluid flow seeks the path of least resistance. In a rotating flow, the difference between the angular velocity of the flow and that of the wetted wall determine the torque. If the wall rotates faster than the fluid, it will increase the flow angular momentum. If the wall angular velocity is less than that of the fluid, the torque produced will decrease the flow angular momentum. Accordingly, the stator torque always reduces the angular momentum of the adjacent fluid flow.

4.2.1   Rotor-Stator Cavity with Radial Outflow

Figure 4.4 shows a rotor-stator cavity with a superimposed radial outflow. For a small outflow rate, shown in Figure 4.4a, the flow streamlines, fully meeting the demand of the induced pumping flow, are along the rotor surface. At a radius where the disk pumping flow rate exceeds the superimposed flow rate, the fluid is entrained from the radially inward flow induced on the stator surface to make up for the difference, as shown in the figure. For a large superimposed radially outflow, exceeding the disk pumping flow, no radially inward flow on the stator surface occurs, as shown in Figure 4.4b.

4.2.2   Rotor-Stator Cavity with Radial Inflow

Figure 4.5 shows a rotor-stator cavity with a superimposed radial inflow. For a small inflow rate with $S_f < 0.5$, shown in Figure 4.5a, the flow enters the cavity along the stator surface. Some of this flow is peeled off by the rotor to satisfy its pumping flow requirement, featuring a flow reversal over a part of the cavity near the rotor surface. At a lower radius, the flow starts swirling faster like a free vortex and preferably migrates to descend down the rotor surface so as to minimize the overall wall shear force opposing it. This part of the rotor disk, where the fluid is flowing radially inward, acts like a bladeless turbine.

In case of a large radial inflow, shown in Figure 4.5b, the flow behaves more like a free vortex and preferentially flows down the rotor surface so as to minimize the overall

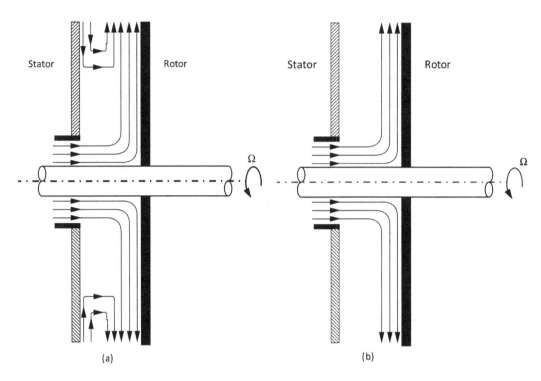

Figure 4.4 Schematic of a rotor-stator cavity with superimposed radial outflow: (a) small outflow rate and (b) large outflow rate.

shear force. The entire rotor disk in this case behaves like a bladeless radial turbine with work transfer from fluid to the rotor.

4.2.3   Rotating Cavity with Radial Outflow

Figure 4.6 shows the complex shear flow streamlines of a radial outflow in the cavity between two rotating disks. The axial flow entering the cavity through the upstream disk undergoes a sudden geometric expansion. The growth of the outer shear layer of the annular jet occurs through entrainment of the pressure-gradient-driven backflow from the downstream stagnation region. This creates the primary recirculation region shown in the figure. The size and strength of this recirculation region are found to depend mainly on the flow rate and rotational speed as discussed in Sultanian and Nealy (1987). The entering axial flow turns 90 degrees over the concave corner and flows radially outward, aided in part by frictional pumping over the downstream disk induced by its rotation. A part of the flow (almost half in this case!) turns back toward the upstream disk and moves radially outward as a result of similar pumping action over that disk.

4.2.4   Rotating Cavity with Radial Inflow

In the rotating cavity shown in Figure 4.7, we have a radial inflow. Like the case of radial outflow, the rotating cavity features essentially four regions. Both the source region at the

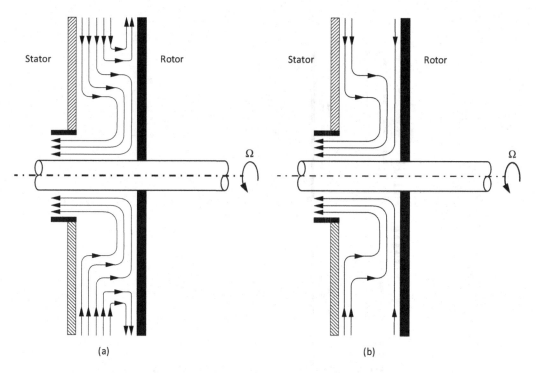

Figure 4.5 Schematic of a rotor-stator cavity with superimposed radial inflow: (a) small inflow rate and (b) large inflow rate.

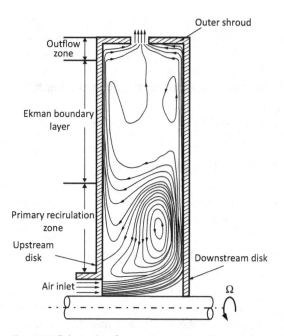

Figure 4.6 Schematic of a rotating cavity with superimposed radial outflow.

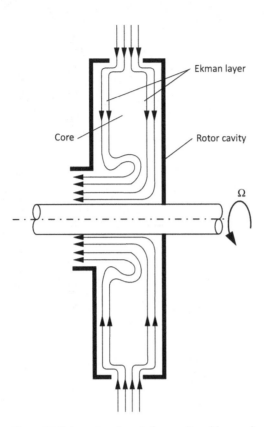

Figure 4.7 Schematic of a rotating cavity with superimposed radial inflow.

inlet and sink region at the outlet are complex shear flows. The core in the mid-section of the cavity features nearly zero axial and radial velocities, and it is bounded by Ekman boundary layers on both disks. These boundary layers are essentially nonentraining. In essence, the flow entering the rotating cavity splits almost in half and flows down radially on each disk with no intermediate entrainment. Because we normally associate disk pumping with a radially outward flow in the disk boundary layer, the flow features shown in Figure 4.7 may appear somewhat counterintuitive to some.

4.3   Windage and Swirl Modeling in a General Cavity

A major task in the design of gas turbines is to compute windage and swirl distributions throughout the path of an internal flow system. These distributions are needed to determine the thermal boundary conditions for structural heat transfer analysis and for establishing static pressure distributions for axial load calculations. Figure 4.8 shows the schematic of a general gas turbine cavity and its key features. This cavity includes multiple axisymmetric surfaces, which may be rotating, co-rotating, counter-rotating, or stationary. Each disk surface may comprise of radial, conical, and horizontal surfaces; for example, shown for surface 2 in the figure, and may feature three-dimensional protrusions, called bolts, which tend to destroy the overall symmetry of the cavity about

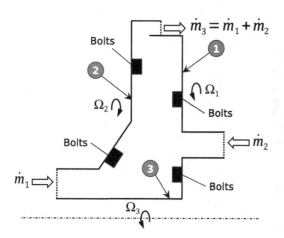

Figure 4.8 Schematic of a general gas turbine cavity and its key features.

the axis of rotation. Additionally, the cavity may have multiple inflows and outflows with different swirl, pressure, and temperature conditions.

Because the flow field in the general cavity shown in Figure 4.8 is highly complex and three dimensional, a 3-D CFD appears to be the only viable analytical method for its analysis and predictions. Such an analysis may not, however, support the shrinking design cycle time and realizing a robust design requiring multiple runs to account of statistical variations in boundary conditions. In the following sections, we present a 1-D flow modeling methodology based on the large control volume analysis for a general gas turbine cavity encountered in design. Because the methodology uses some of the published correlations to compute torque of stator and rotor surfaces, it behooves the readers (designers) to modify them in their design applications based on their design validation studies.

Daily and Nece (1960) studied, both experimentally and theoretically, the fundamental fluid mechanics associated with the rotation of a smooth plane disk enclosed within a right-cylindrical chamber, as shown in Figure 4.9. In this investigation, the torque data were obtained over a range of disk Reynolds numbers from $Re = 10^3$ to $Re = 10^7$ for axial clearance to disk radius ratios from $G = 0.0127$ to $G = 0.217$ for a constant small radial tip clearance; the velocity and pressure data were obtained for both laminar and turbulent flows. The tangential and radial velocity profiles are schematically shown in Figure 4.9a for the case of merged boundary layers and in Figure 4.9b for the case of separate boundary layers.

The study of Daily and Nece (1960) identifies the existence of the following four basic flow regimes, which are delineated in Figure 4.10 for various combinations of Re and G. The rotor disk moment coefficient in each regime is summarized as follows:

- Regime I: Laminar flow with merged boundary layers (small clearance)

$$C_M = \pi G^{-1} Re^{-1} \qquad (4.7)$$

- Regime II: Laminar flow with separate boundary layers (large clearance)

$$C_M = 1.85 G^{1/10} Re^{-0.5} \qquad (4.8)$$

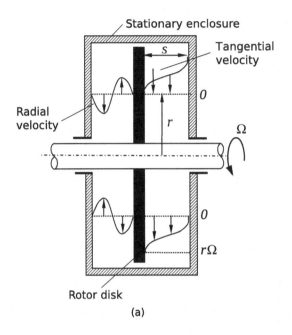

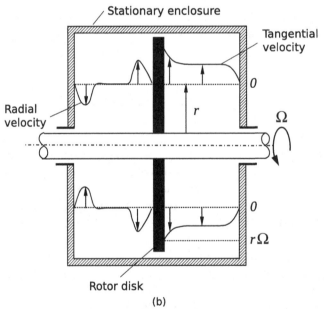

Figure 4.9 Rotor disk in an enclosed cavity: (a) merged boundary layers (Regimes I and III) and (b) separate boundary layers (Regimes II and IV).

- Regime III: Turbulent flow with merged boundary layers (small clearance)

$$C_{\mathrm{M}} = 0.040 G^{-1/6} Re^{-0.25} \tag{4.9}$$

- Regime IV: Turbulent flow with separate boundary layers (large clearance)

$$C_{\mathrm{M}} = 0.051 G^{1/10} Re^{-0.2} \tag{4.10}$$

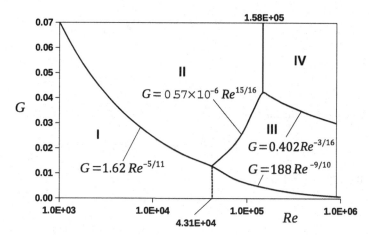

Figure 4.10 Delineation of four flow regimes in the flow of a disk rotating in an enclosed cavity (Daily and Nece, 1960).

Regime IV is generally considered relevant for gas turbine design applications. For our 1-D modeling of a general cavity, we make use of the rotor and stator moment coefficients proposed by Haaser, Jack, and McGreehan (1988) and extend them for partial disks, which may be co-rotating and counter-rotating with arbitrary angular velocities.

Based on the actual gas turbine test experience and the experimental data of Daily and Nece (1960); Haaser, Jack, and McGreehan (1988) proposed the empirical correlation for the shear coefficient on one side of the rotor disk as

$$C_{f_R} = 0.042\left(1 - S_f\right)^{1.35} Re'^{-0.2} \tag{4.11}$$

and that on one side of the stator disk as

$$C_{f_S} = 0.063 S_f^{1.87} Re'^{-0.2} \tag{4.12}$$

where the full disk Reynolds number $(Re = \rho\Omega R_o^2/\mu)$ has been modified to

$$Re' = \frac{\rho\Omega R_o\left(R_o - R_i\right)}{\mu} \tag{4.13}$$

in order to use the correlations for a partial disk $(R_i > 0)$. Note that Equations 4.11 and 4.12 have been deduced from the following moment coefficient correlations assumed for the full disk by assuming a uniform radial distribution of the tangential shear stress. Under this assumption, the shear coefficient and moment coefficient are related as follows:

$$C_M = \frac{\Gamma}{0.5\rho\Omega^2 R_o^5} = \frac{2\pi\tau_\theta \int_0^{R_o} r^2 dr}{0.5\rho\Omega^2 R_o^5} = \frac{2\pi C_f\left(0.5\rho\Omega^2 R_o^2\right)\int_0^{R_o} r^2 dr}{0.5\rho\Omega^2 R_o^2} = \frac{2\pi}{3}C_f \tag{4.14}$$

Using Equation 4.14, we obtain the moment coefficient equation from Equation 4.11 for the partial rotor disk as

$$C_{M_R} = 0.042 \times \left(\frac{2\pi}{3}\right)\left(1 - S_f\right)^{1.35} Re'^{-0.2} = 0.088\left(1 - S_f\right)^{1.35} Re'^{-0.2} \tag{4.15}$$

for the full rotor disk as

$$C_{M_R} = 0.088\left(1 - S_f\right)^{1.35} Re^{-0.2} \tag{4.16}$$

for the partial stator disk as

$$C_{M_S} = 0.063 \times \left(\frac{2\pi}{3}\right)S_f^{1.87} Re'^{-0.2} = 0.132 S_f^{1.87} Re'^{-0.2} \tag{4.17}$$

and for the full stator disk as

$$C_{M_S} = 0.132 S_f^{1.87} Re^{-0.2} \tag{4.18}$$

Equations 4.15 and 4.16 for the rotor disk or Equations 4.17 and 4.18 for the stator disk of outer radius R_o yield the following relation for the ratio of the torque for a partial disk with $R_i > 0$ to that for a full disk with $R_i = 0$:

$$\frac{\Gamma_{\text{partial}}}{\Gamma_{\text{full}}} = \frac{1 - \left(\frac{R_i}{R_o}\right)^3}{\left(1 - \frac{R_i}{R_o}\right)^{0.2}} \tag{4.19}$$

The plot of Equation 4.19 in Figure 4.11 shows that for $R_i/R_o \leq 0.5$, the equation yields $\Gamma_{\text{partial}} > \Gamma_{\text{full}}$, which is physically unacceptable. To mitigate this problem, we make the assumption that, instead of a constant tangential shear stress over the disk, as assumed in Haaser, Jack, and McGreehan (1988), the local shear coefficient of the tangential shear stress is constant, giving

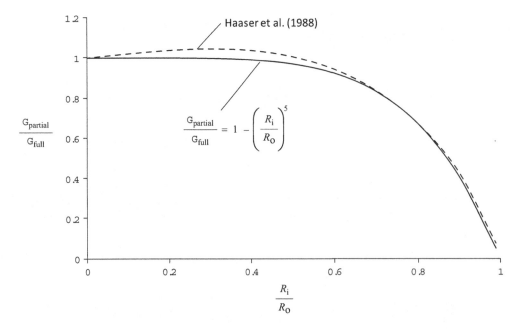

Figure 4.11 Variation of disk torque ratio ($\Gamma_{\text{partial}}/\Gamma_{\text{full}}$) with radius ratio ($R_i/R_o$).

$$C_M = \frac{\Gamma}{0.5\rho\Omega^2 R_o^5} = \frac{2\pi \int_0^{R_o} \tau_\theta r^2 dr}{0.5\rho\Omega^2 R_o^5} = \frac{2\pi \int_0^{R_o} C_f (0.5\rho\Omega^2 r^2) r^2 dr}{0.5\rho\Omega^2 R_o^2} = \frac{2\pi}{5} C_f \qquad (4.20)$$

Thus, from Equations 4.16 and 4.20, we obtain for the rotor disk

$$C_{f_R} = 0.088 \left(\frac{5}{2\pi}\right) (1 - S_f)^{1.35} Re^{-0.2} = 0.070 (1 - S_f)^{1.35} Re^{-0.2} \qquad (4.21)$$

Similarly, from Equations 4.18 and 4.20, we obtain for the stator disk

$$C_{f_s} = 0.132 \left(\frac{5}{2\pi}\right) S_f^{1.87} Re^{-0.2} = 0.105 S_f^{1.87} Re^{-0.2} \qquad (4.22)$$

Based on Equations 4.21 and 4.22, we obtain for rotor and stator disks the following relation

$$\frac{\Gamma_{partial}}{\Gamma_{full}} = 1 - \left(\frac{R_i}{R_o}\right)^5 \qquad (4.23)$$

which is plotted in Figure 4.11. The figure shows that the anomaly associated with Equation 4.19 is absent from Equation 4.23.

In using Equations 4.21 and 4.22 for calculating local tangential shear stress on the rotor surface and stator surface, respectively, one is expected to use the dynamic pressure $0.5\rho\Omega^2 r^2$. For a general 1-D modeling of cavities, it is more appropriate to use the fluid tangential velocity relative to the surface to compute the dynamic pressure, which for the rotor becomes $0.5\rho(1 - S_f)^2\Omega^2 r^2$ and for the stator $0.5\rho S_f^2\Omega^2 r^2$. Accordingly, Equations 4.21 and 4.22 are recast as follows:

$$C_{f_R} = 0.070 (1 - S_f)^{-0.65} Re^{-0.2} \qquad (4.24)$$

$$C_{f_s} = 0.105 S_f^{-0.13} Re^{-0.2} \qquad (4.25)$$

where $Re = \rho R_o^2 \Omega / \mu$.

Let us now consider the 1-D steady adiabatic flow modeling in a simple rotor-stator cavity shown in Figure 4.12a. With $\dot{m}_{in} = \dot{m}_{out} = \dot{m}$, the steady continuity equation in the cavity is satisfied. Because the flow is assumed adiabatic, the change in fluid total temperature occurs entirely as a result of work transfer from the rotor. The stator torque participates in the torque-angular momentum balance only but not directly in the energy transfer with the fluid. To capture accurate variations of flow properties in the cavity, we divide it into a number of control volumes. For the control volume k whose inlet surface is designated by j and the outlet surface by $j+1$, we write the following angular momentum equation:

$$\Gamma_{R_k} - \Gamma_{S_k} = \dot{m}\left(r_{j+1} V_{\theta_{j+1}} - r_j V_{\theta_j}\right) = \dot{m}\left(r_{j+1}^2 S_{f_{j+1}} - r_j^2 S_{f_j}\right)\Omega_{ref} \qquad (4.26)$$

Assuming a forced vortex core with swirl factor S_{f_k} such that $S_{f_{j+1}} = S_{f_k}$ and substituting

$$\Gamma_{R_k} = C_{f_R} \frac{1}{2}\rho\left(1 - S_{f_k}\right)^2 \Omega_{ref}^2 \int_{r_j}^{r_{j+1}} 2\pi r^4 dr = 0.044\rho\left(1 - S_{f_k}\right)^{1.35} \Omega_{ref}^2 \left(r_{j+1}^5 - r_j^5\right) Re^{-0.2}$$

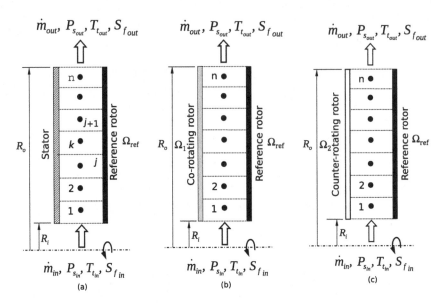

Figure 4.12 Cavity with throughflow: (a) rotor-stator cavity, (b) cavity of co-rotating disks, and (c) cavity of counter-rotating disks.

and

$$\Gamma_{S_k} = C_{f_s} \frac{1}{2} \rho S_{f_k}{}^2 \Omega_{\text{ref}}^2 \int_{r_j}^{r_{j+1}} 2\pi r^4 dr = 0.066 \rho S_{f_k}{}^{1.87} \Omega_{\text{ref}}^2 \left(r_{j+1}^5 - r_j^5 \right) Re^{-0.2}$$

in Equation 4.26, we obtain

$$\left\{ 0.044\rho \left(1 - S_{f_k} \right)^{1.35} - 0.066\rho S_{f_k}{}^{1.87} \right\} \Omega_{\text{ref}}^2 \left(r_{j+1}^5 - r_j^5 \right) \left(\frac{\rho R_o^2 \Omega_{\text{ref}}}{\mu} \right)^{-0.2}$$
$$= \dot{m} \left(r_{j+1}^2 S_{f_k} - r_j^2 S_{f_j} \right) \Omega_{\text{ref}} \tag{4.27}$$

which is a transcendental equation in the unknown S_{f_k}.

Note that in the marching solution from the cavity inlet to outlet we have $S_{f_j} = S_{f_{k-1}}$, which is obtained from the solution for the upstream control volume. One can use the regula falsi method, presented, for example, in Carnahan, Luther, and Wilkes (1969), as a robust and fast iterative solution technique for obtaining S_{f_k} from Equation 4.27; see also Appendix D.

Knowing S_{f_k} in the control volume, the static pressure change from inlet to outlet can be obtained using the radial equilibrium equation

$$\frac{dP_s}{dr} = \frac{\rho V_\theta^2}{r} = \rho r \left(S_{f_k} \Omega_{\text{ref}} \right)^2 \tag{4.28}$$

Using an average value of density $\bar{\rho} = 0.5 \left(\rho_j + \rho_{j+1} \right)$ for the control value, Equation 4.28 can be integrated to yield

$$P_{s_{j+1}} - P_{s_j} = \bar{\rho} \left(S_{f_k} \Omega_{\text{ref}} \right)^2 \left(\frac{r_{j+1}^2 - r_j^2}{2} \right)$$

Thus, the change in fluid total temperature as a result of windage in the control value can be obtained using the equation

$$T_{t_{j+1}} - T_{t_j} = \frac{\Gamma_{R_k} \Omega_{\text{ref}}}{\dot{m} c_{\text{p}}}$$

In the foregoing derivations, we have tacitly assumed that $S_f < 1$ in which case the work transfer occurs from the rotor disk to the fluid. For $S_f > 1$, however, the fluid does work on the rotor. In the following section, we will account for this possibility in the modeling of a general cavity with arbitrary inflow conditions.

Figures 4.12b and 4.12c depict a cavity with two rotor disks, which are either co-rotating or counter-rotating. If we set $\Omega_{\text{ref}} = 0$, these cavities revert to that of Figure 4.12a. Considering the rotor with the highest angular velocity as the reference rotor in a multirotor cavity and using its angular velocity (Ω_{ref}) to normalize other rotor and fluid angular velocity, we can easily extend Equation 4.24 to express the local shear coefficient for any rotor in the cavity as

$$C_{f_R} = 0.070 \, \text{sign} \, (\beta - S_f) |\beta - S_f|^{-0.65} |\beta|^{0.65} Re^{-0.2} \tag{4.29}$$

where

$$\beta = \frac{\Omega}{\Omega_{\text{ref}}}$$

$$\text{sign}(\beta - S_f) \equiv \text{Sign of the term } (\beta - S_f)$$

$$Re = \frac{\rho R^2 |\beta| \Omega_{\text{ref}}}{\mu}$$

$$R \equiv \text{Rotor outer radius}$$

Note that the local dynamic pressure to be used in conjunction with Equation 4.29 equals $0.5\rho(\beta - S_f)^2 \Omega_{\text{ref}}^2 r^2$. Further note that Equation 4.29 is applicable to all rotors in the cavity, including the reference rotor with $\beta = 1$.

For the stator surface, we re-write Equation 4.25 as

$$C_{f_s} = 0.105 |S_f|^{-0.13} Re^{-0.2} \tag{4.30}$$

4.3.1   Arbitrary Cavity Surface Orientation: Conical and Cylindrical Surfaces

A cavity may have a rotor or stator disk comprising conical and cylindrical surfaces. For the conical part of a rotor disk we use the shear stress coefficient correlation given by Equation 4.29, and if the conical surface is a part of a stator disk, we use Equation 4.30 to compute its local shear coefficient. For the conical surface segment of the rotor disk, shown in Figure 4.13a, we express its torque as

$$\Gamma_{R_{\text{cone}}} = C_{f_R} \frac{1}{2} \rho (\beta - S_f)^2 \Omega_{\text{ref}}^2 \int_{r_2}^{r_3} \frac{2\pi r^4}{\sin \alpha} dr$$

$$\Gamma_{R_{\text{cone}}} = \frac{0.044 \, \text{sign} \, (\beta - S_f) |\beta - S_f|^{-0.65} |\beta|^{0.65} \rho (\beta - S_f)^2 \Omega_{\text{ref}}^2 (r_3^5 - r_2^5) Re^{-0.2}}{\sin \alpha} \tag{4.31}$$

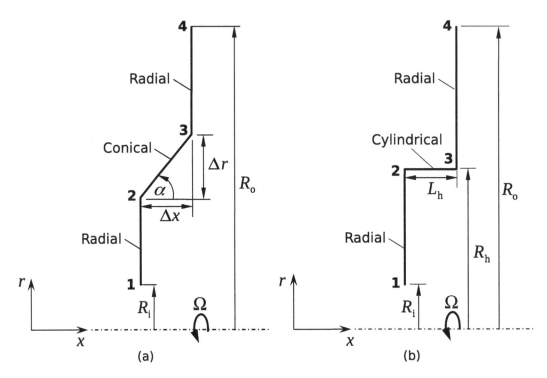

Figure 4.13 (a) Disk with a conical surface and (b) disk with a cylindrical surface.

where

$$\sin \alpha = \sin \left(\tan^{-1}(\Delta r/\Delta x) \right)$$

$$Re = \frac{\rho R_o^2 |\beta| \Omega_{\text{ref}}}{\mu}$$

For the corresponding conical surface segment of a stator disk, we express its torque as

$$\Gamma_{S_{\text{cone}}} = \frac{0.066 |S_f|^{-0.13} \rho S_f^2 \Omega_{\text{ref}}^2 \left(r_3^5 - r_2^5 \right) Re^{-0.2}}{\sin \alpha} \tag{4.32}$$

where

$$Re = \frac{\rho R_o^2 \Omega_{\text{ref}}}{\mu}$$

For the cylindrical rotor surface segment shown in Figure 4.13b, we adopt the shear stress coefficient correlation proposed by Haaser, Jack, and McGreehan (1988), extended for a rotor in addition to the primary rotor

$$C_{f_R} = 0.042 \, \text{sign} \, (\beta - S_f) \, |\beta - S_f|^{-0.65} |\beta|^{0.65} Re^{-0.2} \tag{4.33}$$

where

$$Re = \frac{\rho R_h^2 |\beta| \Omega_{ref}}{\mu}$$

Using Equation 4.33, the torque of the cylindrical surface segment of the rotor can be obtained as follows:

$$\Gamma_{R_{cylinder}} = C_{f_R} \frac{1}{2} \rho (\beta - S_f)^2 \Omega_{ref}^2 R_h^2 \int_{x_2}^{x_3} 2\pi R_h^2 dx$$

$$\Gamma_{R_{cylinder}} = 0.132 \, \text{sign} \, (\beta - S_f) |\beta - S_f|^{-0.65} |\beta|^{0.65} \rho (\beta - S_f)^2 \Omega_{ref}^2 R_h^4 L_h Re^{-0.2} \quad (4.34)$$

For the corresponding cylindrical surface segment of a stator disk, we extend Equation 4.12 for the shear coefficient to the generalized form

$$C_{f_s} = 0.063 |S_f|^{1.87} Re^{-0.2} \quad (4.35)$$

which yields the corresponding torque as

$$\Gamma_{S_{cylinder}} = 0.198 |S_f|^{-0.13} \rho S_f^2 \Omega_{ref}^2 R_h^4 L_h Re^{-0.2} \quad (4.36)$$

where

$$Re = \frac{\rho R_h^2 \Omega_{ref}}{\mu}$$

4.3.2 Bolts on Stator and Rotor Surfaces

Bolts are three-dimensional protrusions on rotor and stator surfaces. They significantly influence both the windage generation and swirl distribution in the cavity. The bolt-to-bolt spacing has a profound effect of its drag force. As the one bolt falls in wake of its upstream bolt, relative to the tangential flow velocity, its drag contribution decreases. Figure 4.14a shows bolts on a disk with small bolt-to-bolt interference, while the increased number of bolts shown in Figure 4.14b result in higher bolt-to-bolt interference.

Following the approach of Haaser, Jack, and McGreehan (1988), the torque as a result of bolts in an axisymmetric cavity is computed as follows:

Bolts on rotor surface:

$$\Gamma_{R_b} = 0.5 N_b h b C_{D_b} I_b R_b^3 \rho \Omega_{ref}^2 (\beta - S_f)^2 \quad (4.37)$$

Bolts on stator surface:

$$\Gamma_{S_b} = 0.5 N_b h b C_{D_b} I_b R_b^3 \rho \Omega_{ref}^2 S_f^2 \quad (4.38)$$

where

$N_b \equiv$ Number of bolts
$h \equiv$ Bolt height from the disk surface
$b \equiv$ Bolt width along the radial direction

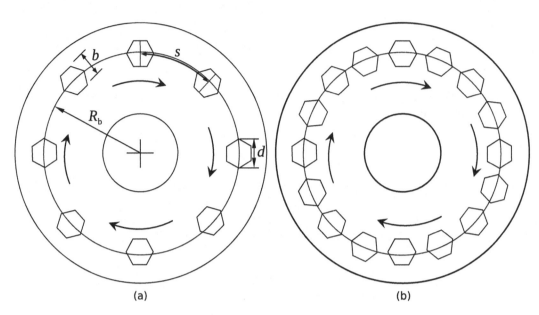

Figure 4.14 (a) Disk with bolts with small interference and (b) disk with bolts with large interference.

$C_{D_b} \equiv$ Baseline drag coefficient of each bolt (≈ 0.6)

$R_b \equiv$ Bolts pitch circle radius

$I_b \equiv$ Bolts interference factor (a function of s/d, see Figure 4.14a)

Based on the empirical data presented in Hoerner (1965), the following approximate correlations are obtained to compute I_b as a function of s/d:

For $1 \leq \frac{s}{d} < 2$

$$I_b = 20.908 - 61.855\left(\frac{s}{d}\right) + 66.616\left(\frac{s}{d}\right)^2 - 30.481\left(\frac{s}{d}\right)^3 + 5.051\left(\frac{s}{d}\right)^4 \qquad (4.39)$$

For $2 \leq \frac{s}{d} < 7$

$$I_b = 5.7185 - 6.5982\left(\frac{s}{d}\right) + 3.1375\left(\frac{s}{d}\right)^2 - 0.6878\left(\frac{s}{d}\right)^3$$
$$+ 0.0717\left(\frac{s}{d}\right)^4 - 0.0029\left(\frac{s}{d}\right)^5 \qquad (4.40)$$

These correlations yield $I_b = 0.25$ for $s/d = 1$ and $I_b = 1.0$ for $s/d > 6$.

4.4   Compressor Rotor Cavity

Axial compressors of gas turbines for aircraft propulsion and power generation feature multiple stages to achieve high compression ratio (20–40) needed in today's high-performance engines. Compressed air temperatures in aft stages of the compressor tend to be very high. As a result of increasing air density along the compressor flow path, its annulus area also decreases with smaller vanes and blades. The thermal growth of compressor rotor disks relative to their outer casing play an important role in determining the blade-tip clearances, which impact compressor aerodynamic performance. During transients of engine acceleration and deceleration, rim-to-bore temperature gradients in each disk determine its low cycle fatigue (LCF) life. This calls for a transient conduction heat transfer analysis of each rotor disk with convection boundary conditions from the air in the cavity (called rotor cavity) formed with the adjacent disk. These rotor cavities are found in all shapes and sizes, some of them are completely closed, while others are cooled with an axial throughflow, called bore flow. Completely closed compressor rotor cavities are generally found in industrial gas turbines. In aircraft engines, except for initial stages of a high-pressure compressor (HPC), the rotor cavities feature a bore flow, which is designed to influence the rotor-tip clearances for better compressor aerodynamic performance and higher LCF life of each rotor disk.

Figure 4.15 schematically shows a compressor rotor cavity with bore flow. Without the bore flow, the rotor cavity may be considered a closed cavity. Because the static pressure at the vane exit is higher than that at its inlet, a reverse flow through each inter-stage seal in the main flow path occurs. Note that the windage generated in these inter-stage seals of an axial-flow compressor significantly change the thermal boundary conditions on rotor surfaces exposed to the main flow path.

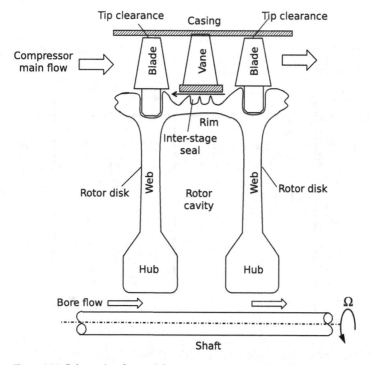

Figure 4.15 Schematic of an axial compressor rotor cavity.

4.4.1   Flow and Heat Transfer Physics

In order to better understand the unsteady and three-dimensional nature of the centrifugally-driven buoyant convection (CDBC) in a typical compressor rotor cavity, schematically shown in Figure 4.15, let us first look at the spin-up and spin-down flow behavior in a closed adiabatic rotor cavity shown in Figure 4.16. When we impulsively rotate a closed cavity to the steady angular velocity Ω, shown in Figure 4.16a, the fluid initially at rest inside the cavity eventually reaches a solid-body rotation with the cavity walls. During the transient, the angular momentum from the walls is transferred to the fluid through the action of viscosity and no-slip boundary condition at the walls. On the cylindrical surface parallel to the axis of rotation, we have negligible axial pressure gradients and the viscous boundary layer (called Stewartson boundary layer) grows radially inward. Each disk surface, which is normal to the axis of rotation, behaves like a free disk rotating adjacent to a nonrotating fluid, discussed earlier in this chapter, and results in the radially-outward pumping secondary flow in the Ekman boundary layer, continuously entraining fluid from the core.

Let us try to physically understand how the radially-outward secondary flow is initiated during spin-up. Being stationary, the core features uniform zero pressure gradient, which is also imposed on the boundary layers on the cavity surfaces. Within the disk boundary layer, the fluid attains angular momentum and causes centrifugal body force, which is balanced by an adverse radial pressure gradient ($dP_s/dr = \rho r \Omega_f^2$). To ensure zero net pressure gradient within the boundary layer, as imposed from the core fluid, a favorable radial pressure gradient is created, which causes increase in radial momentum flow while also overcoming the opposing viscous force from the radial wall

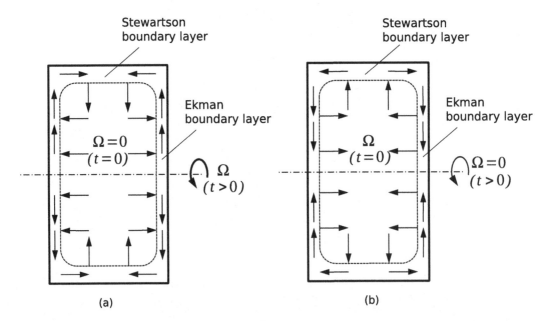

(a)   (b)

Figure 4.16 (a) Fluid spin-up from rest in a rotating closed cavity and (b) spin-down of fluid under solid-body rotation to rest in a nonrotating closed cavity.

shear stresses. This radial disk pumping flow grows by entraining fluid from the core. Thus, to satisfy continuity, circulating flows develop in the meridional plane, as shown in Figure 4.16a.These circulating flows speed up the transient process and the whole system reaches the state of solid-body rotation by an order of magnitude faster than through viscous diffusion alone. If the disks were frictionless, the secondary flow would seize to exist, and the steady state would take place through the growth of the Stewartson boundary layer alone.

During spin-down from solid-body rotation, when the rotor cavity walls are suddenly stopped, as shown in Figure 4.16b, secondary flows in each meridional plane within the Ekman and Stewartson boundary layers circulate in the directions opposite to those in the spin-up process. The occurrence of a radially inward flow in the Ekman boundary layer on each disk can be physically explained as follows. The fluid core initially in solid-body rotation has a radial pressure gradient ($dP_s/dr = \rho r \Omega^2$), which is imposed on each disk boundary layer. Because the fluid in the boundary layer has lower angular velocity than that in the core at the same radius, it experiences a net radially-inward pressure force causing it to flow toward the axis of rotation. Thus, circulating secondary flows develop in the cavity to satisfy continuity.

The spin-down process with its interesting feature of radially-inward flow in the Ekman boundary layer can be easily demonstrated in a simple home experiment. Let us take a cup of water and drop in it a few mustard seeds. With a straw, let us now vigorously stir the water and bring it to a state of solid-body rotation. When we stop stirring and pull the straw out, the water spins down to rest within a minute or so. During the spin-down process, we notice that the mustard seeds, which were mostly rotating away from the axis of rotation, collect at the bottom of the cup, not at the periphery but near the center (around the axis of rotation). The mustard seeds are carried there by the radially-inward flow at the bottom surface.

It may be noted here that, during the spin-up process, the work transfer (windage) from the adiabatic cavity walls will somewhat raise the total temperature of the cavity fluid. In the steady state of solid-body rotation, the fluid features a radial pressure gradient and becomes isothermal. This is true even for a compressible fluid like air. For an adiabatic spin-down, no work transfer occurs, and the fluid total temperature remains constant.

When the annular bore flow passes over the open compressor rotor cavity, shown in Figure 4.15, it undergoes a sudden geometric expansion and may impinge on the web region of the downstream disk. The jet expansion over the cavity results in a toroidal vortex, which recirculates in the axial direction and moves in solid-body rotation with the outer core flow in the cavity under adiabatic conditions. Unlike a driven cavity flow, the radial extent of the toroidal vortex is confined to the web region. For the case of disks temperature being different from that of the bore cooling air flow, CDBC drives a very complex flow structure in the rotor cavity. In their comprehensive review of buoyancy-induced flow in rotor cavities, Owen and Long (2015) note that, for the situation when the temperature of the disks and shroud is higher than that of the air in the cavity, the unstable and unsteady three-dimensional cavity flow features sources (high-pressure regions) and sinks (low-pressure regions), which in the rotating coordinate system (disks rotating counterclockwise, aft looking forward) appear as anti-cyclonic (rotating clockwise) and cyclonic (rotating counterclockwise) flows, respectively. In addition, a part of the axial bore flow enters the cavity through radial arms and return via Ekman layers on the disks. There is little hope of micro-modeling this cavity flow as a means to perform heat transfer calculations for design applications. In the next section, we present a practical approach to 1-D heat transfer modeling of compressor rotor cavities under CDBC.

4.4.2   Heat Transfer Modeling with Bore Flow

In free convection, both the flow and heat transfer are intricately coupled, that is, one depends on the other. Our daily experience involves free convection largely driven by the gravitational force field with a nearly constant acceleration ($g = 9.81$ m/s^2). In the chimney attached to the fireplace in our homes, we know that hot air rises and the cold air sinks to satisfy continuity, generating a free convection flow, which we call gravitationally-driven buoyant convection (GDBC). As we know, water has its maximum density at 4°C. In winter, when a lake starts to freeze, GDBC plays the role in ensuring that the water stays in the liquid phase to save marine life below the frozen top layer, which also acts as a good thermal insulator. When we have cold air next to a hot vertical wall, the free convection flow goes upward in the wall boundary layer. When the vertical wall is colder than the adjacent air, the free-convection flow comes downward in the wall boundary layer. For a downward facing horizontal wall, however, GDBC happens only when the wall is colder than the air; when this wall is hotter than the air, the flow is stably stratified with no GDBC.

The acceleration associated with the centrifugal force in a rotating system is given by $g_c = r\Omega^2$, which varies directly as the radius of rotation and as the square of the angular velocity. At $r = 0.5$ m, for a cavity rotating at 3000 rpm, we have $g_c/g \approx 5000$, which clearly shows the strength of CDBC in a compressor rotor activity vis-à-vis GDBC for similar temperature differences between cavity surface and the surrounding air. In the case of CDBC, cold fluid moves radially outward and warm fluid moves radially inward. Thus, when the wall is hotter than the fluid, the fluid flows radially inward in the disk boundary layer, and, when the wall is colder than the fluid, it flows radially outward in the boundary layer.

For 1-D heat transfer modeling of a compressor rotor cavity with a bore flow, shown in Figure 4.17, we divide the cavity in multiple control volumes, compute the convective heat transfer in each control volume and find the total convective heat transfer rate ($\dot{Q}_c$), which becomes the boundary condition for the steady-flow energy equation over the bore CV to determine the air total temperature change from inlet to outlet. Accordingly, we can write

$$T_{t_{R_{out}}} = T_{t_{R_{in}}} + \frac{\dot{Q}_c}{\dot{m}_{bore}c_p} \tag{4.41}$$

For calculating the convective heat flux $\left(\dot{q}_c = h(T_w - T_{aw})\right)$ from each wall of the cavity CV, we need three quantities: wall temperature (T_w); adiabatic wall temperature (T_{aw}), which acts as the fluid reference temperature; and heat transfer coefficient (h),

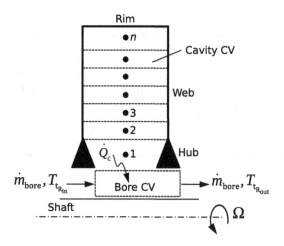

Figure 4.17 1-D Heat transfer modeling of compressor rotor cavity with bore flow.

which is obtained from a specified empirical correlation. In this multitemperature problem, it is not obvious which temperature we should use as the appropriate adiabatic wall temperature. In the present approach, we make use of the fact that, under adiabatic conditions, the rothalpy will remain constant for any excursion of the fluid in solid-body rotation within the rotor cavity. Thus, we can write

$$T_{aw} - \frac{r^2 \Omega^2}{2c_p} = T_{t_{R_{in}}} - \frac{r_{bore}^2 \Omega^2}{2c_p}$$

$$T_{aw} = T_{t_{R_{in}}} + \frac{\Omega^2}{2c_p} \left(r^2 - r_{bore}^2 \right) \tag{4.42}$$

where r is the central radius of a cavity control volume and r_{bore} corresponds to the mean radius of the bore flow. Owen and Tang (2015) also suggest the use of Equation 4.42 for the fluid reference temperature to compute the convective heat transfer between the cavity wall and air in CDBC. They have, however, arrived at this equation using the compressibility effects and invoking the isentropic relationship between the pressure ratio and temperature ratio at two points.

As to the empirical heat transfer correlations for CDBC on the disk and cylindrical rim surfaces, we present here a set of standard correlations for GDBC from McAdams (1954). These correlations conform to the common equation form

$$Nu = C(Ra)^m \tag{4.43}$$

The coefficient C and the exponent m for the Rayleigh number Ra used in Equation 4.43 for various situations are tabulated in Table 4.1. In view of the geometric complexity of a compressor rotor cavity with arbitrary temperature distribution on its bounding surfaces, the empirical correlations suggested by Equation 4.43 are to be treated as nominal correlations for initial modeling purposes only. Such correlations must be later refined and established for a consistent design practice from extensive thermal surveys for various designs. These empirical correlations then become proprietary to the original equipment manufacturer and do not belong to a textbook.

Table 4.1 Constants C and m used in Equation 4.43

Physical situation	Ra range	C	m
Vertical surface	$10^4 - 10^9$	0.59	$\frac{1}{4}$
Vertical surface	$10^9 - 10^{12}$	0.13	$\frac{1}{3}$
Horizontal surface ($T_w > T_{aw}$, CDBC)	$10^5 - 2 \times 10^7$	0.54	$\frac{1}{4}$
Horizontal surface ($T_w > T_{aw}$, CDBC)	$2 \times 10^7 - 3 \times 10^{10}$	0.14	$\frac{1}{3}$
Horizontal surface ($T_w < T_{aw}$, CDBC)	$3 \times 10^5 - 3 \times 10^{10}$	0.27	$\frac{1}{4}$

Note that for the correlations in Table 4.1 with $m = 1/3$, which are to be used for a turbulent boundary layer, the resulting heat transfer coefficients become independent of the characteristic length.

4.4.3 Heat Transfer Modeling of Closed Cavity

The foregoing heat transfer modeling of a compressor rotor cavity with the bore flow can be easily extended for the situation of a closed cavity with no bore flow. When the cavity is closed, it essentially operates under unsteady heat transfer. The mass $\tilde{m}_c$ of air in the cavity remains constant. For each time step Δt we make a quasi-steady-state assumption and use a fictitious bore flow $\dot{m}_{bore} = \tilde{m}/\Delta t$. If $T_{t_R}(t)$ is the air total temperature in the rotor reference frame at the beginning of the time step, then the corresponding air temperature at the end of the time step can be computed as

$$T_{t_R}(t + \Delta t) = T_{t_R}(t) + \frac{\dot{Q}_c}{\dot{m}c_p} \tag{4.44}$$

where the total convective heat transfer $\dot{Q}_c$ for the closed cavity is to be computed just as we did for the cavity with bore flow in the previous section, assuming $T_{t_{R_{in}}} = T_{t_R}(t)$.

4.5 Preswirl System

For internal cooling of turbine blades subjected to high temperatures in the main flow path, the design intent is always to keep the extracted compressor air (used as coolant) as cool as possible at the specified total pressure at the blade root. In its journey from the compressor bleed point to turbine blade inlet, the cooling air will be heated by heat transfer from the stator and rotor surfaces and work transfer from rotor surfaces. Under adiabatic conditions, the cooling air total temperature remains constant over the static structure. When we bring this air onboard the turbine rotor for blade cooling, its total temperature relative to the rotor increases as a result of work transfer from the rotor. This is an undesirable effect on the blade cooling air temperature. A simple technique to avoid this temperature increase is to preswirl the air in the static structure using a number of nozzles, as shown in Figure 4.18. When the coolant air is preswirled to the rotor angular velocity, it enters the holes in the cover plate with no change in its absolute total temperature, as a result, there is no loss in turbine power with simultaneous reduction in the coolant air total temperature relative to the rotor at the same radius.

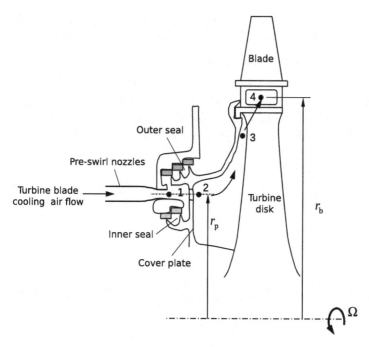

Figure 4.18 Schematic of a preswirl system for turbine blade cooling air.

4.5.1 Flow and Heat Transfer Modeling

As shown in Figure 4.18, the blade cooling air flow from the preswirl nozzles interacts with a complex flow region in the stator-rotor cavity before it enters the rotor cavity between the cover plate and the rotor disk at point 2. At this point, the flow is brought onboard the rotor with solid-body rotation. From point 2 to point 3 in the rotor cavity, the core flow behaves likes a generalized vortex within a complex shear flow. The flow eventually reaches the blade root at point 4 at the design-specified conditions of relative total pressure and temperature for blade internal cooling with the adequate backflow margin so as to prevent any ingestion of hot gases through the blade film cooling holes.

Let us now turn our attention to how the coolant air total temperature changes in a preswirl system from nozzle exit (point 1) to blade inlet (point 4), the first being in the stator reference frame (SRF) and the second in the rotor reference frame (RRF). Under adiabatic conditions with constant c_p for the coolant air, we can invoke the concept of rothalpy to relate total temperatures in SRF and RRF at any point as

$$T_{t_S} - \frac{UV_\theta}{c_p} = T_{t_R} - \frac{U^2}{2c_p}$$

$$T_{t_R} = T_{t_S} - \frac{UV_\theta}{c_p} + \frac{U^2}{2c_p} = T_{t_S} + \frac{U^2}{2c_p}\left(1 - 2S_f\right)$$

$$T_{t_R} = T_{t_S} + \frac{r^2\Omega^2}{2c_p}\left(1 - 2S_f\right) \tag{4.45}$$

where the swirl factor $S_f = V_\theta/(r_p\Omega)$. Equation 4.45 shows that for $S_{f_1} = 0.5$; that is, the air is rotating at half the rotor speed, we obtain $T_{t_R} = T_{t_s}$. For $S_f = 0$, T_{t_R} is higher than T_{t_s} by the dynamic temperature corresponding the solid-body rotation ($r^2\Omega^2/2c_p$), and, for $S_f = 1$, it is lower than T_{t_s} by the same amount and equals the static temperature.

With reference to Figure 4.18, let us now evaluate the reduction in the coolant air total temperature from point 1 to point 4 under no heat transfer. At point 1, Equation 4.45 yields

$$T_{t_{R_1}} = T_{t_{s_1}} + \frac{r_p^2\Omega^2}{2c_p}\left(1 - 2S_{f_1}\right) \tag{4.46}$$

Because the rothalpy remains constant in RRF for an adiabatic flow, we obtain

$$T_{t_{R_4}} - \frac{r_b^2\Omega^2}{2c_p} = T_{t_{R_1}} - \frac{r_p^2\Omega^2}{2c_p} \tag{4.47}$$

Combining Equations 4.46 and 4.47 yields

$$\theta = \frac{T_{t_{s_1}} - T_{t_{R_4}}}{\left(\dfrac{r_b^2\Omega^2}{2c_p}\right)} = 2S_{f_1}\left(\frac{r_p}{r_b}\right)^2 - 1 \tag{4.48}$$

where θ may be interpreted as the blade cooling air temperature reduction coefficient as a result of preswirl nozzles. When we multiply the dynamic temperature of solid-body rotation, $r_b^2\Omega^2/(2c_p)$, at point 4 by θ, we obtain $(T_{t_{s_1}} - T_{t_{R_4}})$.

Equation 4.48 and its plot in Figure 4.19 show that θ varies linearly with the swirl factor S_{f_1} at the preswirl nozzle exit and as the square of the radius ratio r_b/r_p. Negative values of θ imply that $T_{t_{R_4}} > T_{t_{s_1}}$. For $S_{f_1} = 1$, positive values of θ occur only for $r_b/r_p > 0.707$. At first, it appears from Figure 4.19 that over-spinning the blade cooling air flow to $S_{f_1} > 1$ in the preswirl nozzles will have added beneficial effect of reducing

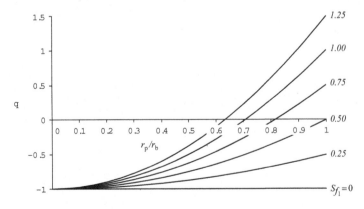

Figure 4.19 Variation of θ with preswirl-to-blade inlet radius ratio r_b/r_p for various exit swirl factor S_{f_1}.

its relative total temperature at the blade inlet. But the gas turbine designer must balance this against the additional reduction in static pressure, which may not be recovered downstream, in order to generate higher dynamic pressure associated with the higher swirl factor at the nozzle exit.

Let us now evaluate the loss in turbine work in the transfer process from preswirl nozzles exit to blade inlet. Because the cooling air attains the solid-body rotation at point 4, we can equate the rothalpy expressed in SRF at point 1 to that at point 4, giving

$$
T_{ts_1} - \frac{r_p^2 \Omega^2 S_{f_1}}{c_p} = T_{ts_4} - \frac{r_b^2 \Omega^2}{c_p}
$$

$$
\tilde{\theta} = \frac{T_{ts_4} - T_{ts_1}}{\left(\dfrac{r_b^2 \Omega^2}{2c_p} \right)} = 1 + \left\{ 1 - 2S_{f_1} \left(\frac{r_p}{r_b} \right)^2 \right\}
\tag{4.49}
$$

where $\tilde{\theta}$ may be interpreted as the turbine work loss coefficient. Combining this equation with Equations 4.48 yields

$$
\tilde{\theta} = 1 - \theta
\tag{4.50}
$$

It is interesting to note from Equation 4.50 that the design goal of higher value of θ and lower value of $\tilde{\theta}$ is achieved simultaneously. For $\tilde{\theta} > 1$, which implies $\theta < 1$, the transfer system exhibits a compressor-like behavior, while for $\tilde{\theta} < 1$, which implies $\theta > 1$, the system behaves like a turbine. In the design of a blade cooling system, it is important to ensure that the required supply pressure at the blade inlet is met even with a slight increase in $\tilde{\theta}$, using the turbine rotor to provide the needed pumping power.

The radial location of preswirl nozzles is an important design consideration. For the same value of S_{f_1}, V_{θ_1} will higher, the higher the radius. This in turn will require higher extraction of dynamic pressure from the total pressure at the preswirl nozzles. Gas turbine designers also need to consider the impact of the preswirl radial location on rotor disk pressure distribution, which determines the axial rotor thrust.

For 1-D flow and heat transfer modeling of blade cooling air from point 1 to point 4, see Figure 4.19, we need to sequentially model the rotor-stator cavity between the inner and outer seals and the rotor cavity formed by the cover plate and the turbine disk up to point 3, using the methodology discussed in Section 4.3. Going from point 3 to point 4, we calculate the change in pressure and temperature using a forced vortex assumption of solid-body rotation of the turbine disk.

4.6 Hot Gas Ingestion: Ingress and Egress

Hot gas ingestion refers to ingress of hot gases from the main flowpath into the cavity, or wheel-space, formed between rotor and stator disks. This problem is most serious in the forward cavity of the first-stage turbine where hot gases exiting the vanes have the highest pressure and temperature. If the stator and rotor parts with no internal cooling, like disks, are exposed to hot gases by way of ingestion, their durability is considerably at risk. The design goal, over considerations of improved turbine efficiency and reduced specific fuel consumption, is always to eliminate or minimize the hot gas ingestion. It, therefore, behooves gas turbine designers to develop a good understanding of the primary factors and basic mechanism behind this phenomenon.

4.6.1   Physics of Hot Gas Ingestion

The phenomenon of hot gas ingestion is schematically shown in Figure 4.20. But for a small pressure loss in the combustor, the maximum total pressure at the trailing edge of the first stage vanes equals the compressor exit pressure. As a result of the dynamic pressure associated with the throughflow and tangential velocities needed for aerodynamic power extraction by the first stage blades, vane-to-vane static pressure distribution, which acts as the gas-path boundary condition for the cavity purge flow, is nonaxisymmetric and circumferentially periodic. As a result of this circumferentially asymmetric pressure distribution, characteristic of all gas turbines, the ingress of hot gases occurs wherever static pressure in the main flowpath is higher than that in the wheel-space. Egress into the main flow path occurs under a favorable pressure gradient over the regions where the gas path acts as the sink boundary condition for the purge air flow.

Unlike the compressor exit air going through the combustor and first stage vanes, where the total pressure decreases downstream, the purge air flow peeled off from the compressor exit loses its total pressure across multiple labyrinth seals but also gains it through work transfer from the rotor surfaces in contact. The cooling and sealing purge air flow typically maintains its swirl factor around 0.5 before reaching the rotor-stator

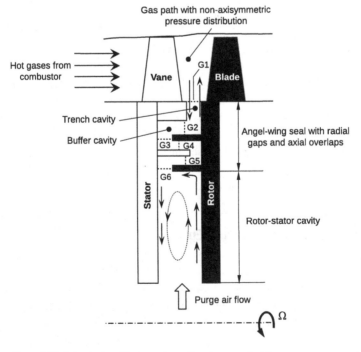

Figure 4.20 Schematic of hot gas ingestion in a gas turbine.

cavity shown in Figure 4.20. The lower the purge air flow, the lower will be the loss in total pressure in the circuit from the compressor exit pressure and higher will be the favorable pressure gradient across the rim-seal opening. This, however, needs to be balanced against the excessive windage temperature rise in this internal flowpath.

If enough purge air flow is not feasible either as a result of source pressure constraint or to achieve the target cycle efficiency, the design intent shifts to restricting the hot gas ingestion to the outer parts of the stator and rotor structures; for example, the trench cavity and the buffer cavity shown in Figure 4.20. This is achieved through an innovative rim-seal design using a system of angel wings with adequate axial overlaps in conjunction with a trench cavity with appreciable radial overlap. For an improved conceptual design of these rim seals, it is important to understand how radial flows occur within boundary layers on the stator and rotor surfaces, separated by an inviscid vortex core.

Let us first consider the rotor boundary with an inviscid vortex core in the rotor-stator cavity shown in Figure 4.20. For the radially outward purge air flow in this cavity, the core is expected to rotate at around half the rotor speed ($S_f \approx 0.5$). The force of adverse pressure gradient in the core flow balances the centrifugal body force as a result of rotation ($P_s/r = \rho r S_f^2 \Omega^2$). At a radial location in the cavity, there is negligible pressure gradient in the axial direction, which is to say that the radial pressure gradient in the inviscid core flow is imposed on the rotor boundary layer. As a result of no-slip boundary condition, the fluid in contact with the rotor surface rotates in solid-body rotation. At the edge of the boundary layer, the fluid rotation equals that of the core flow. Accordingly, the average fluid tangential velocity is higher than that in the core, resulting in higher centrifugal body force and the consequent higher adverse radial pressure gradient in a small control volume encompassing the boundary layer. This difference in the radial pressure gradient in the boundary layer and in the core flow appears as a favorable radial pressure gradient in the rotor boundary layer, causing an increase in radial momentum flow, which is commonly known as disk pumping. Thus, the purge air flow in the rotor-stator cavity gets continuously entrained into disk pumping along the rotor. This pumping flow increases radially outward. If sufficient purge air flow is not available to satisfy the disk pumping flow, the flow will feed on itself through recirculation. From the flow physics behind disk pumping discussed here, it may be noted that, if the core flow rotates faster than the disk, it will induce a radially inward flow in the disk boundary layer, which may be counterintuitive based on our observational experience of free disk pumping.

For the boundary layer on the stator surface in the cavity with a rotating core flow, the radial pressure gradient argument presented in the foregoing will lead us to conclude a radially inward flow along the stator surface. Coming down radially, the flow in the stator boundary layer will continually detrain into the core flow. If we have a significant superimposed favorable pressure gradient in the core flow, we may prevent a radially inward flow in the stator boundary layer. Generally speaking, in the entire wheel space interfacing the main flowpath, the internal flow system of purge air flow features radially outward flows along all rotor surfaces and radially inward flows along all stator surfaces. Thus, the evidence of hot gas ingestion can be first observed on the stator surface in the rim seal area.

If the purge air flow rate equals the rotor disk pumping flow rate in the rotor-stator cavity, it will prevent hot gas ingestion into this cavity. Even with somewhat lower purge air flow, an angel wing on the rotor, see Figure 4.20, will act as an effective discourager and may prevent hot gases from entering the lower cavity. In the process of negotiating the turn around this angel wing, the radially outward boundary layer flow tends to create an aerodynamic stagnation point on the adjacent stator surface forcing the radially downward stator boundary layer flow to turn around rather than enter into the lower cavity. Note that if this angel wing were to be on the stator rather than on the rotor, it will be ineffective in protecting the rotor-stator cavity from the ingress of hot gases.

As shown in Figure 4.20, gas turbine designers use a rim seal system of rotor and stator angel wings with adequate radial and axial gaps and overlaps to restrict hot gas ingestion as close to the rim region as possible. In the rim seal design, the trench cavity with its significant radial overlap plays the key role in attenuating the asymmetry in pressure distribution in the main flowpath. Note that, going from vane exit to blade inlet over the wheel space, the flowpath pressure distribution also tends to become more axisymmetric.

From the foregoing discussion, we can glean the following primary factors that determine hot gas ingestion in a turbine stage:

• Periodic vane/blade pressure field (nonaxisymmetric pressure distribution in the main flowpath of hot gases)
• Disk pumping in the rotor-stator cavity
• Rim seal geometry (radial and axial clearances and overlaps)
• Purge sealing and cooling air flow rate

Some secondary factors, which also influence hot gas ingestion to some extent, include unsteadiness in 3-D flow field, and pressure fluctuations in the wheel space, and turbulent transport in the platform and outer cavity region.

For achieving higher engine performance, gas turbine designers may not have the luxury of minimum purge air flow rate needed to prevent hot gas ingestion by using a simple rim seal design. The competing effect of windage and hot gas ingestion must also be considered in design. When the purge flow is significantly reduced to achieve higher engine performance, the windage generated in the wheel space becomes critical to the creep life of many turbine components that support blades and vanes, which are necessarily protected using internal and film cooling, especially in the initial turbine stages. The design strategy to deal with the hot gas ingestion often follows the following sequence:

• Establish the minimum cavity purge flow needed for acceptable windage temperature rise and heat transfer in the rotor-stator cavity.
• Establish the gas path asymmetric pressure boundary conditions from an appropriate CFD solution.
• Design a seal that will limit the ingress (hot gas ingestion) to trench (the first design target) and buffer cavities (the second design target if we can't meet the first).

4.6.2   1-D Modeling

In view of the flow and heat transfer complexity associated with hot gas ingestion in gas turbines, one may be tempted to use 3-D CFD for its modeling and prediction. Except for providing improved understanding of the flow physics behind hot gas ingestion, as an alternative to expensive and more time-consuming flow visualization through experiments, the use of this technology as a design tool is still limited as a result of the constraint of today's short design cycle time. 1-D modeling methods reinforced by empirical correlations are still the workhorse used by most gas turbine designers to predict hot gas ingestion in all turbine stages.

Judging from the number of papers presented at ASME Turbo Expo in 2016, 2017, and 2018, the hot gas ingestion and rim seals remain the most active area of research to support internal air systems design and technology of modern gas turbines. Scobie et al. (2016) provide a comprehensive review of landmark contributions in this area since 1970. Determined entirely by disk rotation (pumping flow) with axisymmetric flowpath boundary conditions, Bayley and Owen (1970) first proposed an equation to compute the minimum sealing flow rate to prevent hot gas ingestion. For almost a decade, most university-based research considerably simplified the hot gas ingestion flow physics by not simulating the nonaxisymmetric pressure distribution in the annulus to simulate turbine main flowpath, as if in defiance of Albert Einstein's advice, "Make the problem as simple as possible but not simpler." This trend ended when Abe, Kikuchi, and Takeuchi (1979) first presented their experimental investigation and demonstrated that the nonaxisymmetric pressure distribution in the main flowpath is the primary driver for hot gas ingestion in real gas turbines. Hamabe and Ishida (1992) confirmed the importance of the flowpath pressure asymmetry in the ingestion process.

4.6.2.1   Single-Orifice Model

A gas turbine designer may use a simple orifice model to predict ingress and egress flows across a rim seal, such as through the axial gap G1 shown in Figure 4.20. As a way to demonstrate a single-orifice modeling of hot gas ingestion, we present here the model of Scanlon et al. (2004) with its analytical solutions for an assumed parabolic variation of gas path pressure asymmetry and axisymmetric pressure distribution for the sealing coolant flow within the wheel-space as boundary conditions across the rim seal. Phadke and Owen (1988), Chew, Green, and Turner (1994), Reichert and Leiser (1999), Bohn and Wolff (2003), Johnson et al. (2006), Johnson, Wang, and Roy (2008), Owen (2011a, 2011b), Owen et al. (2012), and Owen, Pountney, and Lock (2012) all present variations of a single-orifice model.

For the rim seal control volume shown in Figure 4.21a, the continuity equation yields

$$\dot{m}_{\text{egr}} = \dot{m}_{\text{ing}} + \dot{m}_{\text{cav}} \tag{4.51}$$

Similarly, the energy balance with constant c_{p} gives

$$T_{t_{\text{egr}}} = \frac{\dot{m}_{\text{ing}} T_{t_{\text{ing}}} + \dot{m}_{\text{cav}} T_{t_{\text{cav}}}}{\dot{m}_{\text{egr}}} \tag{4.52}$$

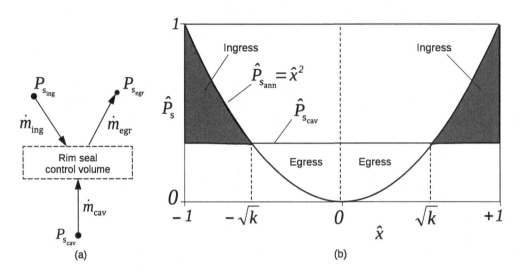

Figure 4.21 (a) Rim seal control volume and (b) parabolic pressure distribution in the gas path annulus.

For calculating $\dot{m}_{ing}$ and $\dot{m}_{egr}$ across the rim seal gap of total area A_{gap}, we make the following simplifying assumptions:

- Incompressible flow with constant density ρ.
- Plenum conditions prevail on either side of the rim seal gap.
- Axisymmetric distribution of static pressure $P_{s_{cav}}$ in the wheel-space at the rim seal gap; for the egress flow, this pressure acts as the total pressure.
- Parabolic distribution of static pressure, see Figure 4.21b, in the gas path annulus at the rim seal gap; for the ingress flow, this pressure acts as the total pressure.

Let us define a dimensionless static pressure in the rim seal system as

$$\hat{P}_s = \left(P_s - P_{s_{min}}\right) / \left(P_{s_{max}} - P_{s_{min}}\right)$$

which yields the annulus static pressure as

$$\hat{P}_{s_{ann}} = \left(P_{s_{ann}} - P_{s_{min}}\right) / \left(P_{s_{max}} - P_{s_{min}}\right)$$

and the cavity static pressure as

$$\hat{P}_{s_{cav}} = \left(P_{s_{cav}} - P_{s_{min}}\right) / \left(P_{s_{max}} - P_{s_{min}}\right)$$

where $P_{s_{max}}$ is the maximum static pressure and $P_{s_{min}}$ the minimum static pressure in the assumed parabolic pressure distribution in the annulus.

For N number of vanes in the flow path, the annular sector per vane equals $N/2\pi$, and the seal area per vane equals $\hat{A}_{gap} = A_{gap}/N$. With $\hat{x} = (N\theta/\pi) - 1$; where, when θ varies from 0 to $N/2\pi$, $\hat{x}$ varies from -1 to $+1$. The dimensionless pressure distributions in the annulus and the cavity are shown in Figure 4.21b. The regions of ingress $(\hat{P}_{s_{ann}} > \hat{P}_{s_{cav}})$ and egress $(\hat{P}_{s_{cav}} > \hat{P}_{s_{ann}})$, which are symmetric across the line $\hat{x} = 0$, are delineated in this figure. The points of intersection between $\hat{P}_{s_{cav}}$ and $\hat{P}_{s_{ann}}$ in the figure correspond to $\hat{x} = -\sqrt{k}$ and $\hat{x} = \sqrt{k}$ where $k = \hat{P}_{s_{cav}}$.

Under the assumption of 1-D incompressible flow through the rim seal, we can write the ingress mass flow rate as

$$\int d\dot{m}_{\text{ing}} = C_{\text{d}}\sqrt{2\rho}\int\left(\sqrt{\left(P_{s_{ann}} - P_{s_{cav}}\right)}\right)dA_{\text{gap}} = NC_{\text{d}}\sqrt{2\rho}\int\left(\sqrt{\left(P_{s_{ann}} - P_{s_{cav}}\right)}\right)d\hat{A}_{\text{gap}}$$
(4.53)

where C_{d} is the discharge coefficient, which is determined experimentally or using a high-fidelity, fully-validated 3-D CFD analysis.

In Equation 4.53, because $\hat{A}_{\text{gap}}$ corresponds to $2\pi/N$, we can write $d\hat{A}_{\text{gap}} = \left(N\hat{A}_{\text{gap}}d\theta\right)/2\pi$. Because $\hat{x} = (N\theta/\pi) - 1$, we can also write $d\theta = \pi d\hat{x}/N$, giving $d\hat{A}_{\text{gap}} = \hat{A}_{\text{gap}}d\hat{x}/2$. Thus, we can write Equation 4.53 as

$$\dot{m}_{\text{ing}} = NC_{\text{d}}\sqrt{2\rho}\int\left(\sqrt{\left(P_{s_{ann}} - P_{s_{cav}}\right)}\right)d\hat{A}_{\text{gap}}$$

$$= 2N\left(\frac{\hat{A}_{\text{gap}}}{2}\right)C_{\text{d}}\sqrt{2\rho}\int_{\sqrt{k}}^{1}\left(\sqrt{\left(P_{s_{ann}} - P_{s_{cav}}\right)}\right)d\hat{x}$$

$$= C_{\text{d}}A_{\text{gap}}\sqrt{2\rho}\int_{\sqrt{k}}^{1}\left(\sqrt{\left(P_{s_{ann}} - P_{s_{cav}}\right)}\right)d\hat{x}$$

which upon substituting $P_{s_{ann}} - P_{s_{cav}} = (x^2 - k)\left(P_{s_{max}} - P_{s_{min}}\right)$ yields

$$\dot{m}_{\text{ing}} = C_{\text{d}}A_{\text{gap}}\sqrt{2\rho\left(P_{s_{max}} - P_{s_{min}}\right)}\int_{\sqrt{k}}^{1}\left(\sqrt{\left(\hat{x}^2 - k\right)}\right)d\hat{x}$$
(4.54)

Let us now evaluate the integral in Equation 4.54. Using the following formula

$$\int\left(\sqrt{(z^2 - a^2)}\right)dz = \frac{z\sqrt{(z^2 - a^2)}}{2} - \frac{a^2}{2}\ln\left(z + \sqrt{(z^2 - a^2)}\right)$$

from the standard table of integrals, we can write

$$\int_{\sqrt{k}}^{1}\left(\sqrt{(\hat{x}^2 - k)}\right)d\hat{x} = \left[\frac{\hat{x}\sqrt{(\hat{x}^2 - k)}}{2} - \frac{k}{2}\ln\left(\hat{x} + \sqrt{(\hat{x}^2 - k)}\right)\right]_{\sqrt{k}}^{1}$$

$$= \frac{1}{2}\left[\sqrt{1 - k} - k\ln\left(\frac{1 + \sqrt{(1 - k)}}{\sqrt{k}}\right)\right]$$

Because $\cosh^{-1}z = \ln\left(z + \sqrt{z^2 - 1}\right)$, we obtain

$$\int_{\sqrt{k}}^{1}\left(\sqrt{(\hat{x}^2 - k)}\right)d\hat{x} = \frac{1}{2}\left[\sqrt{1 - k} - k\cosh^{-1}\frac{1}{\sqrt{k}}\right]$$

whose substitution in Equation 4.54 finally yields

$$\dot{m}_{\text{ing}} = C_d A_{\text{gap}} \sqrt{2\rho \left(P_{s_{\text{max}}} - P_{s_{\text{min}}} \right)} \left[\frac{1}{2} \left(\sqrt{1-k} - k \cosh^{-1} \frac{1}{\sqrt{k}} \right) \right] \tag{4.55}$$

Similarly, we can express the egress mass flow rate through the rim seal as

$$\int d\dot{m}_{\text{egr}} = C_d \sqrt{2\rho} \int \left(\sqrt{\left(P_{s_{\text{cav}}} - P_{s_{\text{ann}}} \right)} \right) dA_{\text{gap}} = N C_d \sqrt{2\rho} \int \left(\sqrt{\left(P_{s_{\text{cav}}} - P_{s_{\text{ann}}} \right)} \right) d\hat{A}_{\text{gap}}$$

which, following the steps we used in the foregoing to simplify the expression for $\dot{m}_{\text{ing}}$, reduces to

$$\dot{m}_{\text{egr}} = C_d A_{\text{gap}} \sqrt{2\rho \left(P_{s_{\text{max}}} - P_{s_{\text{min}}} \right)} \int_0^{\sqrt{k}} \left(\sqrt{(k - \hat{x}^2)} \right) d\hat{x} \tag{4.56}$$

From the standard table of integrals we have the formula

$$\int \left(\sqrt{(a^2 - z^2)} \right) dz = \frac{1}{2} \left[z\sqrt{(a^2 - z^2)} + a^2 \sin^{-1} \left(\frac{z}{a} \right) \right]$$

whose use for the integral in Equation 4.56 yields

$$\int_0^{\sqrt{k}} \left(\sqrt{(k - \hat{x}^2)} \right) d\hat{x} = \frac{1}{2} \left[\hat{x}\sqrt{(k - \hat{x}^2)} + k \sin^{-1} \left(\frac{\hat{x}}{\sqrt{k}} \right) \right]_0^{\sqrt{k}} = \frac{k\pi}{4}$$

giving

$$\dot{m}_{\text{egr}} = C_d A_{\text{gap}} \sqrt{2\rho \left(P_{s_{\text{max}}} - P_{s_{\text{min}}} \right)} \left(\frac{k\pi}{4} \right) \tag{4.57}$$

Note that the discharge coefficient C_d used in Equations 4.55 and 4.57, which have been derived analytically for an assumed periodically parabolic circumferential static pressure distribution in the annulus flowpath and constant axisymmetric static pressure in the wheel-space, may have different numerical values.

4.6.2.2 Multiple-Orifice Spoke Model

For the single-orifice model discussed in the foregoing, we have demonstrated the calculation of ingress and egress flow rates under a number of simplifying assumptions, which enabled analytical equations through integration. In gas turbine design, however, flowpath pressure distribution is generally obtained using a 2-D or 3-D CFD analysis. The discrete results from these analysis favors a numerical solution, offering flexibility needed in the rim seal design for its optimization and robustness. In the framework of 1-D flow network modeling, we generalize the rim seal single-orifice model to a multiple-orifice spoke model, schematically shown in Figure 4.22. In this model, each spoke represents a serially-connected rim seal system of orifices, depicted in Figure 4.20, starting from the axisymmetric boundary conditions at the exit of the

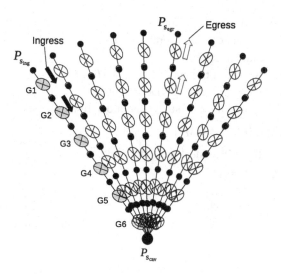

Figure 4.22 Multiple-orifice spoke model.

rotor-stator cavity as the first node and gas path boundary conditions (nonaxisymmetric) as the last node. Any spoke-to-spoke interaction is neglected in this model.

In the multiple-orifice spoke model, the compressible flow of air with all three velocity components is modeled through each orifice to compute mass flow rate using a discharge coefficient, which is determined either empirically or numerically using CFD. For each orifice, part of the wall is rotating and the part stationary. Without any loss of generally, let us assume that V_x is the velocity that determines the mass flow rate through the orifice from the basic equation $\dot{m}_{\text{ideal}} = \rho A V_x$, where ρ is the local air density and A, the exit mechanical area. It is important to note that, for a subsonic air flow, which prevails in the rim-seal system, the static pressure at the orifice exit equals the static pressure of the downstream node. Under ideal (isentropic) conditions, the total pressure and total temperature at the orifice exit correspond to those at the upstream node. We can use either of the two methods presented here to calculate $\dot{m}_{\text{ideal}}$, which then yields $\dot{m}_{\text{real}} = C_d \dot{m}_{\text{ideal}}$. Readers can easily verify that both methods give identical results.

Method 1. This method involves the following calculation steps:

Step 1. Calculate the static temperature: $T_s = T_t - V^2/2c_p$, where V is the total velocity at the orifice exit.

Step 2. Calculate the speed of sound: $C = \sqrt{\kappa R T_s}$

Step 3. Calculate the total-velocity Mach number: $M = V/C$

Step 4. Calculate the static-pressure mass flow function (see Chapter 2):

$$\hat{F}_{f_s} = M\sqrt{\kappa\left(1 + \frac{\kappa - 1}{2}M^2\right)}$$

Step 5. Calculate the orifice ideal mass flow rate: $\dot{m}_{\text{ideal}} = \dfrac{A C_V \hat{F}_{f_s} P_s}{\sqrt{R T_t}}$

where the velocity coefficient $C_V = V_x/V$.

Method 2. This method involves the following calculation steps:

Step 1. Calculate the static temperature: $T_s = T_t - V^2/2c_p$, where V is the total velocity at the orifice exit.

Step 2. Calculate the speed of sound: $C = \sqrt{\kappa R T_s}$

Step 3. Calculate the axial-velocity Mach number: $M_x = V_x/C$

Step 4. Calculate the static-pressure mass flow function (see Chapter 2):

$$\hat{F}_{f_{s,x}} = M_x \sqrt{\kappa \left(1 + \frac{\kappa-1}{2} M_x^2\right)}$$

Step 5. Calculate the orifice ideal mass flow rate: $\dot{m}_{\text{ideal}} = \dfrac{A \hat{F}_{f_{s,x}} P_s}{\sqrt{R T_{t_x}}}$

where $T_{t_x} = T_s + V_x^2/2c_p$.

Note that Method 2 can be thought of as Method 1 under an inertial transformation where the observer is in a coordinate system that is moving with the velocity $V_{r\theta} = \sqrt{V^2 - V_x^2}$.

4.7   Axial Rotor Thrust

When the gas turbine is stationary, it experiences no axial thrust. The primary flow over compressor and turbine blades and cooling and sealing flows (secondary flows) over disk surfaces generate axial rotor thrust from both pressure forces, which are normal to each surface, and shear forces, which are parallel to each surface. Generally, the contribution of shear forces to axial rotor thrust is negligible compared to the contribution from pressure forces. If the rotor surface is along the axial direction, it makes no contribution to axial thrust. Also the pressure distribution in a closed rotor cavity does not contribute to rotor thrust. An accurate calculation of the rotor thrust is critical to the design of the rotor thrust bearing, which plays an important role in ensuring engine operational and structural integrity.

Because compressor main flow occurs under an adverse pressure gradient (pressure increases in the flow direction), the compressor rotor thrust points in the negative flow direction; we call it the forward thrust. In the turbine flowpath, the pressure decreases downstream, causing the rotor thrust in the flow direction; we call it the rearward or aft thrust. Because compressor and turbine are commonly mounted on the same shaft, the net rotor thrust, which the thrust bearing has to withstand, is always lower than the individual contribution either from the compressor or turbine.

4.7.1   Computation of Axial Thrust on Blades

If we know the distribution of static pressure over a rotor surface, for example, from a 3-D CFD analysis, the axial thrust is simply the axial component of the pressure force integrated over the entire rotor surface. In view of both the complex surface geometry and static pressure distribution, the compressor and turbine rotor thrust calculations can be a tedious undertaking. A design-friendly approach to calculate the axial thrust on blades in each compressor stage or turbine stage is by using a large control volume

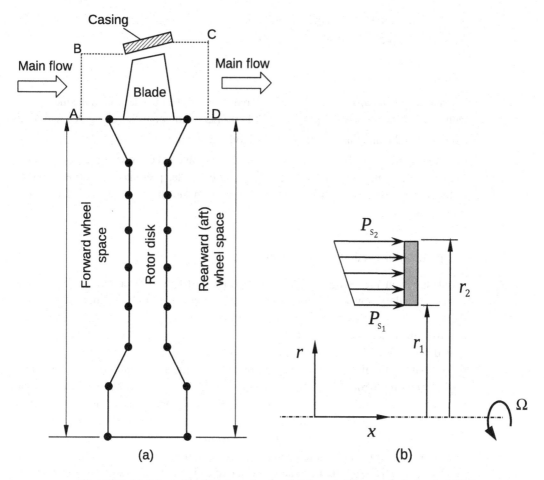

Figure 4.23 (a) Schematic of blade and rotor disk for axial thrust calculation and (b) linear variation of static pressure on the axial projection of an axisymmetric compressor or turbine rotor surface element spanning from r_1 to r_2.

analysis of the linear momentum equation in the axial direction. Such a control volume is shown in Figure 4.23a as ABCD. To further simplify the momentum analysis, we invoke the concept of stream thrust, which is the sum of the inertia force and pressure force at a section, presented in Chapter 2. If $S_{T_{in}}$ is the total stream thrust at inlet AB of the control volume ABCD and $S_{T_{out}}$ the total stream thrust at its outlet CD, we can write

$$S_{T_{out}} = S_{T_{in}} + F_{blades} + F_{casing} \qquad (4.58)$$

where F_{blades} and F_{casing} are the forces from the blades and casing, respectively, assumed to be acting on the fluid control volume in the flow direction. We compute F_{casing} as the product of the average pressure in the blade tip clearance and the projected casing surface area that is normal to the axial direction. If the casing is diverging along the flow direction, F_{casing} is positive, and it is negative for a converging casing. The axial aerodynamic load $\tilde{F}_{\text{blades}}$ on the blades is the reaction force of F_{blades}, giving $\tilde{F}_{\text{blades}} = -F_{\text{blades}}$. Thus, Equation 4.58 yields

$$\tilde{F}_{\text{blades}} = S_{T_{\text{in}}} - S_{T_{\text{out}}} + F_{\text{casing}} \tag{4.59}$$

An accurate calculation of $S_{T_{\text{in}}}$ and $S_{T_{\text{out}}}$ requires a 3-D CFD analysis of the main flow path. We can postprocess the detailed CFD results to compute $S_{T_{\text{in}}}$ along AB and $S_{T_{\text{out}}}$ along CD of the annulus. An alternate method to compute $S_{T_{\text{in}}}$ and $S_{T_{\text{out}}}$ uses the results of a throughflow analysis using circumferentially-averaged flows through multiple stream tubes in the main flow annulus, as for example, presented by Oates (1988). In this case, both at inlet and outlet, we simply sum the stream thrust associated with each stream tube.

4.7.2   Computation of Axial Thrust on Rotor Disks

Axial thrust on the compressor and turbine rotor structure in contact with internal flows (secondary flows) used for cooling and sealing is determined almost entirely by the neighboring vortex structure. We, therefore, first establish the generalized vortex distribution in the rotor-rotor and rotor-stator cavities using the methodology presented in Section 4.3. Using the radial equilibrium equation, we can compute the static pressure at each radial location on the rotor disk. We can also obtain the static pressure distribution on the rotor disk surfaces from a 3-D CFD analysis in the cavities on both forward and rearward (aft) sides. As shown in Figure 4.23a, we use a piece-wise linear distribution of the static pressure along the nodal representation of the axisymmetric rotor disk surfaces. Note that any three-dimensional protrusion, like bolts, on the disk will have negligible additional axial thrust contribution. Between any two nodes on the disk, one can compute the axial thrust contribution either by assuming an average static pressure between the nodes or using a linear variation of static pressure between them.

For the rotor surface element shown in Figure 4.23b, the axial thrust using the average static pressure is computed by

$$\Delta \overline{F}_{\text{rotor}} = \pi(r_2^2 - r_1^2)\left(\frac{P_{s_1} + P_{s_2}}{2}\right) \tag{4.60}$$

The linear variation of static pressure on the rotor surface can be expressed by the equation

$$P_s(r) = Ar + B \tag{4.61}$$

where $A = (P_{s_2} - P_{s_1})/(r_2 - r_1)$ and $B = P_{s_1} - Ar_1$. Using this linear profile equation, the elemental axial thrust can be computed as

$$\Delta \hat{F}_{\text{rotor}} = \int_{r_1}^{r_2} P_s(r)2\pi r dr = \int_{r_1}^{r_2}(Ar+B)r dr \quad \Delta \hat{F}_{\text{rotor}} = \frac{2\pi A}{3}(r_2^3 - r_1^3) + \pi B(r_2^2 - r_1^2) \tag{4.62}$$

Because the axisymmetric rotor surface area varies as square of the radius, the rotor thrust calculated by Equation 4.62 is more accurate than that by Equation 4.60. The difference is given by

$$\Delta \hat{F}_{rotor} - \Delta \overline{F}_{rotor} = \frac{\pi}{6} \left(P_{s_2} - P_{s_1} \right) \left(r_2 - r_1 \right)^2 \qquad (4.63)$$

Because any complex variation in either the rotor surface geometry or the surface pressure can be accurately represented by a piece-wise linear variation, the assumption of the linear pressure variation is not limiting. Equation 4.62, therefore, provides an accurate building block for the calculation of rotor thrust, which can be finally expressed for each rotor in the positive axial direction as

$$\tilde{F}_{rotor} = \sum_{Forward\ side} \Delta \hat{F}_{rotor} - \sum_{Rearward\ side} \Delta \hat{F}_{rotor} \qquad (4.64)$$

4.8 Concluding Remarks

We have presented in this chapter some unique thermofluids concepts that are at the heart of designing gas turbine internal flows and their 1-D modeling. These concepts include generalized vortex structure in rotor-stator and rotor-rotor cavities, windage, and induced rotor disk pumping flow for which a simple physics-based argument is presented to determine its direction (radially inward or outward) in the disk boundary layer (Ekman layer) based on the swirl factor of the adjacent vortex core outside the boundary layer. The discussion of the centrifugally-driven buoyant convection in compressor rotor cavities, with and without the bore flow, highlights the nonequilibrium and complex flow and heat transfer physics present in these cavities. However, a simplified 1-D heat transfer modeling methodology for these cavities is presented for design implementation and continuous product validation. The case without the bore flow is treated like a closed rotor cavity. We also presented in this chapter a design-oriented methodology to compute windage and swirl distributions in a purged rotor cavity formed by surfaces with arbitrary rotation, counter-rotation, and no rotation.

Hot gas ingestion and rim seals currently remain the areas of most active research. In a gas turbine design, the objective of any reduction in the flow through a rim-seal to prevent hot gas ingestion must be weighed against the competing increase in the windage temperature rise in the wheel-space. Both the single- and multiple-orifice models for hot gas ingestion presented here will provide the validation flexibility and modeling simplicity necessary to develop an effective rim-seal design in gas turbines.

The preswirl system, while it helps to reduce the temperature of the blade cooling air, must simultaneously ensure that the supply pressure at the blade root is adequate to provide an acceptable backflow margin for the film cooling holes.

We have also presented in this chapter a method to calculate the rotor axial thrust, the accuracy of which depends on the accuracy of the pressure force, which is to be determined from the static pressure distribution (in the generalized vortex core) integrated with area over each disk face.

Worked Examples

Example 4.1 In Section 4.3, for a rotor-stator cavity, we formulated the calculation of windage power and the related change in fluid total temperature in an inertial reference frame. This formulation essentially uses the angular momentum and windage equations as follows:

Angular momentum equation: $\Gamma_R - \Gamma_S = \dot{m} r_{out}^2 S_{f_{out}} \Omega - \dot{m} r_{in}^2 S_{f_{in}} \Omega$

Windage equation: $T_{t_{s_{out}}} - T_{t_{s_{in}}} = \frac{\Gamma_R \Omega}{\dot{m} c_p}$

After replacing the rotor torque Γ_R by the stator torque Γ_S from the angular momentum equation, express the windage equation in terms of total temperatures in the rotor reference frame and interpret the terms in the resulting equation.

Solution
Using the angular momentum equation to substitute for Γ_R in the windage equation, we obtain

$$T_{t_{s_{out}}} - T_{t_{s_{in}}} = \frac{r_{out}^2 S_{f_{out}} \Omega^2 - r_{in}^2 S_{f_{in}} \Omega^2}{c_p} + \frac{\Gamma_S \Omega}{\dot{m} c_p}$$

$$c_p T_{t_{s_{out}}} - c_p T_{t_{s_{in}}} = U_{out} V_{\theta_{out}} - U_{in} V_{\theta_{in}} + \frac{\Gamma_S \Omega}{\dot{m}}$$

$$\left(c_p T_{t_{s_{out}}} - U_{out} V_{\theta_{out}} \right) - \left(c_p T_{t_{s_{in}}} - U_{in} V_{\theta_{in}} \right) = \frac{\Gamma_S \Omega}{\dot{m}}$$

$$\dot{m} I_{out} - \dot{m} I_{in} = \Gamma_S \Omega$$

which is the windage equation in terms of rothalpy where I_{in} and I_{out} are the fluid rothalpy at the cavity inlet and outlet, respectively. This equation shows that the increase in rothalpy outflow over its inflow equals $\Gamma_S \Omega$, as if the stator becomes a rotor in the rotor reference frame and imparts the equivalent windage power to the fluid. When the stator torque is zero, implying an all-rotor cavity, the fluid rothalpy remains constant for an adiabatic flow, which is consistent with the Euler's turbomachinery equation presented in Chapter 2.

In the aforementioned windage equation, expressing I_{in} and I_{out} in the rotor reference frame by $I_{in} = c_p T_{t_{R_{in}}} - r_{in}^2 \Omega^2 / 2$ and $I_{in} = c_p T_{t_{R_{in}}} - r_{in}^2 \Omega^2 / 2$, respectively, yields the following windage equation in terms of the total temperature in this reference frame:

$$T_{t_{R_{out}}} - T_{t_{R_{in}}} = \frac{(r_{out}^2 - r_{in}^2) \Omega^2}{2 c_p} + \frac{\Gamma_S \Omega}{\dot{m} c_p}$$

This equation shows that the change in relative total temperature as a result of windage consists of two parts. The first part represents the change in the relative total temperature from radius r_{in} to r_{out} in the flow through an adiabatic duct rotating with the angular velocity Ω. The second part is the change in the relative total temperature as a result of work done by an equivalent rotor whose torque equals that of the stator.

Example 4.2 For a small single-stage centrifugal air compressor, schematically shown in Figure 4.24, the key impeller dimensions are $r_{sh} = 20$ mm, $r_1 = 50$ mm, and $r_2 = 75$ mm. At $\Omega = 60000$ rpm test speed and inlet air mass flow rate $\dot{m} = 1$ kg/s at $P_{t_1} = 1$ bar and $T_{t_1} = 290$ K, the compressor discharge occurs at $P_{t_2} = 4.5$ bar and $T_{t_2} = 524.5$ K. The slip coefficient at the discharge is 0.94. Assuming that the air in the gap between the stationary casing and impeller, both on forward and aft sides, behaves like a forced vortex with the swirl factor $S_f = 0.50$ and constant density corresponding to the compressor discharge conditions, calculate the net axial thrust on the impeller. Neglect the contribution of axial thrust as a result of the change in stream thrust of the impeller flow.

Solution

The radial equilibrium equation for a forced vortex with swirl factor S_f becomes

$$\frac{dP_s}{dr} = \rho r S_f^2 \Omega^2$$

which for constant density yields the following radial variation of static pressure in the gap between the impeller and casing

$$P_s = P_{s_2} - \frac{\rho S_f^2 \Omega^2}{2}\left(r_2^2 - r^2\right)$$

Figure 4.24 shows that the axial thrust exerted by the fluid on surface AB is balanced by the equal and opposite thrust on surface CD. Thus, the net contribution to rotor axial thrust is from the fluid pressure force exerted on surface DE. This rotor thrust is directed to the left and is equal to

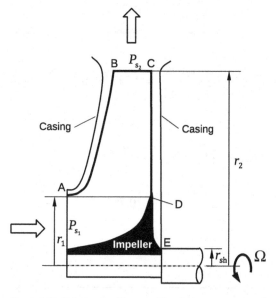

Figure 4.24 Schematic of a centrifugal air compressor for axial thrust calculation (Example 4.2).

$$F_{\text{rotor}} = 2\pi \int_{r_{\text{sh}}}^{r_1} P_s r \, dr$$

$$F_{\text{rotor}} = 2\pi \int_{r_{\text{sh}}}^{r_1} \left\{ P_{s_2} - \frac{\rho S_f^2 \Omega^2}{2} \left(r_2^2 - r^2 \right) \right\} r \, dr$$

$$F_{\text{rotor}} = \pi \left(r_1^2 - r_{\text{sh}}^2 \right) \left\{ P_{s_2} - \frac{\rho S_f^2 \Omega^2}{2} \left(r_2^2 - \frac{r_1^2 + r_{\text{sh}}^2}{2} \right) \right\}$$

In the aforementioned equation to calculate rotor axial thrust, we need to first calculate the values of P_{s_2} and ρ as follows:

$$\Omega = \frac{600000 \times \pi}{30} = 6283.185 \text{ rad/s}$$

$$V_{\theta_2} = 0.94 \times r_2 \times \Omega = 0.94 \times \left(\frac{75}{1000} \right) \times 6283.185 = 442.965 \text{ m/s}$$

Neglecting the contribution of dynamic temperature associated with the flow radial velocity at compressor discharge, we write

$$T_{s_2} = T_{t_2} - \frac{V_{\theta_2}^2}{2c_p} = 524.5 - \frac{(442.965)^2}{2 \times 1004} = 427 \text{ K}$$

From isentropic relations we obtain

$$\frac{P_{t_2}}{P_{s_2}} = \left(\frac{T_{t_2}}{T_{s_2}} \right)^{\frac{\kappa}{\kappa-1}} = \left(\frac{524.5}{427} \right)^{3.5} = 2.058$$

$$P_{s_2} = \frac{450000}{2.058} = 218688 \text{ N}$$

and

$$\rho = \frac{P_{s_2}}{R T_{s_2}} = \frac{218688}{287 \times 427} = 1.785 \text{ kg/m}^3$$

Thus, the rotor axial thrust can now be calculated as

$$F_{\text{rotor}} = \pi \left((0.05)^2 - (0.02)^2 \right) \times$$

$$\left\{ 218688 - \frac{1.785 \times (0.5)^2 \times (6283.185)^2}{2} \left((0.75)^2 - \frac{(0.05)^2 + (0.02)^2}{2} \right) \right\}$$

$$F_{\text{rotor}} = 1200 \text{ N}$$

Problems

4.1   An adiabatic, constant-area pipe is rotating at angular velocity Ω about an axis normal to the pipe axis. For the pipe inlet at $r = r_1$ and outlet at $r = r_2$, where r is the radial distance from the axis of rotation, compute for an air mass flow rate of 1 kg/s the change in total pressure and total temperature (in an inertial reference frame) from the pipe inlet to outlet. Neglect any frictional pressure loss in the pipe. How will your answers change if the pipe outlet is closed, resulting in zero mass flow rate?

4.2   In an isentropic forced vortex with constant angular velocity Ω_f, show that, between radii r_1 and r_1, the change in static temperature equals the change in dynamic pressure associated with the rotational velocity. Thus, the change in total temperature in this vortex equals twice the change in static temperature or the change in dynamic temperature, which is easy to calculate.

4.3   In an isentropic free vortex with constant angular momentum H_f, show that, between radii r_1 and r_1, the change in static temperature is equal and opposite to the change in dynamic temperature associated with the rotational velocity. Thus, the total temperature in this vortex remains constant everywhere.

4.4   In some gas turbines, radial pipes are used in the compressor rotor cavity to bring a part of the compressor primary air flow from rim to bore for downstream cooling of turbine parts. One such design is schematically shown in Figure 4.25. The primary design intent is to have the coolant air at location A with a minimum drop in pressure from its value at the extraction point. For this design, the following representative data are given:

$P_{t_s} = 5$ bar; $T_{t_s} = 500$ K; $\dot{m} = 10$ kg/s; $\Omega = 3600$ rpm
Number of radial pipes = 20
Pipe ID = 40 mm; Pipe OD = 45 mm; $L = 1$ m; $r_1 = 0.15$ m; $r_2 = 0.2$ m;
Darcy friction factor $f = 0.022$.

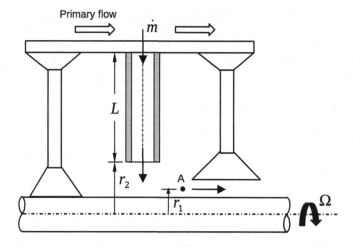

Figure 4.25 Schematic of a compressor rotor cavity with radial pipes for supplying compressor air for downstream cooling of turbine parts (Problem 4.4).

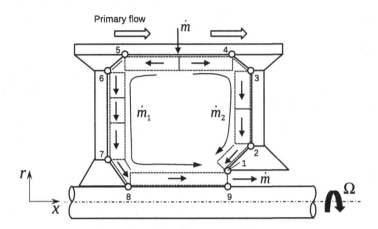

Figure 4.26 Schematic of radially inward flow of air in a compressor rotor cavity extracted from the primary flow for downstream cooling of turbine parts (Problem 4.5).

Note that the compressor air enters each pipe in solid-body rotation with the rotor ($S_f = 1.0$), and the flow is assumed to be adiabatic.

Compute at location A:

(a)   Static pressure and the corresponding total pressure in both stator and rotor reference frames.

(b)   Total temperature in both stator and rotor reference frames.

(c)   Air swirl factor.

4.5   Figure 4.26 schematically shows radially inward flow of air in a compressor rotor cavity extracted for downstream cooling of turbine parts. If one assumes that the cooling air reaches the bore region via a free vortex, its static pressure will drop significantly – an undesirable outcome that may result in some backflow. Fortunately, the air that enters the rim region flows radially inward in the Ekman boundary layers on the rotor surface. In general, as shown in the figure, the total mass flow rate $\dot{m}$ splits into two unequal mass flow rates $\dot{m}_1$ and $\dot{m}_2$, where $\dot{m}_1$ is for the flow along the left-hand side rotor surface, and $\dot{m}_2$ along the right-hand side rotor surface.

The coordinates of nodes for a piece-wise linear profile of the axisymmetric compressor rotor surface shown in Figure 4.26 are given as follows:

Node	x (m)	r (m)
1	1.90	0.25
2	2.10	0.40
3	2.10	1.05
4	2.00	1.20
5	1.05	1.20
6	0.90	1.05
7	0.90	0.30
8	1.10	0.15
9	1.90	0.15

The compressor rotor cavity rotates at $\Omega = 3600$ rpm. The coolant air enters the cavity at the rim at $P_{t_s} = 5$ bar, $T_{t_s} = 500$ K, at a mass flow rate of $\dot{m} = 10$ kg/s. Using multiple control volumes, which may be refined from the ones shown in the figure, along each rotor surface and carrying out the torque (based on empirical correlations presented in this chapter) and angular momentum balance in each control volume, compute $\dot{m}_1$, $\dot{m}_2$, and the common static pressure at the location in the bore region where the two flows join together.

4.6 Repeat the axial thrust calculation of Example 4.1 for a centrifugal air compressor, assuming the swirl factor of 0.45 between the casing and impeller and treating the forced vortex as isentropic.

4.7 A design tool, such as developed under Project 4.1, to compute the distribution of windage and swirl velocity in a general rotor cavity is hardwired for air as the working fluid. Develop a set of scaling equations to convert the results obtained from this design tool to be applicable for another working fluid with known thermophysical properties.

4.8 What role does the windage heating in a wheel space cavity play in the turbine rim seal design for maximum sealing effectiveness while minimizing the sealing air flow rate?

4.9 Repeat the calculations of Example 4.2 under both isentropic and isothermal assumptions for the forced vortex (see Chapter 2) in the gaps between the impeller and casing. Assume that the static pressure and temperature obtained in Example 4.2 at the impeller exit prevail at the gap outer radius. Compare and comment on the rotor axial thrust values obtained under different forced vortex assumptions.

4.10 Assuming that the centrifugal compressor in Example 2.4 discharges radially with no axial velocity, compute the contribution to rotor axial thrust from the given flow through the compressor. Will it increase of decrease the rotor axial thrust calculated in this example?

Projects

4.1 Develop a general purpose stand-alone computer code in FORTRAN, or the language of your choice, to solve for swirl distribution and total temperature change as a result of windage in a general rotor-stator cavity, which is schematically shown in Figure 4.8. For calculating the torque of stator and rotor surfaces in the angular momentum equation, you may use the empirical equations presented in this chapter or others of your preference. For solving the resulting transcendental equation to find the fluid swirl factor in each cavity control volume, you may use the regula falsi routine listed in Appendix D. Assume that all the cavity surfaces are adiabatic. How will you verify your computer code, and how will you validate it?

4.2 Develop a general purpose stand-alone computer code in FORTRAN, or the language of your choice, for handing hot gas ingestion using the multiorifice spoke model presented in Section 4.6.2.2 and schematically shown in Figure 4.22. For the numerical solution of the two-point boundary value problem for each spoke represented by serially-connected orifices, use the Thomas Algorithm presented in Appendix E.

4.3 In the rim seal design shown in Figure 4.20 and the corresponding multispoke model shown in Figure 4.22, we have assumed that the flow happens in each

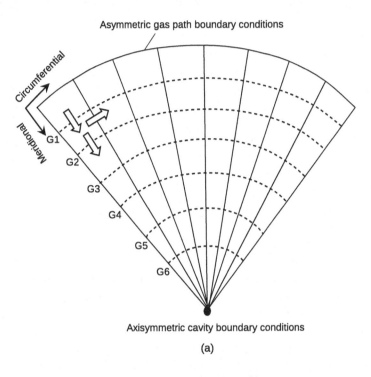

(a)

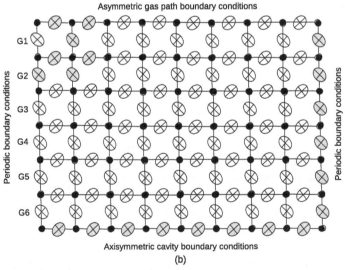

(b)

Figure 4.27 (a) Grid system for the multiorifice fan model in the meridional-tangential plane and (b) Flow network representation of multiorifice fan model for hot gas ingestion simulation (Project 4.3).

spoke in the meridional plane only, with no spoke-to-spoke interactions in the tangential direction. To include such interactions we can easily extend the spoke model into a fan model for which the grid system in the meridional-tangential plane is shown in Figure 4.27a and the corresponding flow network model in Figure 4.27b. In this project, you are asked to extend the computer code developed under Project 4.2 to solve for various flow properties in the network shown in Figure 4.27b. The problem is akin to 2-D numerical heat conduction solution requiring row-by-row solution using TDMA first in the meridional direction followed by column-by-column TDMA-based solution in the tangential direction until convergence.

References

Abe, T., J. Kikuchi, and H. Takeuchi. 1979. An investigation of turbine disk cooling: experimental investigation and observation of hot gas flow into a wheel space. 13th International Congress on Combustion Engines (CIMAC), Vienna, Austria, May 7–10, Paper No. GT30.

Bayley, F. J., and J. M. Owen. 1970. Fluid dynamics of a shrouded disk system with a radial outflow of coolant. *ASME J. Eng. Power.* 92(3): 335–341.

Bohn, D., and M. Wolff. 2003. Improved formulation to determine minimum sealing flow – cw, min – for different sealing configurations. ASME Paper No. GT2003–38465.

Carnahan, B., H. A. Luther, and J. O. Wilkes. 1969. *Applied Numerical Methods.* New York: John Wiley & Sons.

Chew, J. W., T. Green, and A. B. Turner. 1994. Rim sealing of rotor-stator wheelspaces in the presence of external flow. ASME Paper No. 94-GT-126.

Childs, P. R. N. 2011. *Rotating Flow.* New York: Elsevier.

Daily, J. W., and R. E. Nece. 1960. Chamber dimension effects on induced flow and frictional resistance of enclosed rotating disks. *J. Basic Eng. Trans. ASME.* 82(1): 217–232.

Haaser, F., J. Jack, and W. F. McGreehan. 1988. Windage rise and flowpath gas ingestion in turbine rim cavities. *ASME J. Eng. Gas Turbines Power.* 110 (1): 78–85.

Hamabe, K., and K. Ishida. 1992. Rim seal experiments and analysis of a rotor-stator system with non-axisymmetric main flow. ASME Paper No. 92-GT-160.

Hoerner, S. F. 1965. *Fluid-Dynamic Drag: Theoretical, Experimental, and Statistical Information.* Author-published.

Johnson, B. V., R. Jakoby, D. Bohn, and D. Cunat. 2006. A method for estimating the influence of time-dependent vane and blade pressure fields on turbine rim seal ingestion. ASME Paper No. GT2006–90853.

Johnson, B. V., C. Z. Wang, and R. Roy. 2008. A rim seal orifice model with two cds and effects of swirl in seals. ASME Paper No.GT2008–50650.

McAdams, W. H. 1954. *Heat Transmission*, 3rd edn. New York: McGraw-Hill.

Newman, B. G. 1983. Flow and heat transfer on a disk rotating beneath a forced vortex. *AIAA J.* 22 (8):1066–1070.

Oates, G. C. 1988. *Aerothermodynamics of Gas Turbines and Rocket Propulsion: Revised and Enlarged.* Washington, DC: AIAA.

Owen, J. M. 2011a. Prediction of ingestion through turbine rim seals–part1: rotationally induced ingress. *ASME J. Turbomach.* 133(3): 031005.1–031005.9.

Owen, J. M. 2011b. Prediction of ingestion through turbine rim seals–part 2: externally induced and combined ingress. *ASME J. Turbomach.* 133(3):031006.1–031006.9.

Owen, J. M., and R. H. Rogers. 1989. *Flow and Heat Transfer in Rotating-Disc System. Vol. 1 Rotor-Stator Systems.* Taunton, England: Research Studies Press.

Owen, J. M., and R. H. Rogers. 1995. *Flow and Heat Transfer in Rotating-Disc System: Vol. 2 Rotating Cavities.* Taunton,England: Research Studies Press.

Owen, J., O. Pountney, and G. Lock. 2012. Prediction of ingress through turbine rim seals – part 2: combined ingress. *ASME J.Turbomach.* 134(3): 031013.1–031013.7.

Owen, J. M., K. Zhou, O. Pountney, M. Wilson, and G. Lock, 2012. Prediction of ingress through turbine rim seals–part 1: externally induced ingress. *ASME J. Turbomach.* 134(3): 031012.1–031012.1 3.

Owen, J. M., and C. A. Long. 2015. Review of buoyancy-induced flow in rotating cavities. *ASME J. Turbomach.* 137(11): 111001.1–111001.13.

Owen, J. M., and H. Tang. 2015. Theoretical model of buoyancy-induced flow in rotating cavities. *ASME J. Turbomach.* 137(11): 111005.1–111005.7.

Phadke, U., and J. Owen. 1988. Aerodynamic aspects of the sealing of gas-turbine rotor-stator systems: part 3: the effect of non-axisymmetric external flow on seal performance. *Inter. J. Heat and Fluid Flow.* 9(2): 113–117.

Reichert, A. W., and D. Lieser. 1999. Efficiency of air-purged rotor-stator seals in combustion turbine engines. ASME Paper No. 99-GT-250.

Scanlon, T., J. Wilkes, D. Bohn, and O. Gentilhomme. 2004. A simple method of estimating ingestion of annulus gas into a turbine rotor stator cavity in the presence of external pressure gradients. ASME Paper No.GT2004–53097.

Schlichting, H. 1979. *Boundary Layer Theory,* 7th edn. New York: McGraw-Hill.

Scobie, J. A., C. M. Sangan, J. M. Owen, and G. D. Lock. 2016. Review of ingress in gas turbines. *ASME J. Eng. Gas Turbines & Power.* 138:120801.1–120801.16.

Sultanian, B. K., and D. A. Nealy. 1987. Numerical modeling of heat transfer in the flow through a rotor cavity. In D. E. Metzger, ed., *Heat Transfer in Gas Turbines,* HTD-Vol. 87, 11–24, New York: ASME.

Bibliography

Atkins, N. R., and V. Kanjirakkad. 2014. Flow in a rotating cavity with axial throughflow at engine representative conditions. ASME Paper No. GT2014–27174.

Bayley, F. J., and J. M. Owen. 1969. Flow between a rotating and stationary disc. *Aeronautical Quarterly.* 20:333–354.

Bayley, F. J., C. A. Long, and A. B. Turner. 1993. Discs and drums: the thermo-fluid dynamics of rotating surfaces. *Proc. IMechE. Part C: J. Mech. Engineering Science.* 207: 73–81.

Benim, A. C., D. Brillert, and M. Cagan. 2004. Investigation into the computational analysis of direct-transfer pre-swirl system for gas turbine blade cooling. ASME Paper No. GT2004–54151.

Benra, F.-K., H. J. Dohmen, and O. Schneider. 2008. Application of an enhanced 1D network model to calculate the flow properties of a pre-swirl secondary air system. ASME Paper No. GT2008–50442.

Bohn, D., E. Deuker, R. Emunds, and V. Gorzelite. 1995. Experimental and theoretical investigations of heat transfer in closed gas-filled rotating annuli. *ASME J. Turbomach.* 117(1): 175–183.

Bricaud, C., T. Geis, K. Dullenkopf, and H.-J. Bauer. 2007. Measurement and analysis of aerodynamic and thermodynamic losses in pre-swirl system arrangements. ASME Paper No. GT2007–27191.

Case, P. 1966. Measurements of entrainment by a free rotating disk. *J. Roy. Aero. Soc.* 71: 124–129.

Cham, T.-S. and M. R. Head. 1969. Turbulent boundary layer flow on a rotating disk. *J. Fluid Mech.* 37(1): 129–149.

Chew, J. W., N. J. Hills, S. Khalatov, T. Scanlon, and A. B. Turner. 2003. Measurement and analysis of flow in a pre-swirled cooling air delivery system. ASME Paper No. GT2003–38084.

Chew, J. W., F. Ciampoli, N. J. Hills, and T. Scanlon. 2005. Pre-swirled cooling air delivery system performance. ASME Paper No. GT2005–68323.

Ciampoli, F., J. W. Chew, S. Shahpar, and E. Willocq. 2007. Automatic optimization of pre-swirl nozzle design. *ASME J. Eng. Gas Turbines & Power.*129: 387–393.

Didenko, R. A., D. V. Karelin, D. G. Levlev, Y. N. Shmotin, and G. P. Nagoga. 2012. Pre-swirl cooling air delivery system performance study. ASME Paper No. GT2012–68342.

Dittmann, M., T. Geis, V. Schramm, S. Kim, and S. Wittig. 2002. Discharge coefficients of a pre-swirl system in secondary air systems. *ASME J. Turbomach.* 124: 119–124.

Dittmann, M., K. Dullenkopf, and S. Wittig. 2004. Discharge coefficients of rotating short orifices with radiused and chamfered inlets. *ASME J. Eng. Gas Turbine & Power.* 126: 803–808.

Dittmann, M., K. Dullenkopf, and S. Wittig. 2005. Direct-transfer pre-swirl system: A one-dimensional modular characterization of the flow. *ASME J. Eng. Gas Turbine & Power.* 127: 383–388.

Dweik, Z., R. Briley, T. Swafford, and B. Hunt. 2009. Computational study of the heat transfer of the buoyancy-driven rotating cavity with axial throughflow of cooling air. ASME Paper No. GT2009–59978.

Dweik, Z., R. Briley, T. Swafford, and B. Hunt. 2009. Computational study of the unsteady flow structure of the buoyancy-driven rotating cavity with axial throughflow of cooling air. ASME Paper No. GT2009–59969.

El-Oun, Z. B., and J. M. Owen. 1989. Pre-swirl blade-cooling effectiveness in an adiabatic rotor-stator system. *ASME J. Turbomach.* 111(4):522–529.

Farthing, P. R., C. A. Long, J. M. Owen, and J. R. Pincombe. 1992. Rotating cavity with axial throughflow of cooling air: Flow structure. *ASME J. Turbomach.* 114(1): 237–246.

Farthing, P. R., C. A. Long, J. M. Owen, and J. R. Pincombe. 1992. Rotating cavity with axial throughflow of cooling air: Heat transfer. *ASME J. Turbomach.* 114(1): 229–236.

Gartner, W. 1997. A prediction method for the frictional torque of a rotating disk in a stationary housing with superimposed radial outflow. ASME Paper No. 97-GT-204.

Geis, T., M. Dittmann, and K. Dullenkopf. 2004. Cooling air temperature reduction in a direct transfer pre-swirl system. *ASME J. Eng. Gas Turbines & Power.* 126: 809–815.

Glicksman, L. R. and Lienhard, J. H. 2016. *Modeling and Approximation in Heat Transfer.* New York: Cambridge University Press.

Gord, M. F., M. Wilson, and J. M. Owen. 2005. Numerical and theoretical study of flow and heat transfer in a pre-swirl rotor-stator system. ASME Paper No. GT2005– 68135.

Green, B. R., R. M. Mathison, and M. G. Dunn. 2012. Time-averaged and time-accurate aerodynamic effects of forward rotor cavity purge flow for a high-pressure turbine – part I: Analytical and experimental comparisons. ASME Paper No. GT2012–69937.

Green, B. R., R. M. Mathison, and M. G. Dunn. 2012. Time-averaged and time-accurate aerodynamic effects of forward rotor cavity purge flow for a high-pressure turbine – part II: Analytical flow field analysis. ASME Paper No. GT2012–69939.

Guida, R., D. Lengani, D. Simoni, M. Ubaldi, and P. Zunino. 2018. New facility setup for the investigation of cooling flow, viscous and rotational effects on the interstage seal flow behavior of a gas turbine. ASME Paper No. GT2018–75630.

Huening, M. 2010. Parametric single gap turbine rim seal model with boundary generation for asymmetric external flow. ASME Paper No. GT2010–22434.

Imayama, S., P. H. Alfredsson, and R. J. Lingwood. 2014. On the laminar-turbulent transition of the rotating-disk flow: The role of absolute instability. *J. Fluid Mech.* 745(2):132–163.

Idelchik, I. E. 2005. *Handbook of Hydraulic Resistance*, 3rd edn. Delhi: Jaico Publishing House.

Jarzombek, K., H. J. Dohmen, F.-K. Benra, and O. Schneider. 2006. Flow analysis in gas turbine pre-swirl cooling air systems – variation of geometric parameters. ASME Paper No. GT2006–90445.

Jarzonbek, K., F.-K. Benra, H. J. Dohmen, and O. Schneider. 2007. CFD analysis of flow in high-radius pre-swirl systems. ASME Paper No. GT2007–27404.

Javiya, U., J. Chew, and N. Hills. 2011. A comparative study of cascade vanes and drilled nozzle design for pre-swirl. ASME Paper No. GT2011–46006.

Karnahl, J., J. V. Wolfersdorf, K. M. Tham, M. Wilson, and G. Lock. 2011. CFD simulations of flow and heat transfer in a pre-swirl system: influence of rotating-stationary domain interface. ASME Paper No. GT2011–45085.

Karabay, H., J. X. Chen, R. Pilbrow, M. Wilson, and J. M. Owen. 1999. Flow in a "cover-plate" pre-swirl rotor-stator system. *ASME J. Turbomach.* 121: 160–166.

Karabay, H., R. Pilbrow, M. Wilson, and J. M. Owen. 2000. Performance of pre-swirl rotating-disc systems. *J. Eng. Gas Turbines & Power.* 122: 442–450.

Karabay, H., M. Wilson, and J. M. Owen. 2001. Approximate solutions for flow and heat transfer in pre-swirl rotating-disc systems. ASME Paper No. 2001-GT-0200.

Kim, Y. W., M. N. Okpara, H. Hamm, R. P. Roy, and H.-K. Moon. 2018. Hot gas ingestion model employing flow network with axisymmetric solvers. ASME Paper No. GT2018–77055.

Laurello, V., M. Yuri, and K. Fujii. 2004. Measurement and analysis of an efficient turbine rotor pump work reduction system incorporating pre-swirl nozzles and a free vortex pressure augmentation chamber. ASME Paper No. GT2004–53090.

Lee, H. J. Lee, S. Kim, and J. Cho. 2018. Pre-swirl system design including inlet duct shape by using CFD analysis. ASME Paper No. GT2018–76323.

Lewis, P. 2008. Pre-swirl rotor-stator systems: Flow and heat transfer. PhD thesis, University of Bath.

Lewis, P., M. Wilson, G. Lock, and J. M. Owen. 2008. Effect of radial location of nozzles on performance of pre-swirl systems. ASME Paper No. GT2008–50296.

Liu, G., B. Li, Z. Jiang, and L. Zhang. 2012. Influences of pre-swirl angle on the flow characteristics of pre-swirl nozzle. *J. Propulsion Technology.* 33(5):740–746.

Liu, G., H. Wu, Q. Feng, and S. Liu. 2016. Theoretical and numerical analysis on the temperature drop and power consumption of a pre-swirl system. ASME Paper No. GT2016–56742.

Liu, Y., G. Liu, X. Kong, and Q. Feng. 2016. Design and numerical analysis of a new type of pre-swirl nozzle. ASME Paper No. GT2016–56738.

Lock, G. D., M. Wilson, and J. M. Owen. 2005. Influence of fluid dynamics on heat transfer in a pre-swirl rotating-disc system. *ASME J. Eng. Gas Turbines & Power.* 127:791–797.

Long, C. A., N. D. D. Miche, and P. R. N. Childs. 2007. Flow measurements inside a heated multiple rotating cavity with axial throughflow. *Int. J. Heat Fluid Flow.* 28(6): 1391–1404.

Lugt, H. J. 1995. *Vortex Flow in Nature and Technology*. Malabar: Krieger Publishing Company.

Miller, D. S. 1990. *Internal Flow Systems*, 2nd edn. Houston: Gulf Publishing Company

Mirzamoghadam, A. V., S. Kanjiyani, A. Riahi, R. Vishnumolakala, and L. Gundeti. 2014. Unsteady 360 computational fluid dynamics validation of a turbine stage mainstream/disk cavity interaction. *ASME J. Turbomach.* 137: 011008.1– 011008.9.

Mott, R. L. 2006. *Applied Fluid Mechanics*, 6th edn. Upper Saddle River, NJ: Pearson Prentice Hall.

Nickol, J., M. Tomko, R. Mathison, J. S. Liu, M. Morris, and M. F. Malak. 2018. Heat transfer and pressure measurements for the forward purge cavity, inner endwall, and rotor platform of a cooled transonic turbine stage. ASME Paper No. GT2018–76978.

Ong, C. L., and J. M. Owen. 1989. Boundary-layer flows in rotating cavities. Trans. ASME, J. *Turbomachinery.* 111(3): 341–348.

Owen, J. M. 1988. Air-cooled gas turbine discs: A review of recent research. *Int. J. Heat and Fluid Flow.* 9(4): 354–365.

Owen, J. M. 1989. An approximate solution for the flow between a rotating and a stationary disk. Trans. ASME, J. *Turbomachinery.* 111(3): 323–332.

Owen, J. M. 2010. Thermodynamic analysis of buoyancy-induced flow in rotating cavities. *ASME, J. Turbomachinery.* 132(3). 031006-1-031006–7.

Owen, J. M., and J. Powell. 2006. Buoyancy-induced flow in heated rotating cavities. *ASME J. Eng. Gas Turbines Power.* 128(1): 128–134.

Owen, J. M., H. Abrahamsson, and K. Linblad. 2007. Buoyancy-induced flow in open rotating cavities. *ASME J. Eng. Gas Turbines Power.* 129(4): 893–900.

Patinios, M., I. L. Ong, J. A. Scobie, G. D. Lock, and C. M. Sangan. 2018. Influence of leakage flows on hot gas ingress. ASME Paper No. GT2018–75071.

Pitz, D. B., J. W. Chew, and O. Marksen. 2018. Large-eddy simulation of buoyancy-induced flow in a sealed rotating cavity. ASME Paper No. GT2018–75111.

Popp, O., H. Zimmerman, and J. Kutz. 1998. CFD analysis of cover-plate receiver flow. *ASME J. Turbomach.* 120: 43–49.

Puttock-Brown, M. R., M. G. Rose, and C. A. Long. 2017. Experimental and computational investigation of Rayleigh-Benard flow in the rotating cavities of a core compressor. ASME Paper No. GT2017–64884.

Puttock-Brown, M. R., and M. G. Rose. 2018. Formation and evolution of Rayleigh-Benard streaks in rotating cavities. ASME Paper No. GT2018–75497.

Richardson, L. F. 1922. *Weather Prediction by Numerical Process.* Cambridge: Cambridge University Press.

Scobie, J. A., R. Teuber, L. Y. Sheng, C. M. Sangan, M. Wilson, and G. D. Lock. 2015. Design of an improved turbine rim-seal. *ASME. J. Eng. Gas Turbines & Power.* 138: 022503.1– 022503.10.

Smout, P. D., J. W. Chew, and P. R. N. Childs. 2002. ICAS-GT: a european collaborative research programme on internal cooling air systems for gas turbines. ASME Paper No. GT-2002–30479.

Snowsill, G. D., and C. Young. 2008. Application of CFD to assess the performance of a novel pre-swirl configuration. ASME Paper No. GT2008–50684.

Soghe, R. D., C. Bianchini, and J. D'Errico. 2017. Numerical characterization of flow and heat transfer in pre-swirl systems. ASME Paper No. GT2017–64503.

Sultanian, B. K. 2015. *Fluid Mechanics: An Intermediate Approach.* Boca Raton, FL: Taylor & Francis.

Sun, Z., A. Kilfoil, J. W. Chew, and N. J. Hills. 2004. Numerical simulation of natural convection in stationary and rotating cavities. ASME Paper No. GT2004–53528.

Sun, X., K. Linbald, J. W. Chew, and C. Young, 2007. LES and RANS investigations into buoyancy-affected convection in a rotating cavity with a central axial throughflow. *ASME J. Eng. Gas Turbines Power.* 129(2): 318–325.

Tang, H., S. Tony, and J. M. Owen. 2015. Use of fin equation to calculate numbers for rotating disks. *ASME J. Turbomach.* 137(12): 121003.1–121003.10.

Tang, H., and J. M. Owen. 2017. Effect of buoyancy-induced rotating flow on temperatures of compressor discs. *ASME J. Eng. Gas Turbines & Power.* 139(06): 062506.1–062506.10.

Tian, S., Z. Tao, S. Ding, and G. Xu. 2004. Investigation of flow and heat transfer in a rotating cavity with axial throughflow of cooling air. ASME Paper No. GT2004–53525.

Tian, S., Q. Zhang, and H. Liu. 2013. CFD investigation of vane nozzle and impeller design for HPT blade cooling air delivery system. ASME Paper No. GT2013–95396.

Wilson, M., R. Pilbrow, and J.M. Owen. 1997. Flow and heat transfer in a pre-swirl rotor-stator system. *ASME J.Turbomach.*119: 364–373.

Yan, Y., M. F. Gord, G. D. Lock, M. Wilson, and J. M. Owen. 2003. Fluid dynamics of a pre-swirl rotor-stator system. *ASME J. Turbomach.* 125: 641–647.

Zhu, X., G. Liu, and S. Liu. 2010. Numerical study on the temperature drop and pressure loss characteristic of pre-swirl system with cover plate. *J. Aerospace Power.* 25(11): 2489–2506.

Zografos, A. T., W. A. Martin, and J. E. Sunderland. 1987. Equations of properties as a function of temperature for seven fluids. *Computer Methods in Applied Mechanics and Engineering.* 61: 177–187.

Nomenclature

A	Area
$\hat{A}_{gap}$	Seal area per vane ($\hat{A}_{gap} = A_{gap}/N$)
b	Width of bolt head
c_p	Specific heat at constant pressure
C	Speed of sound
C_d	Discharge coefficient
C_D	Drag coefficient
C_f	Shear coefficient
C_M	Moment coefficient
C_V	Velocity coefficient ($C_V = V_x/V$)
CV	Control volume
CFD	Computational fluid dynamics
CDBC	Centrifugally-driven buoyant convection
F	Force
$\hat{F}_{f_s}$	Dimensionless static-pressure mass flow function
g	Acceleration as a result of gravity
G	Ratio of axial clearance to cavity outer radius
GDBC	Gravitationally-driven buoyant convection
h	Heat transfer coefficient
I_b	Interference factor of adjacent bolts on a disk
k	Dimensionless cavity static pressure ($k = \hat{P}_{s_{cav}}$)
L	Length
$\dot{m}$	Mass flow rate
M	Mach number
N	Number of turbine vanes in the primary flow path

N_b	Number of bolts
Nu	Nusselt number
P	Pressure
$\hat{P}_s$	Dimensionless static pressure $\left(\hat{P}_s = \left(P_s - P_{s_{min}}\right) \middle/ \left(P_{s_{max}} - P_{s_{min}}\right)\right)$
Pr	Prandtl number
$\dot{q}_c$	Convective heat flux
$\dot{Q}_c$	Convective heat transfer rate
r	Cylindrical polar coordinate r
R	Gas constant; disk outer radius
Ra	Rayleigh number
Re	Disk Reynolds number $(Re = \rho R^2 \Omega / \mu)$
Re'	Modified disk Reynolds number $(Re' = \rho \Omega R_o (R_o - R_i)/\mu$
RRF	Rotor (noninertial) reference frame
s	Gap between rotor and stator in a cavity; tangential distance between bolds
SRF	Stator (inertial) reference frame
t	Time
S_f	Swirl factor
S_T	Stream thrust
T	Temperature
T_{aw}	Adiabatic wall temperature
U	Rotor tangential velocity
V	Velocity
x	Cartesian coordinate x
$\hat{x}$	Dimensionless θ coordinate $(\hat{x} = (N\theta/\pi) - 1)$

Subscripts and Superscripts

1	Location 1; section 1
2	Location 2; section 2
3	Location 3; section 3
4	Location 4; section 4
ann	Annulus
b	Bolt
bore	Compressor bore region
cav	Cavity
cone	Conical surface
cylinder	Cylindrical surface
disk pump	Disk pumping mass flow rate
egr	Egress
free disk	Rotor disk adjacent to stagnant air core
gap	Gap
h	Horizontal
i	Inner
in	Inlet

ing	Ingress
max	Maximum value
min	Minimum value
o	Outer
out	Outlet
r	Radial direction
$r\theta$	In radial-tangential $(r - \theta)$ plane
R	Rotor; rotor reference frame
ref	Reference rotor
s	Static
sh	Shaft
S	Stator; stator reference frame
t	Total (stagnation)
w	Wall
x	Axial direction; flow direction
θ	Tangential direction
$(^-)$	Average

Greek Symbols

α	Angle of the conical surface from horizontal
β	Ratio disk angular velocity to reference rotor angular velocity $(\beta = \Omega/\Omega_{ref})$
Γ	Torque
Δ	Small change in a quantity
ζ	Fraction of free disk pumping mass flow rate
θ	Blade cooling air temperature reduction coefficient
$\tilde{\theta}$	Turbine work loss coefficient
κ	Ratio of specific heats
μ	Dynamic viscosity
ρ	Density
τ	Shear stress
Ω	Angular velocity

5 Labyrinth Seals

5.0 Introduction

Sealing technology is critical to modern high-performance gas turbine designs whose internal flow systems involve a variety of static (between stator surfaces) and dynamic (between stator and rotor surfaces) seals. Some of these seals are designed to minimize parasitic leakage flows, while others simultaneously meter the cooling flows as needed for the downstream components under various operating conditions. They endure severe operating and boundary conditions. The key design goals for seals used in gas turbines are: (1) to achieve pressure sealing between stationary and rotating surfaces with minimum leakage flow, (2) to minimally impact turbine and compressor efficiencies, (3) to generate acceptable windage temperature rise across them, (4) to achieve acceptable rotor thrust across them, and (5) to impact rotor dynamics within acceptable limits. Key design parameters that influence each seal performance include seal geometry, drag characteristics of the stator and rotor surfaces; and operating parameters such as inlet pressure, temperature, swirl, seal pressure ratio, fluid viscosity, and leakage flow compressibility.

During a gas turbine operation, as a result of the radial thermal and centrifugal growth of the rotor and the radial thermal growth of the adjacent stator, the seal clearance between them decreases and may lead to interference and excessive material loss through rubbing. These rubs typically result in the permanent removal of effective sealing material, increasing leakage. To avoid a rub, dynamic seals with rigid geometry; for example, straight-through labyrinth seals, inevitably have a larger build clearance. To better accommodate rotor excursions, the performance of other legacy labyrinth seals is improved by having teeth on the rotor and a layer of abradable or honeycomb material on the stator. In contrast, compliant seals track rotor movement with minimal loss of effective sealing material when a rub is experienced. The gas turbine returns to steady-state performance with minimal increase in leakage. Because brush seals can follow rotor movement with minimal wear, they fall into the category of contacting compliant seals, reducing leakage gaps that occur from large thermal differences between rotating and stationary components. The sealing technology used in modern gas turbines is trending toward the development of noncontacting compliant seals.

Chupp et al. (2006, 2014) provide an extensive and authoritative review of standard static and dynamics seals and advanced seal designs used in turbomachinery, including a rich bibliography for those who are interested in further advancing the seal technology. Rotordynamics considerations of seals are extensively presented in Childs (1993). The main thrust of this chapter is on the thermofluids modeling of standard dynamic seals used in gas turbines, enabling the calculation of seal leakage flow rate and the associated windage temperature rise for 1-D flow network simulation of secondary cooling and sealing flows. Toward this objective, we limit our discussion in this chapter to the labyrinth seals. For the other types of seals, the interested readers may refer to the references and bibliography at the end of this chapter.

5.1   Straight-Through and Stepped-Tooth Designs

As the name suggests, a labyrinth seal creates a complex flow passage to increase pressure loss across it and reduces leakage flow rate for a given clearance. Being the most widely used in gas turbines, these seals are inexpensive, noncontacting, and they operate under a wide range of pressures, temperatures, and rotor speeds. Some of their disadvantages include inevitable wear that enlarges clearance and worsens leakage.

Three most commonly used design configurations of a labyrinth seal are schematically shown in Figure 5.1. The straight-through design with nominal labyrinth is the most basic one with nearly equal inlet and outlet radii. Stepped designs are used to handle leakage through an inclined rotor-stator interface, and they offer higher flow resistance than a straight-through design.

5.1.1   Flow Physics

Figure 5.2a shows typical streamlines in a straight-through labyrinth seal. As the flow jets out the clearance gap between the rotating tooth and stator, it undergoes a sudden radially inward expansion in the downstream pocket formed between two adjacent teeth. The entrainment of the fluid for jet expansion creates flow recirculation in the

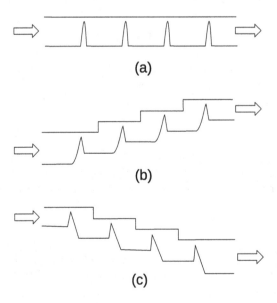

(a)

(b)

(c)

Figure 5.1 (a) Straight-through design, (b) step-up design, and (c) step-down design.

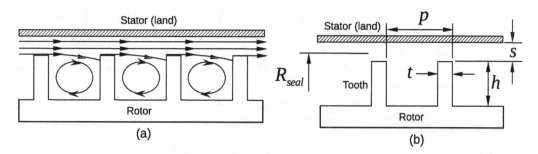

Figure 5.2 (a) Expected streamlines in a straight-through labyrinth seal and (b) seal geometric parameters.

pocket. Exceptfor rotation, the flow field in each pocket is similar to that of a driven cavity flow. Rotation produces radial pumping along each tooth, generating in each pocket additional shear layers, which are primarily responsible to destroy most of the incoming dynamic pressure, resulting in excess entropy production and loss in total pressure. The wall jet flow along the stator surface (land) suffers minimum loss in dynamic pressure and mainly contributes to the "kinetic energy carry-over factor" for the next tooth passage. An ideal labyrinth seal aims at total loss in dynamic pressure associated with its through-flow into each pocket, enabling a 1-D tube-and-tank modeling of the seal. This makes the seal more predicable in terms of its leakage flow characteristics. For an accurate assessment of the carry-over factor for each seal design, one must resort to either experimental data or high-fidelity 3-D CFD simulation.

An inevitable consequence of rotation is the generation of windage in the seal as a result of work transfer from the rotor surface. Windage is an undesirable feature of each seal and must be accurately quantified to calculate the rise in coolant air temperature. Air enters each seal pocket at a swirl factor close to 0.5. As a result of the preponderance of rotor surface over the stator surface in each pocket, the air is expected to exit the pocket at a swirl factor greater than 0.5.

5.1.2 Leakage Mass Flow Rate

The well-known formula of Martin (1908) to compute mass flow rate through a multi-tooth straight-through seal is given by

$$\dot{m} = 5.68 C_d A \beta \frac{P_{t(0)}}{\sqrt{RT_{t(0)}}} \tag{5.1}$$

where C_d is the discharge coefficient, A the geometric area available for the seal flow, and β the gland factor, which is given by

$$\beta = \left[\frac{1 - \left(\dfrac{P_{s(n)}}{P_{t(0)}}\right)^2}{n - \ln\left(\dfrac{P_{s(n)}}{P_{t(0)}}\right)} \right]^{\frac{1}{2}} \tag{5.2}$$

In Equations 5.1 and 5.2, $P_{t(0)}$ and $T_{t(0)}$ are the total pressure and total temperature, respectively, at the inlet, and $P_{s(n)}$ is the static pressure at the exit of the seal with n teeth. Martin's formula assumes that the flow throughout the seal is subsonic, the dynamic pressure through each tooth is completely lost in the downstream cavity, and the total temperature remains nearly constant.

For the seal geometry shown in Figure 5.2b, Egli (1935) proposed the following equation to estimate leakage mass flow rate:

$$\dot{m} = C_t C_c C_r \left(2\pi R_{seal} s\right) \frac{P_{t(0)}}{\sqrt{RT_{t(0)}}} \qquad (5.3)$$

where

$C_t \equiv$ Seal throttling coefficient
$C_c \equiv$ Seal carry-over coefficient
$C_r \equiv$ Seal contraction coefficient

For C_t, C_c, and C_r, Aungier (2000) converted the graphical representations in Egli (1935) into the following easy-to-use equations:

$$C_r = 1 - \frac{1}{3 + \left[\dfrac{54.3}{1 + 100s/t}\right]^{3.45}} \qquad (5.4)$$

$$C_t = \frac{2.143[\ln(n) - 1.464]}{n - 4.322}\left[1 - \left(P_{s(n)}/P_{t(0)}\right)\right]^{0.375\left(P_{s(n)}/P_{t(0)}\right)} \qquad (5.5)$$

$$C_c = 1 + \frac{X_1\left[\dfrac{s}{p} - X_2 \ln\left(1 + \dfrac{s}{p}\right)\right]}{1 - X_2} \qquad (5.6)$$

where

$$X_1 = 15.1 - 0.05255 \exp[0.507(12 - n)]; \quad n \le 12$$

$$X_1 = 13.15 + 0.1625n; \quad n > 12$$

$$X_2 = 1.058 + 0.0218n; \quad n \le 12$$

$$X_2 = 1.32; \quad n > 12$$

Equations 5.4, 5.5, and 5.6 are shown plotted in Figure 5.3.

Vermes (1961) further extended Martin's work and modified Equation 5.1 to compute leakage flow rate for straight, stepped, and combination seals. In particular, to account for the change in total temperature through the seal as a result of windage, he augmented the numerical coefficient from 5.68 to 5.76 in Equation 5.1. Furthermore, using Zabriskie and Sternlicht (1959), he proposed the following modified leakage equation:

$$\dot{m} = 5.76 C_d A\beta \frac{P_{t(0)}}{\sqrt{RT_{t(0)}(1 - \alpha)}} \qquad (5.7)$$

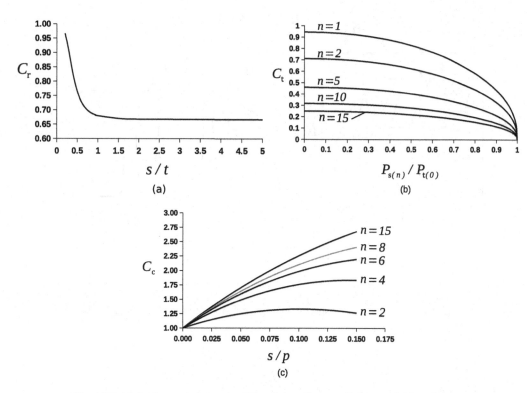

Figure 5.3 (a) Seal contraction ratio, (b) seal throttling coefficient, and (c) seal carry-over coefficient.

where α is called the kinetic energy carry-over factor, which is related to seal geometry. With the aid of the boundary layer analysis, he proposed the following equation to compute α:

$$\alpha = \frac{8.52}{\dfrac{(p-b)}{s} + 7.23} \qquad (5.8)$$

where p is the seal pitch, b the tooth tip width, and s the seal clearance. Note that the seal flow total temperature can only be changed either by heat transfer or windage (work transfer), and it should not depend upon the kinetic energy carry-over factor. Only for an incompressible flow, the dynamic pressure, which equals $0.5\rho V^2$, can be interpreted as the flow kinetic energy per unit volume.

For the seal geometry shown in Figure 5.4a, McGreehan and Ko (1989) proposed the following leakage mass flow rate equation, which is a modified form of Equation 5.7:

$$\dot{m} = K_{L} A \beta \frac{P_{t(0)}}{\sqrt{RT_{t(0)}(1-\alpha)}} \qquad (5.9)$$

where K_{L} is the seal configuration factor, which is like an overall discharge coefficient for the seal.

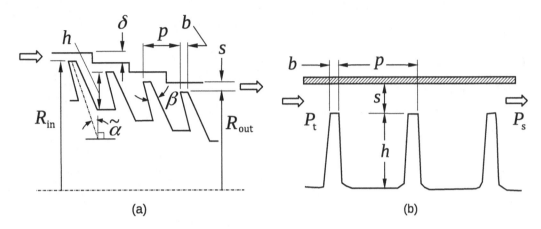

Figure 5.4 (a) Seal geometric parameters in McGreehan and Ko (1989) (b) Seal geometric parameters in Zimmermann and Wolff (1998).

For the seal geometry shown in Figure 5.4b, Zimmermann and Wolff (1998) proposed the following seal mass flow rate equation:

$$\dot{m} = k_2 C_d \beta \frac{P_{t(0)}}{\sqrt{RT_{t(0)}}} \tag{5.10}$$

where C_d is the discharge coefficient, and k_2, the corrected carry-over coefficient given by $k_2 = kk_1$, where the carry-over coefficient k, first proposed by Hodkinson (1939), is expressed in terms of seal geometry as

$$k = \sqrt{\frac{1}{1 - \left(\frac{n-1}{n}\right)\left(\frac{s/p}{s/p + 0.02}\right)}} \tag{5.11}$$

and the correction factor k_1 is given by

$$k_1 = \sqrt{\frac{n}{n-1}} \tag{5.12}$$

Equations 5.11 and 5.12 are shown plotted in Figure 5.5a and 5.5b, respectively. While the variations of the carry-over coefficient k shown in Figure 5.11a exhibit acceptable tends for a seal with the number of teeth ranging from 2 to 15, the corrected carry-over coefficient k_2 shown in Figure 5.5c seems suspect for $0 < s/p < 0.08$. Note that both k and k_2 are akin to C_c used in Equation 5.3, which is shown in Figure 5.3c.

5.2   Tooth-by-Tooth Modeling

Various empirical correlations proposed in the foregoing section generally compute significantly different values (see Problem 5.1) for the seal leakage mass flow rate. This

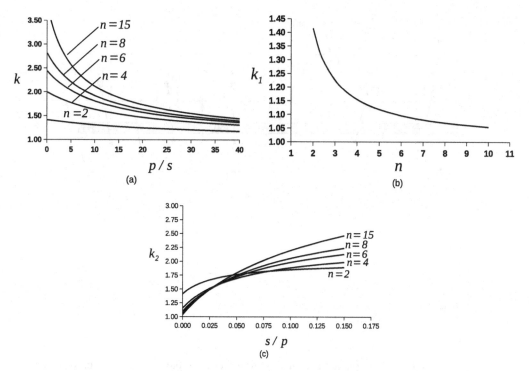

Figure 5.5 (a) Seal carry-over coefficient k_1, (b) correction factor k, and (c) corrected carry-over coefficient k_2.

is to be expected as the leakage mass flow rate depends on both the seal operating conditions and various geometric features, which are difficult to accurately capture in a universal empirical correlation. Furthermore, these correlations do not explicitly account for the swirl velocity variation and windage temperature rise in these seals, somewhat impacting their leakage mass flow rate. Nevertheless, these empirical correlations are still valuable in providing an initial estimate of the leakage mass flow rate to enable an iterative solution method based on the tooth-by-tooth distributed modeling of the labyrinth seal discussed next in this section.

The tooth-by-tooth distributed modeling of a six-tooth labyrinth seal is schematically shown in Figure 5.6. This modeling methodology from seal inlet to outlet allows us to incorporate seal geometry variations, changes in total temperature as a result of both heat transfer and windage, and forced vortex pressure changes as a result of the change in radius between two adjacent teeth, as found in a stepped labyrinth seal design. The prediction results in this case include not only the seal leakage mass flow rate but also the total temperature and swirl factor at the seal exit.

For an incompressible flow, the mass flow rate at a section depends on the difference between the total pressure and static pressure at the section. In case of a compressible flow, however, the mass flow rate depends upon both the total pressure and total-to-static pressure ratio, and the seal may feature a choked flow condition if the Mach number over a tooth becomes unity. For a labyrinth seal with uniform clearance across

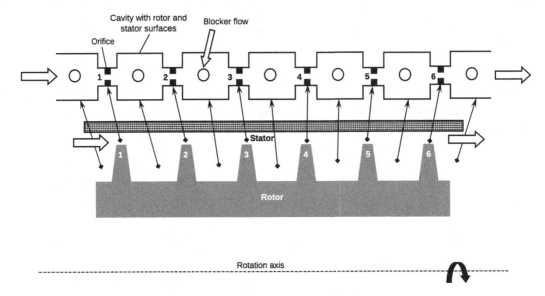

Figure 5.6 Tooth-by-tooth modeling of a six-tooth labyrinth seal.

each tooth, we can make the following argument that the seal can only choke at the last tooth. In steady state, the mass flow rate remains constant throughout the seal. Over each seal tooth with constant leakage flow area, discharge coefficient, swirl factor, and total temperature, the mass flow rate will be proportional to the total pressure and total-pressure mass flow function, which is a function of Mach number, which in turn is a function of total-to-static pressure ratio (see Chapter 2). When the flow chokes at a section, its Mach number becomes unity, yielding a constant value of the total-pressure mass flow function. Because the total pressure decreases monotonically across the labyrinth seal, the maximum leakage mass flow rate through it must correspond to the minimum total pressure, which occurs at the last tooth.

Let us assume that the six-tooth labyrinth seal shown in Figure 5.6 is operating under ideal conditions such that the entire dynamic pressure generated at each tooth is dissipated in the downstream cavity, implying zero kinetic energy carry-over factor ($\alpha = 0$). In addition, under adiabatic conditions with no rotation, the total air temperature remains constant throughout the seal. Accordingly, the static pressure in each cavity becomes the total pressure for the flow over the downstream tooth. This implies that the static-pressure mass flow function over one tooth must equal the total-pressure mass flow function over the next tooth, as shown in Figure 5.7a for a choked air flow through the seal. This figure also shows how Mach number increases from seal inlet to outlet. While the air flow remains subsonic from tooth 1 to tooth 5, it increases rapidly to unity over tooth 6, the last tooth. The corresponding variation in total-to-static pressure ratio from tooth 1 to tooth 6 is shown in Figure 5.7b.

The overall total-to-static pressure ratio for a choked air flow through an ideal labyrinth seal varies with the number of teeth. This variation is depicted in Figure 5.8. We can represent the curve in this figure by the following cubic polynomial:

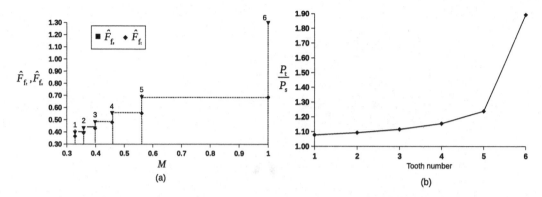

Figure 5.7 (a) Mach number variation across a six-tooth choked ideal labyrinth seal and (b) tooth-wise pressure ratio variation across the seal.

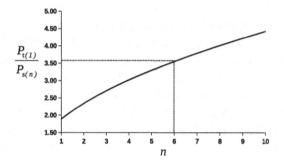

Figure 5.8 Overall total-to-static pressure ratio variation with n for the choked flow through an ideal labyrinth seal.

$$\frac{P_{t(1)}}{P_{s(n)}} = 1.4115 + 0.5261n - 0.0363n^2 + 0.0014n^3 \qquad (5.13)$$

For a labyrinth seal with n number of teeth, if the overall pressure ratio is equal or higher than that computed by Equation 5.13, the seal flow will choke at the last tooth, and if it is lower, the flow will remain subsonic throughout the seal. For the six-tooth labyrinth seal, the overall critical pressure ratio is exactly calculated to be 3.556, while the curve-fit Equation 5.13 yields 3.564, which is within an acceptable error of 0.23 percent.

5.2.1 Orifice-Cavity Model

For a given labyrinth seal with known geometry, the typical design objective is to calculate the seal leakage mass flow rate when the total pressure, total temperature, and swirl factor are specified at the seal inlet, and the static pressure is specified at the seal outlet. The orifice-cavity model shown in Figure 5.9 forms the basis for the tooth-by-tooth modeling of a multitooth labyrinth seal shown in Figure 5.6. For the rotor-stator cavity control volume, which is formed between two consecutive teeth together with the

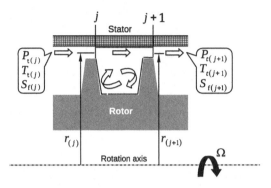

Figure 5.9 Orifice-cavity model control volume formed between tooth j and tooth $j + 1$ in a multitooth labyrinth seal.

stator surface, shown in Figure 5.9, we present here a methodology to compute the compressible flow properties at outlet $j + 1$, knowing their values at inlet j. Note that, if the seal operates under the choked flow condition, which typically occurs at its last tooth, the specified outlet static pressure boundary condition becomes irrelevant. In this case, we should use $M = 1$ as the new boundary condition at the last tooth. In steady state, the air leakage mass flow rate over each seal tooth remains constant.

With an initial estimate of the seal leakage mass flow rate $\dot{m}$, which may be obtained using any of the empirical equations presented in the previous section, we summarize here an iterative solution method for the orifice-cavity model shown in Figure 5.9.

1.   Assume Mach number $M_{(j)}$ and calculate the total-pressure mass flow function at section j

$$\hat{F}_{f,(j)} = M_{(j)} \sqrt{\frac{\kappa}{\left(1 + \dfrac{\kappa - 1}{2} M_{(j)}^2\right)^{\frac{\kappa+1}{\kappa-1}}}}$$

2.   Knowing the empirical discharge coefficient $C_{d(j)}$, calculate the velocity coefficient $C_{V(j)} = V_{x(j)}/V_{(j)}$ from the equation

$$C_{V(j)} = \frac{\dot{m}\sqrt{RT_{t(j)}}}{A_{(j)} C_{d(j)} \hat{F}_{f,(j)}} P_{t(j)}$$

where $A_{(j)}$ is the mechanical flow area at j.

3.   Obtain $V_{(j)} = \sqrt{V_{\theta(j)}^2 / \left(1 - C_{V(j)}^2\right)}$ where $V_{\theta(j)} = r_{(j)} S_{f(j)} \Omega$.

4.   Calculate the static temperature $T_{s(j)}$

$$T_{s(j)} = T_{t(j)} - \frac{V_{(j)}^2}{2c_p}$$

5.   Calculate the Mach number $\tilde{M}_{(j)}$

$$\tilde{M}_{(j)} = \frac{V_{(j)}}{\sqrt{\kappa R T_{s(j)}}}$$

6.   Repeat steps from 1 to 5 until $M_{(j)}$ equals $\tilde{M}_{(j)}$ within an acceptable tolerance.
7.   Calculate the static pressure $P_{s(j)}$

$$P_{s(j)} = \frac{P_{t(j)}}{\left(\frac{T_{t(j)}}{T_{s(j)}}\right)^{\frac{\kappa}{\kappa-1}}}$$

Thus, using the foregoing seven calculation steps, we now have established at section j all the flow properties: $P_{t(j)}$, $P_{s(j)}$, $T_{t(j)}$, $T_{s(j)}$, $M_{(j)}$, $S_{f(j)}$, $V_{x(j)}$, and $V_{\theta(j)}$. Within the cavity control volume, the leakage air flow experiences heat transfer from the cavity stator and rotor surfaces, work transfer (windage) from the cavity rotor surfaces, and the change in its swirl factor as a result of the net torque from the stator and rotor surfaces. For nonzero kinetic energy carry-over factor, the cavity static pressure must be modified to become the total pressure at section $j+1$. If $r_{(j)} \neq r_{(j+1)}$, we must also account for the change in static pressure in the forced vortex associated with the seal leakage flow.

Additional calculation steps to obtain $P_{t(j+1)}$, $T_{t(j+1)}$, and $S_{f(j+1)}$ are summarized as follows:

1.   Using the methodology presented in Chapter 4, calculate the windage temperature rise $\Delta T_{windage}$ and the swirl factor $S_{f(j+1)}$.
2.   From the structural heat transfer analysis of the labyrinth seal, calculate the change in temperature (ΔT_{HT}) of the leakage air flow.
3.   Calculate the total temperature $T_{t(j+1)}$

$$T_{t(j+1)} = T_{t(j)} + \Delta T_{windage} + \Delta T_{HT}$$

4.   Calculate the static temperature $T_{s(j+1)}$

$$T_{s(j+1)} = T_{t(j+1)} - \frac{r_{(j+1)}^2 \Omega^2 S_{f(j+1)}^2}{2c_p} - \frac{\alpha V_{x(j)}^2}{2c_p}$$

where α is the kinetic energy carry-over factor given by Equation 5.8.
5.   Calculate the static pressure $P_{s(j+1)}$

$$P_{s(j+1)} = P_{s(j)} \left\{ 1 + \left(\frac{\kappa-1}{2}\right) M_\theta^2 \left(\frac{r_{(j+1)}^2}{r_{(j)}^2} - 1\right) \right\}^{\frac{\kappa}{\kappa-1}}$$

where

$$M_\theta = \frac{r_{(j)} S_{f(j+1)} \Omega}{\sqrt{\kappa R T_{s(j)}}}$$

6.   Calculate the total pressure $P_{t(j+1)}$

$$P_{t(j+1)} = P_{s(j+1)} \left(\frac{T_{t(j+1)}}{T_{s(j+1)}} \right)^{\frac{\kappa}{\kappa-1}}$$

The foregoing six steps to calculate the flow properties at section $j+1$ and the previous seven steps for properties at section j constitute a methodology to march the solution from inlet to outlet of a multitooth labyrinth seal. We iterate this tooth-by-tooth calculation procedure by adjusting the seal leakage mass flow rate $\dot{m}$ until the computed discharge static pressure for the last tooth equals the specified seal discharge pressure within an acceptable tolerance.

The tooth-by-tooth modeling technique presented here offers significant flexibility to simulate variable seal clearances and other features, including blocker flows if present.

5.3   Concluding Remarks

In this chapter, we have presented comprehensive thermofluids modeling of labyrinth seals, which are ubiquitous in internal flow systems of gas turbines. The tooth-by-tooth modeling combined with the orifice-cavity model presented here will allow a gas turbine designer to handle a general labyrinth seal to accurately compute the leakage mass flow rate, windage temperature rise, and change in flow swirl factor. Considering the labyrinth seal as super-element, this modeling methodology is ideally suited for integration into a 1-D flow network model used for simulating gas turbine internal flow systems. It is important to note that, for the transient operation of a labyrinth seal, one must account for both thermal and centrifugal growth of its structural parts, requiring their multiphysics (mechanical-thermofluids) analysis for which the general methodology is presented in Chapter 6.

For many other static seals such as C-seals, E-seals, W-seals, and spline seals, and dynamic seals such as brush seals, leaf and cloth seals, hybrid brush seals, finger seals, carbon segmented seals, and others, interested readers may find the references and bibliography at the end of this chapter helpful.

Worked Examples

Example 5.1 Consider a two-tooth labyrinth seal whose mean radius under each tooth is $r_m = 0.5$ m with a clearance of $s = 3$ mm. The rotor rotates at 3000 rpm, and the swirl factor of the leakage air flow remains constant at $S_f = 0.5$. The seal initially operates under the inlet total pressure and total temperature of $P_{t_{in}} = 10$ bar and $T_{t_{in}} = 500$ K. The outlet static pressure (back pressure) remains constant at $P_{s_{out}} = 5$ bar. Assume that the kinetic energy carry-over factor is zero, and discharge coefficient $C_d = 0.8$. Neglect any change in air total temperature as a result of heat transfer and rotational work transfer (windage).

(a)   Compute the leakage air mass flow rate through the seal.

(b)   Use a noniterative method to compute the minimum leakage air mass flow rate and inlet total pressure such that the seal just chokes ($M_2 = 1$) at the second tooth.

Solution

Because the swirl factor remains constant, we may solve this problem in the relative reference frame rotating with the air swirl velocity, using the total pressure and total temperature based only on the axial throughflow velocity. These quantities are computed as follows:

Air tangential velocity:

$$V_\theta = S_f r_m \Omega = 0.5 \times 0.5 \times \left(\frac{3000 \times \pi}{30}\right) = 78.540 \text{ m/s}$$

Relative total temperature at the first tooth inlet:

$$T_{t_{x_1}} = 500 - \frac{78.540 \times 78.540}{2 \times 1004.5} = 496.930 \text{ K}$$

Relative total pressure at the first tooth inlet:

$$\frac{P_{t_{in}}}{P_{t_{x_1}}} = \left(\frac{T_{t_{in}}}{T_{t_{x_1}}}\right)^{\frac{\kappa}{\kappa-1}} = \left(\frac{500}{496.30}\right)^{3.5} = 1.02179$$

$$P_{t_{x_1}} = \frac{P_{t_{in}}}{1.02179} = \frac{10 \times 10^5}{1.02179} = 9.7867 \times 10^5 \text{ Pa}$$

(a)   Leakage air mass flow rate through the seal
     Leakage flow area over each tooth: $A = 2\pi r_m s = 2 \times \pi \times 0.5 \times \left(\frac{3}{1000}\right)$
     $= 0.009425 \text{ m}^2$

Tooth 1: $\dot{m}_1 = \dfrac{A C_d \hat{F}_{f_{t_1}} P_{t_{x_1}}}{\sqrt{RT_{t_{x_1}}}}$

Tooth 2: $\dot{m}_2 = \dfrac{A C_d \hat{F}_{f_{t_2}} P_{t_{x_2}}}{\sqrt{RT_{t_{x_2}}}} = \dfrac{A C_d \hat{F}_{f_{t_2}} P_{s_1}}{\sqrt{RT_{t_{x_1}}}}$

Note that in this problem we have $\dot{m}_1 = \dot{m}_2$, $T_{t_{x_1}} = T_{t_{x_2}}$, and $P_{t_{x_2}} = P_{s_1}$.

The leakage air mass flow rate can be computed using the following iterative method where the converged values are shown within parentheses:

1.   Assume M_1 (0.5496)

2.   Compute $\hat{F}_{f_{t_1}} = M_1 \sqrt{\dfrac{\kappa}{(1 + \frac{\kappa-1}{2}M_1^2)^{\frac{\kappa+1}{\kappa-1}}}}$ (0.5454)

3. Compute $\dot{m}_1 = \dfrac{AC_d \hat{F}_{f_{t_1}} P_{t_{x_1}}}{\sqrt{RT_{t_{x_1}}}}$ (10.657)

4. Compute $P_{s_1} = \dfrac{P_{t_{x_1}}}{\left(1 + \frac{\kappa-1}{2} M_1^2\right)^{\frac{\kappa}{\kappa-1}}}$ (7.9701×10^5)

5. Compute pressure ratio P_{s_1}/P_{s_2} and the Mach number M_2

$$M_2 = \left[\frac{2}{\kappa-1}\left\{\left(\frac{P_{s_1}}{P_{s_{out}}}\right)^{\frac{\kappa-1}{\kappa}} - 1\right\}\right]^{\frac{1}{2}}$$ (0.8441)

6. Compute $\hat{F}_{f_{t_2}} = M_2 \sqrt{\dfrac{\kappa}{\left(1 + \frac{\kappa-1}{2} M_2^2\right)^{\frac{\kappa+1}{\kappa-1}}}}$ (0.6697)

7. Compute $\dot{m}_2 = \dfrac{AC_d \hat{F}_{f_{t_2}} P_{s_1}}{\sqrt{RT_{t_{x_1}}}}$ (10.657)

8. Repeat steps from 1 to 7 until $\dot{m}_2 = \dot{m}_1$.

Thus, the seal leakage air mass flow rate computed in this case is 10.657 kg/s.

(b)   A noniterative method to compute $\dot{m}$ and $P_{t_{in}}$ for $M_2 = 1$ and $P_{s_{out}} = 5$ bar

When the leakage flow through the labyrinth seal is entirely subsonic, the exit static pressure must be equal to the specified static back pressure. When the last tooth chokes, the exit static pressure could be equal to or higher than the back pressure. In the present case, the inlet total pressure is increased with concurrent increase in the seal leakage air mass flow rate until $M_2 = 1$, and the exit static pressure equals the specified back pressure of $P_{s_{out}} = 5$ bar.

To solve for both the mass flow rate and total inlet pressure, we can certainly use an iterative method, similar to the one used in (a). We present here a simple direct solution method, which is based on backward marching from tooth 2 to tooth 1.

Tooth 2:

For $M_2 = 1$ at tooth 2, we compute the following quantities:

$$\frac{P_{s_1}}{P_{s_{out}}} = \left(\frac{\kappa+1}{2}\right)^{\frac{\kappa}{\kappa-1}} = (1.2)^{3.5} = 1.8929$$

which yields

$$P_{s_1} = \left(\frac{P_{s_1}}{P_{s_{out}}}\right) \times P_{s_{out}} = 1.8929 \times 5 \times 10^5 = 9.4646 \times 10^5$$

and

$$\hat{F}_{f_{s_2}} = M_2 \sqrt{\kappa\left(1 + \frac{\kappa-1}{2} M_2^2\right)} = \sqrt{1.4 \times \frac{1.4+1}{2}} = 1.2961$$

which directly yields the seal air leakage mass flow rate as

$$\dot{m} = \frac{AC_d \hat{F}_{f_{s_2}} P_{s_{out}}}{\sqrt{RT_{t_{x_1}}}} = \frac{0.009425 \times 0.8 \times 1.2961 \times 5 \times 10^5}{\sqrt{287 \times 496.930}} = 12.939 \text{ kg/s}$$

and

$$\hat{F}_{f_{t_2}} = \frac{\hat{F}_{f_{s_2}}}{\dfrac{P_{s_1}}{P_{s_{out}}}} = \frac{1.2961}{1.8929} = 0.6847$$

For directly computing the inlet total pressure at tooth 1, we make use of the following facts:

(1) Static pressure at tooth1 exit equals the total pressure at tooth 2 inlet ($\alpha = 0$)
(2) For constant total temperature and seal clearance (equal flow area under both tooth 1 and tooth 2), the static-pressure mass flow function of tooth 1 equals the total-pressure mass function of tooth 2.

Therefore, we obtain $\hat{F}_{f_{s_1}} = \hat{F}_{f_{t_2}} = 0.6847$, which directly yields M_1 as

$$M_1 = \sqrt{\frac{-\kappa + \sqrt{k^2 + 2\kappa(\kappa - 1)\hat{F}^2_{f_{s_1}}}}{\kappa(\kappa - 1)}}$$

$$= \sqrt{\frac{-1.4 + \sqrt{(1.4)^2 + 2 \times 1.4 \times (1.4 - 1) \times (0.6847)^2}}{1.4 \times (1.4 - 1)}}$$

$$M_1 = 0.5613$$

which from isentropic relations gives the total-to-static pressure ratio

$$\frac{P_{t_{x_1}}}{P_{s_1}} = \left(1 + \frac{\kappa - 1}{2}M_1^2\right)^{\frac{\kappa}{\kappa - 1}} = \left(1 + 0.2 \times (0.5613)^2\right)^{3.5} = 1.2385$$

and

$$P_{t_{x_1}} = 1.2385 \times 9.4646 \times 10^5 = 1.1722 \times 10^5 \text{ Pa}$$

which finally yields

$$P_{t_1} = P_{t_{x_1}} \times 1.02179 = 1.1722 \times 10^5 \times 1.02179$$

$$P_{t_1} = 1.1977 \times 10^5 \text{ Pa}$$

Therefore, for the choked second tooth, the direct solution method yields the leakage mass flow rate of 12.939 kg/s at the required inlet total pressure of 1.1977×10^5 Pa. It may be verified that, once the seal is choked, the leakage air mass rate varies linearly with the total inlet pressure while the total-to-static pressure ratio across the seal remains constant. The resulting exit static pressure, therefore, becomes higher than the specified back pressure. This behavior of the choked two-tooth labyrinth seal in this example has significant practical implication. For a given back pressure, we can first compute its leakage mass flow rate and total pressure by using the direct solution method and then scale the values to other inlet and outlet pressure conditions while the seal remains choked.

Example 5.2 For the two-tooth seal of Example 5.1, show that, in order for the leakage air flow to choke under both teeth, the clearance under the second tooth must be increased to satisfy the following relation:

$$\frac{s_2}{s_1} = \left(\frac{\kappa+1}{2}\right)^{\frac{\kappa}{\kappa-1}} = 1.8929$$

where s_1 is the seal clearance under tooth 1 and s_2 under tooth 2.

Solution

From the flow physics of the two-tooth seal of Problem 5.1, we can write $T_{t_{x_1}} = T_{t_{x_2}}$, $P_{t_{x_2}} = P_{s_1}$, and $\dot{m}_1 = \dot{m}_2$. We can express the choked mass flow rate under each tooth as follows:

$$\dot{m}_1 = \frac{A_1 C_d \hat{F}_{f_{s_1}} P_{s_1}}{\sqrt{RT_{t_{x_1}}}}$$

$$\dot{m}_2 = \frac{A_2 C_d \hat{F}_{f_{t_2}} P_{t_{x_2}}}{\sqrt{RT_{t_{x_2}}}} = \frac{A_2 C_d \hat{F}_{f_{t_2}} P_{s_1}}{\sqrt{RT_{t_{x_1}}}}$$

where A_1 and A_2 are seal clearances under teeth 1 and 2, respectively, and $\hat{F}_{f_{s_1}}$ is the static-pressure mass flow function evaluated at $M_1 = 1$ and $\hat{F}_{f_{t_2}}$ the total-pressure mass flow function evaluated at $M_2 = 1$.

Equating $\dot{m}_1$ and $\dot{m}_2$ and simplifying the resulting expression yields

$$A_1 \hat{F}_{f_{s_1}} = A_2 \hat{F}_{f_{t_2}}$$

which for air with $\kappa = 1.4$ results in the following relation:

$$\frac{A_2}{A_1} = \frac{s_2}{s_1} = \frac{\hat{F}_{f_{s_1}}}{\hat{F}_{f_{t_2}}} = \left(\frac{\kappa+1}{2}\right)^{\frac{\kappa}{\kappa-1}} = 1.8929$$

Example 5.3 During the air flow testing of a two-tooth labyrinth seal in a nonrotating test rig, the measurements made correspond to $P_{t_{in}} = 122500$ Pa, $P_{s_{out}} = 100000$ Pa, $T_{t_{in}} = 300$ K, and $\dot{m} = 0.5680$ kg/s. The mechanical clearance area under each tooth equals $A_1 = A_2 = 0.005$ m^2. Assuming the leakage flow to be adiabatic with a discharge coefficient $C_d = 0.65$ under each tooth, compute the kinetic energy carry-over factor α from the first tooth exit to the second tooth inlet.

Solution

Because the flow is adiabatic, with no rotational work transfer (windage) in the test rig, the total temperature remains constant from seal inlet to outlet. For the given seal leakage air mass flow rate, we can compute additional flow properties at tooth 1 and tooth 2 as follows:

Tooth 1

From the mass flow rate equation expressed in terms of total-pressure mass flow function, we can write

$$\hat{F}_{f_{t_1}} = \frac{\dot{m}_1 \sqrt{RT_{t_{in}}}}{A_1 C_d P_{t_{in}}} = \frac{0.5680 \times \sqrt{287 \times 300}}{0.005 \times 0.65 \times 122500} = 0.4186$$

which yields $M_1 = 0.3865$ using either an iterative method or table look-up with interpolation.

Static pressure at tooth 1 exit:

$$\frac{P_{t_{in}}}{P_{s_1}} = \left(1 + \frac{\kappa - 1}{2}M_1^2\right)^{\frac{\kappa}{\kappa - 1}} = \left(1 + 0.2 \times (0.3865)^2\right)^{3.5} = 1.1085$$

$$P_{s_1} = \frac{122500}{1.1085} = 110507 \text{ Pa}$$

Mach number and static-to-total temperature ratio at tooth 1 exit:

$$\hat{F}''_{f_{s_1}} = \frac{\dot{m}_1 \sqrt{RT_{t_{in}}}}{A_1 P_{s_1}} = \frac{0.5680 \times \sqrt{287 \times 300}}{0.005 \times 110507} = 0.3017$$

$$M''_1 = \sqrt{\frac{-\kappa + \sqrt{k^2 + 2\kappa(\kappa - 1)\hat{F}''^2_{f_{s_1}}}}{\kappa(\kappa - 1)}}$$

$$= \sqrt{\frac{-1.4 + \sqrt{(1.4)^2 + 2 \times 1.4 \times (1.4 - 1) \times (0.3017)^2}}{1.4 \times (1.4 - 1)}}$$

$$M''_1 = 0.2533$$

$$\frac{T_{t_{in}}}{T''_{s_1}} = 1 + \left(\frac{\kappa - 1}{2}\right)M''^2_1 = 1 + \left(\frac{1.4 - 1}{2}\right) \times (0.2533)^2 = 1.0128$$

$$\frac{T''_{s_1}}{T_{t_{in}}} = \frac{1}{1.0128} = 0.9873$$

Tooth 2

Using equal mass flow rate through each tooth and the given static pressure at tooth 2 exit, we compute the inlet total pressure P_{t_2} as follows:

$$\hat{F}_{f_{s_1}} = \frac{\dot{m}_2 \sqrt{RT_{t_{in}}}}{A_2 C_d P_{s_{out}}} = \frac{0.5680 \times \sqrt{287 \times 300}}{0.005 \times 0.65 \times 100000} = 0.5128$$

$$M_2 = \sqrt{\frac{-\kappa + \sqrt{k^2 + 2\kappa(\kappa - 1)\hat{F}_{f_{s_1}}^2}}{\kappa(\kappa - 1)}}$$

$$= \sqrt{\frac{-1.4 + \sqrt{(1.4)^2 + 2 \times 1.4 \times (1.4 - 1) \times (0.5128)^2}}{1.4 \times (1.4 - 1)}}$$

$$M_2 = 0.4258$$

$$\frac{P_{t_2}}{P_{s_2}} = \left(1 + \frac{\kappa - 1}{2} M_2^2\right)^{\frac{\kappa}{\kappa - 1}} = \left(1 + 0.2 \times (0.4258)^2\right)^{3.5} = 1.13276$$

$$P_{t_2} = 1.1328 \times P_{s_2} = 1.1328 \times 100000 = 113276 \text{ Pa}$$

$$\frac{P_{s_1}}{P_{t_2}} = \frac{110507}{113276} = 0.9755$$

The kinetic energy carry-over factor α can now be computed as

$$\alpha = \frac{1 - \left(\dfrac{P_{s_1}}{P_{t_2}}\right)^{\frac{\kappa - 1}{\kappa}}}{1 - \dfrac{T''_{s_1}}{T_{t_{in}}}} = \frac{1 - (0.9755)^{\frac{0.4}{1.4}}}{1 - 0.9873} = 0.556$$

Project

5.1   Develop a general purpose stand-alone computer code in FORTRAN, or the language of your choice, to solve for swirl distribution, total temperature change as a result of windage and heat transfer, and leakage mass flow rate through a general labyrinth seal. Use the tooth-by-tooth modeling scheme presented in this chapter along with an orifice-cavity model between two adjacent teeth. Include in this compute code the flexibility to input for each cavity a blocker flow and the associated loss in total pressure.

References

Aungier, R. H. 2000. *Centrifugal Compressors: A Strategy for Aerodynamic Design and Analysis*. New York: ASME Press.

Childs, D. W. 1993. *Turbomachinery Rotordynamics: Phenomena, Modeling and Analysis*. New York: John Wiley & Sons.

Chupp, R. E., R. C. Hendricks, S. B. Lattime, and B. M. Steinetz. 2006. *Sealing in Turbomachinery*. NASA/TM–2006–214341.

Chupp, R. E., R. C. Hendricks, S. B. Lattime, B. M. Steinetz, and M. F. Aksit. 2014. Turbomachinery clearance control. In T. I. -P. Shih and V. Yang, eds., *Turbine Aerodynamics, Heat Transfer, Materials, and Mechanics*. Reston, VA: AIAA.

Egli, A. 1935. The leakage of steam through labyrinth glands. *Trans. ASME*. 57: 115–122.

Hodkinson, B. 1939. Estimation of leakage through a labyrinth gland. *Proceedings of the Institute of Mechanical Engineers*. 141: 283–288.

McGreehan, W. F., and S. H. Ko.1989. Power dissipation in smooth and honeycomb labyrinth seals. ASME Paper No. 89-GT-220.

Martin, H. 1908. Labyrinth packings: Engineering. *January* 10. pp. 35–36.

Vermes, G. 1961. A fluid mechanics approach to the labyrinth seal leakage problem. *ASME J. Eng. Power*. 83: 161–169.

Zabriskie, W., and B. Sternlicht. 1959. Labyrinth seal leakage analysis. *J. Basic Eng*. 81: 332–340.

Zimmermann, H., and K. H. Wolff. 1998. Air system correlations: part 1- labyrinth seals. ASME Paper No. 98-GT-206.

Bibliography

Andrés, L. S., and Z. Ashton. 2010. Comparison of leakage performance in three types of gas annular seals operating at a high temperature (300°C). *Tribol. Trans*. 53(3): 463–471.

Andrés, L. S., and A. Anderson. 2015. An all-metal compliant seal versus a labyrinth seal: a comparison of gas leakage at high temperatures. *ASME J. Eng. Gas Turbines Power*. 137 (5): 052504.1– 052504.8.

Chen, Y., L. Zhigang, Y. Xin, and L. Jun. 2018. Effects of mushroom-shaped tooth wear on the leakage performance and rotordynamic coefficients of labyrinth seals. ASME Paper No. GT2018–75147.

Chupp, R. E., F. H. Glenn, and E. S. Thomas. 1986. Labyrinth seal analysis, Vol. IV – user's manual for the labyrinth seal design model. AFWAL-TR-85–2103.

Chupp, R. E., R. P. Johnson, and R. G. Loewenthal. 1995. Brush seal development for large industrial gas turbines. AIAA Paper No. 1995–3146.

Chupp, R. E., R. C. Hendricks, S. B. Lattime, and B. M. Steinetz. 2006. Sealing in turbomachinery. *AIAA J. Propulsion and Power*. 22(2): 313–349.

Collins, D. 2007. The effects of wear on abradable honeycomb labyrinth seals. EngD Thesis. Cranfield University.

El-Gamal, H. A., T. H. Awad, and E. Saber. 1996. Leakage from labyrinth seals under stationary and rotating conditions. *Tribology International*. 29(4): 291–297.

Eser, D., and J. Y. Kazakia. 1995. Air flow in cavities of labyrinth seals. *Int. J. Eng. Science*. 33 (15): 2309–2326.

Ferguson, J. G. 1988. Brushes as high performance gas turbine seals. ASME Paper No. 88-GT-182.

Justak, J. F., and P. F. Crudgington. 2006. Evaluation of a film riding hybrid seal. AIAA Paper No. 2006–4932.

Kim, T. S., and K. S. Cha. 2009. Comparative analysis of the influence of labyrinth seal configuration on leakage behavior. *J. Mech. Sci. Tech.* 23: 2830–2838.

Meyer, C. A., and J. A. Lowrie. 1975. The leakage thru straight and slant labyrinths and honeycomb seals. *J. Eng. Power.* 97(4): 495–501

Millward, J. A., and M. F. Edwards. 1994. Windage heating of air passing through labyrinth seals. ASME Paper No. 94-GT-56.

Stocker, H. L., D. M. Cox, and, G. F. Holle. 1977. Aerodynamic performance of conventional and advanced design labyrinth seals with solid-smooth, abradable, and honeycomb lands. NASA CR-135307.

Sultanian, B. K. 2015. *Fluid Mechanics: An Intermediate Approach.* Boca Raton, FL: Taylor & Francis.

Xia, Q., D. Gillespie, A. Owen, and G. Franceschini. 2008. Quasi-static thermal modelling of multi-scale sliding contact for unlubricated brush seal materials. ASME Paper No. GT2018–75920.

Yan, X., X. Dai, K. Zhang, J. Li, and K. He. 2018. Influence of hole-pattern stator on leakage performance of labyrinth seals. ASME Paper No. GT2018–75349.

Zhang, Y., J. Li, X. Yan, and Z. Li. 2018. Investigations on the leakage flow characteristics of the brush seal with high pressure-resistance. ASME Paper No. GT2018–76235.

Zuk, J. 1976. *Fundamentals of Fluid Sealing.* NASA TN-D8151, Lewis Research Center, Cleveland, OH.

Nomenclature

A	Mechanical area for seal flow
C_c	Seal carry-over coefficient
C_d	Discharge coefficient
C_r	Seal contraction coefficient
C_t	Seal throttling coefficient
C_V	Velocity coefficient
$\hat{F}_{f_s}$	Dimensionless static-pressure mass flow function
$\hat{F}_{f_t}$	Dimensionless total-pressure mass flow function
h	Tooth height
k	Carry-over coefficient
k_1	Correction factor
k_2	Corrected carry-over coefficient
$\dot{m}$	Mass flow rate
M	Mach number
n	Number of teeth
p	Pitch
P	Pressure
r	Coordinate r of cylindrical polar coordinate system; radius
R	Gas constant

R_{seal}	Mean radius of a labyrinth seal
s	Seal clearance
S_{f}	Swirl factor
t	Tooth thickness
T	Temperature
V	Total velocity
x	Cartesian coordinate x

Subscripts and Superscripts

(0)	Upstream of tooth number 1
HT	Heat transfer
m	Mean
(n)	Downstream of tooth number n
s	Static
t	Total (stagnation)
x	Axial direction

Greek Symbols

α	Kinetic energy carry-over factor
$\tilde{\alpha}$	Tooth slant angle from vertical
β	Gland factor
θ	Tangential direction
κ	Ratio of specific heats
ρ	Density
Ω	Angular velocity

6   Whole Engine Modeling

6.0   Introduction

Gas turbines in all the applications for aircraft propulsion, electric power generation, and mechanical drive have many of critical components that limit life. The performance of a gas turbine, therefore, must not only include its thermodynamic cycle efficiency but also the acceptable durability of its structural members. If the most efficient gas turbine blade fails in operation because of creep, oxidation, low cycle fatigue (LCF), or high cycle fatigue (HCF), the damage to the rest of engine could be catastrophic. For safe and reliable operation, the gas turbine component durability becomes critically important. Gas turbine manufacturers handle the durability problem by actively managing the life of critical components of all their engines through scheduled maintenance, refurbishment, and replacement. The life management department at these OEMs (original equipment manufacturers) is often as big a department as the one that performs the original design of these components. An important objective behind the life management of engine critical components is to explore their life extension to reduce the frequency of their replacement with significant cost saving. More recently, airlines enter into a "power by the hour" contract with OEMs. In this arrangement OEMs retain ownership of the engines and thus responsibility for them. They actively monitor engines and charge customers per hour of use. Additionally, they own and manage engine maintenance facilities and perform various engine maintenance and repair. Thus, the overall design and development of an advanced gas turbine calls for a balanced delivery of cycle efficiency, engine durability and reliability, and cost. Multiphysics-based accurate whole engine modeling (WEM) is foundational to both the initial design and operational life management of all gas turbines. The Internet of Things (IoT) revolution, feeding actual service data into the predictive models, help GT OEMs perform more reliable life assessments.

In the previous chapters our emphasis has been on accurate 1-D compressible flow modeling of various components of gas turbine internal flow systems. In this chapter we discuss the integration of the flow modeling with conduction heat transfer in the solid, including mechanical displacement analysis, to discern distributions of mechanical and thermal stresses under transient operations. These results become the basis of life assessment of gas turbine components. As discussed in this chapter, the transient multiphysics WEM is also used to feed engine design, e.g., setting of cold clearances for blade tips in the gas path and also for seals in the secondary air system.

6.1   Multiphysics Modeling of Engine Transients

Because the material yield strength is temperature dependent, both the temperatures and their gradients in the structure play the key role in determining its durability under a given mechanical loading, including centrifugal stresses as a result of rotation if present. Both the mechanical and thermal stress distributions in each component must be combined for identifying critical areas for the component life assessment. Typically, the maximum stress in the components of a gas turbine is found during an engine transient, not in its steady state operation. Transients for commercial aircraft engines, for example, arise from taxiing to take-off, take-off to cruise, and cruise to decent and landing. Military engines operate under harsher transients. For power generation gas turbines, transients include many cold and hot restarts not only to meet variable seasonal demand of a power grid but also to support the renewable energy contribution. When the sun does not shine or the wind stops blowing, the gas turbine helps to support the grid requirements. One way to avoid restarts is to "park" the engine overnight at a 20- to 30-percent load, typically called turndown.

The whole engine modeling and analysis becomes a necessity to realize improved component life prediction, better performance of various seals for higher engine efficiency through the reduction in cooling and sealing flows. Accurate prediction of metal temperatures throughout the engine transients requires the knowledge of local coolant air temperatures and heat transfer coefficients to simulate the convective boundary conditions for the conduction solution within the stator and rotor components. Although the coolant air properties are known at the point of compressor bleed, its conditions at any location can only be obtained through an accurate energy balance along its preceding internal flow path. Thus, the thermal analysis problem in gas turbines uniquely involves convective boundary conditions that themselves depend on the thermal solution elsewhere and cannot be specified a priori. Note that the gas turbine internal flow systems respond orders of magnitude faster (convection time constant) than the thermal response (diffusion time constant) of the structural members in contact.

The whole engine modeling broadly consists of two types of modeling: thermal modeling and thermomechanical modeling. The first leads to thermal FEA and the second to mechanical FEA. Both are carried out using one or more commercial and in-house computer codes. As shown in Figure 6.1, the results of thermal FEA form an essential input to the mechanical FEA. Both analyses are by and large linear in nature, and the accuracy of their grid-independent solutions largely depends upon the accuracy of their boundary conditions. Generally, for the stress analysis we need a model with an order of magnitude more finite elements than are needed in the model for the thermal FEA. Accordingly, transferring temperatures from the thermal FEA model to the mechanical FEA model requires mapping from a coarse mesh to a fine mesh. Nowadays, in a "pushbutton" design software, there are smoothing functions that help interpolate data from coarse to fine mesh.

The conduction solution to predict temperatures within the gas turbine structure requires accurate convective boundary conditions at a fluid-solid interface. The CFD modeling in principle is capable of simulating both the fluid and solid domain through a

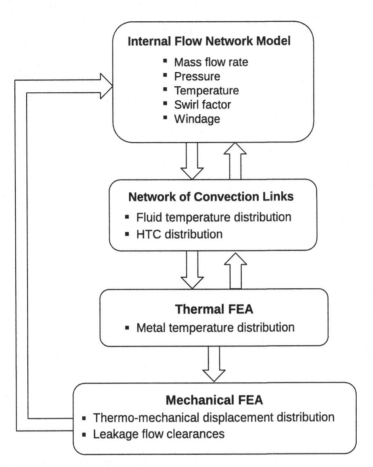

Figure 6.1 Multiphysics (aero-thermal-mechanical) whole engine modeling and analysis scheme.

conjugate heat transfer (CHT) modeling, thus eliminating the need for explicit specifications of convective boundary conditions in terms of heat transfer coefficients and reference fluid temperatures. For WEM, the use of CHT is generally not used as a result of the preponderance of the solid domain in the whole model. In addition, the heat transfer modeling with CFD at a fluid-solid interface becomes unreliable in the presence of rotation, which causes a highly nonisotropic turbulence field, and the variable turbulent Prandtl number in the thermal boundary layer. The current gas turbine design practice by and large involves iterative integration of separate solutions in the fluid and solid domains. Most of the uncertainty in prediction accuracy is associated with the fluid domain solution, which uses empirical correlations that are proprietary to each OEM and are fine-tuned based on extensive and continuous product validation studies and engine thermal surveys.

Figure 6.1 shows the multiphysics (aero-thermal-mechanical) modeling and analysis scheme used in gas turbine WEM. As discussed in the foregoing, the overall accuracy of results from WEM largely depends on the physics-based generation of the convective boundary conditions, which are specified in terms of fluid temperatures (adiabatic wall temperatures) and the associated heat transfer coefficients (HTC) for the thermal FEA. The flow network model (FNM) of cooling and sealing flows, which we discussed in Chapter 3, is the starting point for a WEM. This model yields the distributions of mass flow rate, pressure, temperature, swirl factor and windage (rotor work transfer). For this model to effectively communicate with the solid model for thermal FEA, we propose here a buffer model comprising the network of convection links, as shown in the figure. This buffer model offers two key advantages. First, it frees the internal flow network model from the energy equation allowing its robust and fast solution convergence. Second, it handles the steady flow energy equation directly in concert with the conduction solution in the solid domain. In this approach, we take advantage of the fact that the coupling between the internal FNM and the network of convection links is one-way stronger. This is because the mass flow rates computed in FNM significantly influence the fluid temperatures computed in the convection links, but these temperatures in turn have a weak influence on the computed mass flow rates. The coupling among various models shown in Figure 6.1 is handled through their iterative solutions, ensuring the continuity of heat flux across the entire fluid-solid interface.

Figure 6.1 further shows a one-way coupling between the thermal FEA and mechanical FEA. The results from the thermo-mechanical FEA in terms of changes in the seal clearances and leakage gaps are fed back to FNM, which in turn influences the thermal FEA, resulting in a fully integrated WEM methodology.

6.2   Nonlinear Convection Links

In this section we develop a general multisurface forced vortex convection link with windage, which forms the basis for the buffer model (network of convection links) shown in Figure 6.1.

6.2.1   Linear versus Nonlinear Convection Links

The primary function of each convection link is to compute from inlet to outlet the change in fluid total temperature resulting from both convective heat transfer and rotational work transfer (windage) from the linked solid surfaces. To facilitate computation of convective heat transfer from an assumed isothermal wall at temperature T_w with a constant heat transfer coefficient h, a linear variation of fluid total temperature is often assumed. Let us assume a recovery factor of unity, which makes the adiabatic wall temperature equal to the local fluid total temperature. Accordingly, for the convection link associated with the segment of a pipe flow shown in Figure 6.2a, we can write the net convection heat transfer rate as

$$\dot{Q}_c = hA_w \left(T_w - \frac{T_{t_{in}} + T_{t_{out}}}{2} \right) = \frac{hA_w}{2} \left\{ \left(T_w - T_{t_{in}} \right) + \left(T_w - T_{t_{out}} \right) \right\} \qquad (6.1)$$

where A_w is the total heat transfer surface area, which varies linearly over the pipe segment from its inlet to outlet.

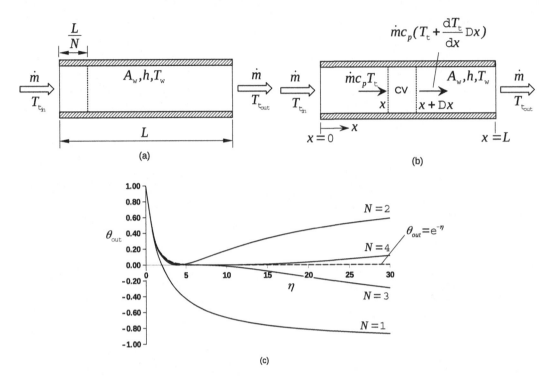

Figure 6.2 (a) Linear convection link, (b) nonlinear convection link, and (c) variation of fluid total temperature at outlet (θ_{out}) with η for linear and nonlinear convection links.

From the steady flow energy equation over this convection link we can write

$$\dot{m}c_p\left(T_{t_{out}} - T_{t_{in}}\right) = \dot{Q}_c = \frac{hA_w}{2}\left\{\left(T_w - T_{t_{in}}\right) + \left(T_w - T_{t_{out}}\right)\right\}$$

$$\left\{\left(T_w - T_{t_{in}}\right) - \left(T_w - T_{t_{out}}\right)\right\} = \frac{\eta}{2}\left\{\left(T_w - T_{t_{in}}\right) + \left(T_w - T_{t_{out}}\right)\right\}$$

$$1 - \theta_{out} = \frac{\eta}{2}\left(1 + \theta_{out}\right)$$

$$\theta_{out} = \frac{2 - \eta}{2 + \eta} \tag{6.2}$$

where

$$\eta = \frac{hA_w}{\dot{m}c_p} \text{ and } \theta_{out} = \frac{T_w - T_{t_{out}}}{T_w - T_{t_{in}}}$$

Note that in the heat exchanger literature η is known as the number of transfer units (NTU).

Equation 6.2 shows that for $\eta > 2$, θ_{out} becomes negative, which means that the fluid outlet total temperature is higher than the wall temperature. This is in violation of the second law of thermodynamics and physically unrealizable. How can a wall heat the fluid to exceed its own temperature? One often tries to mitigate this numerical anomaly by dividing the convection link into smaller links. When we divide the initial link into N

equal sublinks, we can write the new solution with linear variation in the fluid total temperature in each sublink as

$$\theta_{\text{out}} = \left(\frac{2 - \dfrac{\eta}{N}}{2 + \dfrac{\eta}{N}} \right)^N \tag{6.3}$$

which is plotted in Figure 6.1c. As can be seen from this figure, Equation 6.3 does not exhibit a physically acceptable behavior for all values of η.

For the derivation of a nonlinear convection link in this case, we use a small control volume (CV) shown in Figure 6.2b. The steady flow energy balance over this CV yields

$$\dot{m}c_p \left(T_t + \frac{dT_t}{dx}\Delta x \right) - \dot{m}c_p T_t = h\frac{A_w}{L}(T_w - T_t)\Delta x$$

$$\frac{dT_t}{d\xi} = \eta(T_w - T_t) \tag{6.4}$$

where $\xi = x/L$. Using the dimensionless temperature $\theta = (T_w - T_t)\big/\left(T_w - T_{t_{\text{in}}}\right)$, we can rewrite Equation 6.4 as

$$\frac{d\theta}{d\xi} + \eta\theta = 0 \tag{6.5}$$

which with the boundary condition $\theta = 1$ at $\xi = 0$ yields the classical solution

$$\theta = e^{-\eta\xi} \tag{6.6}$$

which, for $\xi = 1$, yields $\theta_{\text{out}} = e^{-\eta}$ shown plotted in Figure 6.2c. This solution yields physically realistic values of θ_{out} for all values of $0 \leq \eta \leq \infty$ and is independent of the convection link length as long as the heat transfer coefficient and wall temperature remain constant over the link. Note that the closed formed analytical solution given by Equation 6.6 shows that the fluid total temperature in an isothermal duct with constant heat transfer coefficient varies nonlinearly such that the difference between the wall temperature and fluid total temperature decays exponentially from inlet to outlet with the exponent η, which equals the number of transfer units (NTU). For the situation depicted in Figure 6.2a, one must therefore use the nonlinear convection link with the solution $\theta_{\text{out}} = e^{-\eta}$, which ensures both accuracy and robustness in the related numerical calculations.

6.2.2 Nonlinear Convection Link in a Multisided Duct

In many design applications, the convective heat transfer occurs in a multisided duct where the wall traversed by each side in the flow direction exhibits its own constant wall temperature and heat transfer coefficient. One such example is the heat transfer in the serpentine passage used for internal cooling of turbine airfoils. In the following, we develop solutions for the fluid outlet total temperature of a nonlinear convection link in a multisided duct with and without internal heat generation.

6.2.2.1   Without Internal Heat Generation

For the multisided duct shown in Figure 6.3a, let us first consider the nonlinear convection link without internal heat generation. Each wall of this duct is isothermal at a known wall temperature and has a constant heat transfer coefficient. The wall heat transfer area varies linearly along the duct. As shown in the figure, for the duct wall i, the heat transfer area, the wall temperature, and the heat transfer coefficient are denoted by A_{w_i}, T_{w_i}, and h_i, respectively. We can write the steady flow energy equation for a small control volume (CV) bounded by N distinct walls as

$$\dot{m}c_p\left(T_t + \frac{dT_t}{dx}\Delta x\right) - \dot{m}c_p T_t = \sum_{i=1}^{i=N} h_i \frac{A_{w_i}}{L}\left(T_{w_i} - T_t\right)\Delta x$$

$$\dot{m}c_p\frac{dT_t}{d\xi} = \sum_{i=1}^{i=N} h_i A_{w_i} T_{w_i} - \sum_{i=1}^{i=N} h_i A_{w_i} T_t$$

which further simplifies to

$$\frac{d\theta}{d\xi} + \overline{\eta}\theta \tag{6.7}$$

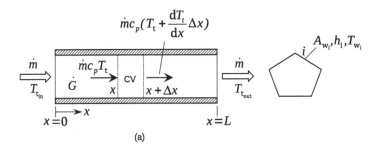

(a)

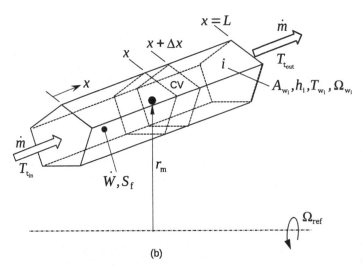

(b)

Figure 6.3 (a) Multisided duct with internal heat generation and (b) multisurface forced vortex convection link with windage.

where

$$\bar{\eta} = \frac{\sum\limits_{i=1}^{i=N} h_i A_{w_i}}{\dot{m}c_p} \qquad \bar{T}_w = \frac{\sum\limits_{i=1}^{i=N} h_i A_{w_i} T_{w_i}}{\sum\limits_{i=1}^{i=N} h_i A_{w_i}} \qquad \theta = \frac{\bar{T}_w - T_t}{\bar{T}_w - T_{t_{in}}}$$

Equation 6.7 has the same form as Equation 6.5. With the boundary condition $\theta = 1$ at $\xi = 0$, we can write its solution as

$$\theta = e^{-\bar{\eta}\xi} \tag{6.8}$$

which yields $\theta_{out} = e^{-\bar{\eta}}$ at $\xi = 1$.

6.2.2.2 With Uniform Internal Heat Generation

When we include the uniform heat generation term $\dot{G}$ in the steady flow energy balance over the small control volume shown in Figure 6.3a, we obtain

$$\dot{m}c_p\left(T_t + \frac{dT_t}{dx}\Delta x\right) - \dot{m}c_p T_t = \sum_{i=1}^{i=N} h_i \frac{A_{w_i}}{L}\left(T_{w_i} - T_t\right)\Delta x + \dot{G}\frac{\Delta x}{L}$$

$$\dot{m}c_p \frac{dT_t}{d\xi} = \sum_{i=1}^{i=N} h_i A_{w_i} T_{w_i} - \sum_{i=1}^{i=N} h_i A_{w_i} T_t + \dot{G}$$

$$\frac{d\theta}{d\xi} + \bar{\eta}\theta = -\tilde{\varepsilon} \tag{6.9}$$

where

$$\tilde{\varepsilon} = \frac{\dot{G}}{\dot{m}c_p\left(T_w - T_{t_{in}}\right)}$$

Equation 6.9 is a first-order linear nonhomogeneous differential equation with its solution given by

$$\theta = C\theta_h + \theta_P$$

where θ_h is the solution of the homogeneous part given by $\theta_h = e^{-\bar{\eta}\xi}$, θ_P is a particular solution given by $\theta_P = \tilde{\varepsilon}/\bar{\eta}$, and C is a constant to be determined by the boundary condition $\theta = 1$ at $\xi = 0$. Finally, we obtain the solution of Equation 6.9 as

$$\theta = \left(1 + \frac{\tilde{\varepsilon}}{\bar{\eta}}\right)e^{-\bar{\eta}\xi} - \frac{\tilde{\varepsilon}}{\bar{\eta}} \tag{6.10}$$

Note that for $\tilde{\varepsilon} = 0$, which represents the case with no internal heat generation, Equation 6.10 reduces to Equation 6.8. Equation 6.10 for the nonlinear convection link yields the temperature at the outlet ($\xi = 1$) as

$$\theta_{out} = \left(1 + \frac{\tilde{\varepsilon}}{\bar{\eta}}\right)e^{-\bar{\eta}} - \frac{\tilde{\varepsilon}}{\bar{\eta}} \tag{6.11}$$

6.2.3   Multisurface Forced Vortex Convection Link with Windage

Let us now obtain the closed-form analytical solution for a general convection link as a forced vortex with convective heat transfer from solid surfaces with arbitrary rotation. This link also features windage as a result of work transfer from the associated rotor surfaces. Such a convection link of length L is schematically shown in Figure 6.3b. In the current formulation, we assume that each wall area varies linearly along the convection length. In addition, the wall temperature, the heat transfer coefficient, and the angular velocity for each wall are constant. The windage energy generation $\dot{W}$ is uniformly distributed over the convection link.

If, for example, a forced vortex flow with the swirl factor S_f passes over two surfaces, one rotating at Ω_{ref} and the other stationary, except for $S_f = 0.5$, the adiabatic wall temperature for each surface will be different for the same fluid total temperature. Assuming the predominance of the swirl velocity for the convection link, we will use the recovery factor of $Pr^{1/3}$, which has been empirically established for a turbulent boundary layer, to calculate the adiabatic wall temperature for each surface. The methodology to compute swirl factor distribution along with windage in a general rotor-stator cavity is presented in Chapter 4.

For the small control volume shown in Figure 6.3b, we can write the steady flow energy equation as

$$\dot{m}c_p\left(T_t + \frac{dT_t}{dx}\Delta x\right) - \dot{m}c_p T_t = \sum_{i=1}^{i=N} h_i \frac{A_{w_i}}{L}\left(T_{w_i} - T_t\right)\Delta x + \dot{W}\frac{\Delta x}{L}$$

$$\frac{dT_t}{d\xi} = \overline{\eta}\overline{T}_w - \sum_{i=1}^{i=N}\frac{h_i A_{w_i} T_{aw_i}}{\dot{m}c_p} + \frac{\dot{W}}{\dot{m}c_p} \qquad (6.12)$$

In Equation 6.12, we can express the adiabatic wall temperature T_{aw_i} for surface i as

$$T_{aw_i} = T_t - \frac{r_m^2 S_f^2 \Omega_{ref}^2}{2c_p} + \frac{r_m^2 Pr^{1/3}}{2c_p}\left(S_f\Omega_{ref} - \Omega_{w_i}\right)^2$$

$$T_{aw_i} = T_t - \frac{r_m^2 S_f^2 \Omega_{ref}^2}{2c_p} + \frac{r_m^2 Pr^{1/3}}{2c_p}\left(S_f^2\Omega_{ref}^2 - 2S_f\Omega_{ref}\Omega_{w_i} + \Omega_{w_i}^2\right)$$

$$T_{aw_i} = T_t - \frac{r_m^2\left(1 - Pr^{1/3}\right)S_f^2\Omega_{ref}^2}{2c_p} - \frac{r_m^2 Pr^{1/3}S_f\Omega_{ref}\Omega_{w_i}}{c_p} + \frac{r_m^2 Pr^{1/3}\Omega_{w_i}^2}{2c_p} \qquad (6.13)$$

Substituting T_{aw_i} from Equation 6.13 into Equation 6.12 and simplifying the resulting expression yields

$$\frac{d\theta}{d\xi} + \overline{\eta}\theta = -\hat{\varepsilon} \qquad (6.14)$$

where

$$\hat{\varepsilon} = \frac{\overline{\eta} r_m^2\left(1 - Pr^{1/3}\right)S_f^2\Omega_{ref}^2}{2c_p\left(\overline{T}_w - T_{t_{in}}\right)} + \frac{r_m^2 Pr^{1/3}S_f\Omega_{ref}\overline{\eta\Omega_w}}{c_p\left(\overline{T}_w - T_{t_{in}}\right)} - \frac{r_m^2 Pr^{1/3}\overline{\eta\Omega_w^2}}{2c_p\left(\overline{T}_w - T_{t_{in}}\right)} + \frac{\dot{W}}{\dot{m}c_p\left(\overline{T}_w - T_{t_{in}}\right)}$$

$$\overline{\Omega_{w}} = \frac{\sum\limits_{i=1}^{i=N} h_i A_{w_i} \Omega_{w_i}}{\sum\limits_{i=1}^{i=N} h_i A_{w_i}}$$

$$\overline{\Omega_{w}^{2}} = \frac{\sum\limits_{i=1}^{i=N} h_i A_{w_i} \Omega_{w_i}^{2}}{\sum\limits_{i=1}^{i=N} h_i A_{w_i}}$$

Because the linear nonhomogeneous ordinary differential equation given by Equation 6.14 is of the same form as Equation 6.9, we can write its solution under the boundary condition $\tilde{\theta} = 1$ at $\xi = 0$ as

$$\theta = \left(1 + \frac{\hat{\varepsilon}}{\overline{\eta}}\right) e^{-\overline{\eta}\xi} - \frac{\hat{\varepsilon}}{\overline{\eta}} \tag{6.15}$$

which yields the dimensionless fluid total temperature of the convection link at its outlet $(\xi = 0)$

$$\theta_{out} = \left(1 + \frac{\hat{\varepsilon}}{\overline{\eta}}\right) e^{-\overline{\eta}} - \frac{\hat{\varepsilon}}{\overline{\eta}}$$

In terms of the physical variables, we thus obtain the following equation to compute the total fluid temperature at the convection link outlet:

$$T_{t_{out}} = \overline{T}_{w} - \left(\overline{T}_{w} - T_{t_{in}}\right) e^{-\overline{\eta}} + \left(1 - e^{-\overline{\eta}}\right) \tilde{E} \tag{6.16}$$

where

$$\tilde{E} = \frac{r_{m}^{2}}{2c_{p}} \left[\left(1 - Pr^{1/3}\right) S_{f}^{2} \Omega_{ref}^{2} + 2Pr^{1/3} S_{f} \Omega_{ref} \overline{\Omega_{w}} - Pr^{1/3} \overline{\Omega_{w}^{2}}\right] + \frac{\dot{W}}{\sum\limits_{i=1}^{i=N} h_i A_{w_i}}$$

The convection link associated with a forced vortex flow and represented by Equation 6.16 computes convective heat transfer using the difference between the wall (surface) temperature and adiabatic wall temperature (fluid recovery temperature at the wall), which accounts for the relative angular velocity between the fluid and each bounding surface, be it co-rotating, counter-rotating, or stationary. Further note that this equation yields $T_{t_{out}} \approx T_{t_{in}}$ when $\overline{\eta} \approx 0$ (large mass flow rate) and $T_{t_{out}} \rightarrow \left(\overline{T}_{w} + \tilde{E}\right)$ as $\overline{\eta} \rightarrow \infty$ (nearly zero mass flow rate).

According to Equation 6.16, the accuracy of fluid temperatures computed in the network of convection links largely depends on $\overline{\eta}$, which in turn depends on the mass flow rate in each convection link and the associated heat transfer coefficient, which is mostly obtained from one or more empirical correlations. Note that the mass flow rate in each convection link is an iterative input from the internal flow network, as shown in Figure 6.1.

6.2.4   Junction of Convection Links

The network of convection links features junctions. Each junction is connected to the outlet of one or more convection links of incoming flows and to the inlet of one or more convection links of outgoing flows. In the steady state, the sum of all mass flow rates entering the junction must be equal to the sum of all mass flow rates leaving the junction. Under adiabatic conditions with zero torque associated with the junction, we can use the following equation to compute the junction total temperature as the mixed mean total temperature of k convection links of incoming flows, denoted by i, which varies from $i = 1$ to $i = k$

$$T_{t_j} = \frac{\sum\limits_{i=1}^{i=k} \dot{m}_i T_{t_{out_i}}}{\sum\limits_{i=1}^{i=k} \dot{m}_i} \tag{6.17}$$

Similarly, the mixed mean angular momentum of k convection links whose flows enter the junction yields the fractional swirl velocity of the junction as

$$S_{f_j} = \frac{\sum\limits_{i=1}^{i=k} \dot{m}_i S_{f_i}}{\sum\limits_{i=1}^{i=k} \dot{m}_i} \tag{6.18}$$

The junction properties computed by Equations 6.17 and 6.18 determine how the total temperature at the inlet of each convection link, denoted by j, with the flow leaving the junction J needs to be evaluated. If the convection link with the outgoing flow is associated with a cavity of rotor and stator surfaces, where we perform a separate set of calculations to compute the force vortex swirl factor S_{f_j} and the windage power $\dot{W}_j$, we simply use S_{f_j} and T_{t_j} as its inlet conditions. If, by contrast, the outgoing convection link is a forced vortex with the specified swirl factor S_{f_j}, its inlet total temperature changes as a result of the energy transfer in the process of the junction fluid getting aboard this convection link. Using the Euler's turbomachinery equation for this energy transfer, we can write

$$T_{t_{j_{in}}} = T_{t_j} + \frac{r_j^2 \Omega_{ref}^2 \left(S_{f_j} - S_{f_j}\right) S_{f_j}}{c_p} \tag{6.19}$$

6.3   Role of Computational Fluid Dynamics (CFD)

Computational fluid dynamics (CFD) provides numerical prediction of the distributions of velocity, pressure, temperature, and other relevant properties throughout the calculation domain. Leveraging concurrent advances in computing power and technology, CFD has made impressive strides since its infancy in the 1960s and remains the only method for computing time-dependent three-dimensional flows and heat transfer. Gas turbine internal flows are essentially turbulent and dominated by rotation, which results in complex shear layers with a highly nonisotropic turbulence field. Even with the highest quality grid and the most accurate numerical solution, the accuracy of CFD results depends on the predictive capability of the selected turbulence model, which continues to be the major weakness of the state-of-the-art CFD technology. Even today, the following assessment of Peter Bradshaw (1997) on the progress in turbulence modeling holds true: "There are signs of real progress in modeling, but industry still awaits a model that is both reliable and cheap."

The governing conservation equations of mass, momentum, and energy used in CFD are universally applicable and, therefore, need no further validation. The same, however, is not true about the turbulence models, whose development is based on several modeling approximations, including the use of measurements from simple shear flows (equilibrium wall boundary layers and free shear flows). A comprehensive application-specific validation of CFD is the key to its successful design application. This undisputed proposal, however, faces three major challenges. First, there is general lack of benchmark quality data representative of the state-of-the-art gas turbine internal flow systems. Second, most of currently available data sets are missing the required measurements at the inflow, outflow, and wall boundaries to enable a meaningful CFD analysis and detailed comparison. Third, some of the experimental measurements, which are regarded as the ultimate truth, may contain substantial errors. For example, Sultanian, Neitzel, and Metzger (1987) ended up invalidating the detailed LDV measurements published in the AIAA journal. These comments are intended to emphasize that good experiments, like good CFD, are hard to do and find in the open literature. Both have their errors and their problems. Nevertheless, CFD technology still offers the best available analysis method to fully understand the existing design within the available database by way of validation, and then provide its meaningful exploration in new operating regimes.

In a modern gas turbine design, the CFD technology plays multiple key roles for the modeling of internal flow systems. First, the CFD-based numerical flow visualization helps to understand the key features of the flow field, which is not obviously discernible. This understanding guides the development of a reduced-order model such as a flow network model. Second, where the empirical flow data are unavailable, CFD provides a quantitative assessment of the friction factor, discharge coefficient, and loss coefficient, which then become input to the reduced-order model. The accuracy of these numerical data depends on the capability of the turbulence model to accurately predict the entire flow field, both near-wall and far-wall. Third, CFD helps to generate heat transfer coefficients whose accuracy depends in large part on the accurate near-wall turbulence modeling for heat transfer. This perhaps is the most desirable but least reliable role of CFD in the design of internal flow systems. Additional roles of CFD include predicting circumferential pressure asymmetry in the gas path as an input to the ingestion model discussed in Chapter 4, ranking different SAS schemes or designs on the basis of their relative performance, and guiding the instrumentation definition to probe the test at the right places.

The greatest strength of CFD technology is to provide quick and inexpensive flow visualization, which is often better than an unrealistic guess. Just knowing the flow behavior in a design can be extremely valuable to perform a reduced-order analysis with higher speed and direct use of adjusted empirical correlations to yield results in better agreement with the test results.

6.3.1   Radially Inward Flow in a Compressor Rotor Drum Cavity

In the way of an example, let us consider an axial-flow compressor rotor drum cavity with radially inward cooling flow, which is bled from the primary flow path, as shown schematically in Figure 6.4. This cooling flow, which is intended for cooling turbine parts operating at higher temperatures, has a mass flow rate of $\dot{m}$ and passes through the bore of the right rotor disc. The quantities of design interest are the drop in static pressure from Point A to Point B and swirl factor of the cooling flow at Point B. Higher the swirl velocity at B, higher the strength of the "vortex whistle," which must be avoided in a good design. The static pressure at Point B should be large enough to prevent any backflow from the downstream turbine flow path.

One may intuitively expect the cooling flow in the rotor cavity of Figure 6.4 to flow from point A to Point B as a complex shear flow dominated by a free vortex. A CFD analysis of the flow, however, shows an unexpectedly interesting and simple behavior. The coolant essentially flows along two paths hugging the rotor walls, one along ACB and another along ADB. The total mass flow rate $\dot{m}$ is split at Point A into $\dot{m}_1$ and $\dot{m}_2$ flowing along these paths and joining together at point B. One finds from the CFD results that there is little flow activity in the core of this rotor drum cavity. This understanding of the flow field obtained from CFD can be used to alternatively model the flow using two one-dimensional control volume analyses, one along ACB and the other along ADB, with the common static pressure boundary conditions at Points A and B. In fact, such a constraint can be used to establish the flow split along two paths, obtaining mixed-mean swirl and total temperature at Point B. This example clearly demonstrates a novel application of CFD in design that potentially leads to a reduced-order model that promises higher speed and accuracy.

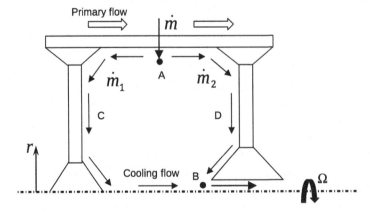

Figure 6.4 Radially inward cooling flow through a compressor rotor drum cavity.

6.4 CFD Methodology

In this section, we present a brief overview of the CFD methodology, including the governing conservation equations, which are nonlinearly-coupled partial differential equations. The overview also includes widely used turbulence models and physics-based postprocessing of 3-D CFD results for their meaningful interpretation, design applications, and comparison with the results obtained from a reduced-order model. Detailed reviews of turbulent flow simulation using direct numerical simulation (DNS) and large eddy simulation (LES) may be found in the literature listed at the end of this chapter.

As shown in Figure 6.5, the flow and heat transfer physics of design is mathematically modeled using the conservation equations of mass, momentum, and energy. For the applicable boundary conditions relevant to the design physics, these equations are numerically solved by the CFD methods. The numerical results are then postprocessed, properly interpreted, and used to make necessary changes in design for better performance.

Until the mid-1980s, most industries were engaged in the development and maintenance of their own in-house CFD computer codes. Universities were also engaged in developing their own CFD codes for their CFD-related research activities, including more accurate and robust numerical methods and improved turbulence models. With the emergence of several general purpose commercial codes, most CFD development activities in the industries and universities have dramatically subsided over the next thirty years.

When one uses a leading commercial CFD code, the task of CFD investigation of an industrial design becomes routine, as shown in Figure 6.6. A CFD analyst essentially focuses on generating a high-quality grid in the calculation domain, setting up the right boundary conditions, and selecting the appropriate turbulence model, which is consistent with the design physics. The commercial CFD code used by the analyst takes care of the rest. The postprocessing of CFD results for their design applications remains a major critical task.

For details on numerical aspects of CFD, including derivations of the discretization equations and their iterative solution methods, interested engineers may want to study the work by Patankar (1980) and Pletcher, Tannehill, and Anderson (2012). Thompson, Soni, and Weatherill (1998) provide an excellent coverage of the CFD grid generation

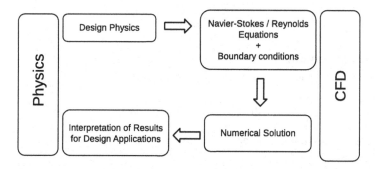

Figure 6.5 CFD methodology.

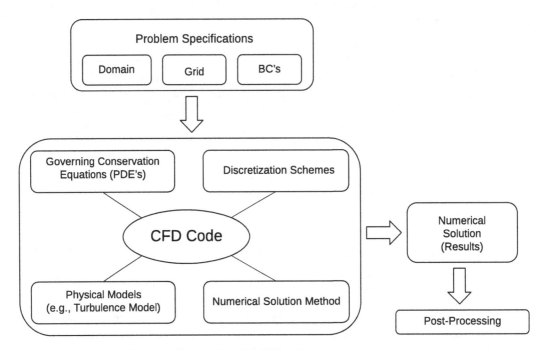

Figure 6.6 Solution with a commercial CFD code.

technology. One may find comprehensive details of turbulence modeling in Wilcox (1993) and Leschziner (2016). Durbin and Shih (2005) present a state-of-the-art review of turbulence modeling.

6.4.1   Governing Conservation Equations

Continuity equation:
Equation in vector form:

$$\frac{\partial \rho}{\partial t} + \text{div}\left(\rho \vec{V}\right) = 0 \tag{6.20}$$

Equation in tensor notation:

$$\frac{\partial \rho}{\partial t} + \frac{\partial}{\partial x_j}\left(\rho U_j\right) = 0 \tag{6.21}$$

Momentum equations:
The x-component momentum equation in vector form:

$$\frac{\partial}{\partial t}\left(\rho U\right) + \text{div}\left(\rho \vec{V} U\right) = \text{div}\left(\mu \, \text{grad} \, U\right) - \frac{\partial P_s}{\partial x} + \tilde{S}_U \tag{6.22}$$

The y-component momentum equation in vector form:

$$\frac{\partial}{\partial t}\left(\rho V\right) + \text{div}\left(\rho \vec{V} V\right) = \text{div}\left(\mu \, \text{grad} \, V\right) - \frac{\partial P_s}{\partial y} + \tilde{S}_V \tag{6.23}$$

The z-component momentum equation in vector form:

$$\frac{\partial}{\partial t}(\rho W) + \text{div}\left(\rho \vec{V} W\right) = \text{div}(\mu \text{ grad } W) - \frac{\partial P_s}{\partial z} + \tilde{S}_W \tag{6.24}$$

Momentum equations in tensor notation:

$$\frac{\partial}{\partial t}(\rho U_i) + \frac{\partial}{\partial x_j}\left(\rho U_j U_i\right) = \frac{\partial}{\partial x_j}\left(\mu \frac{\partial U_i}{\partial x_j}\right) + S_{U_i} \tag{6.25}$$

Energy equation:

In terms of static enthalpy (h_s) in vector form:

$$\frac{\partial}{\partial t}(\rho h_s) + \text{div}\left(\rho \vec{V} h_s\right) = \text{div}\left(\tilde{k} \text{ grad } T_s\right) + S_h \tag{6.26}$$

Using $h_s = c_p T_s$ to replace T_s on the right-hand side of Equation (6.26) yields

$$\frac{\partial}{\partial t}(\rho h_s) + \text{div}\left(\rho \vec{V} h_s\right) = \text{div}\left(\frac{\tilde{k}}{c_p} \text{ grad } h_s\right) + S_h \tag{6.27}$$

Equation 6.27 in tensor notation becomes

$$\frac{\partial}{\partial t}(\rho h_s) + \frac{\partial}{\partial x_j}\left(\rho U_j h_s\right) = \frac{\partial}{\partial x_j}\left(\frac{\tilde{k}}{c_p} \frac{\partial h_s}{\partial x_j}\right) + S_h \tag{6.28}$$

Substituting $h_s = c_p T_s$ in Equation 6.27 yields the energy equation in vector form in terms of the static temperature as follows:

$$\frac{\partial}{\partial t}(\rho T_s) + \text{div}\left(\rho \vec{V} T_s\right) = \text{div}\left(\frac{\tilde{k}}{c_p} \text{ grad } T_s\right) + \frac{S_h}{c_p} \tag{6.29}$$

which in tensor notation becomes

$$\frac{\partial}{\partial t}(\rho T_s) + \frac{\partial}{\partial x_j}\left(\rho U_j T_s\right) = \frac{\partial}{\partial x_j}\left(\frac{\tilde{k}}{c_p} \frac{\partial T_s}{\partial x_j}\right) + \frac{S_h}{c_p} \tag{6.30}$$

Chemical species conservation equation:

In vector form:

$$\frac{\partial}{\partial t}(\rho m_\ell) + \text{div}\left(\rho \vec{V} m_\ell\right) = \text{div}\left(\Gamma_\ell \text{ grad } m_\ell\right) + S_\ell \tag{6.31}$$

where

$m_\ell \equiv$ Mass fraction of chemical species ℓ

$\Gamma_\ell \equiv$ Diffusion coefficient for chemical species ℓ

$_\ell \equiv$ Generation rate per unit volume (from chemical reaction)

In tensor notation:

$$\frac{\partial}{\partial t}\left(\rho m_\ell\right) + \frac{\partial}{\partial x_j}\left(\rho U_j m_\ell\right) = \frac{\partial}{\partial x_j}\left(\Gamma_\ell \frac{\partial m_\ell}{\partial x_j}\right) + S_{m_\ell} \tag{6.32}$$

The common equation form:

In tensor notations, the common form of the conservation equations suitable for a common numerical solution method can be expressed as

$$\frac{\partial}{\partial t}\left(\rho \Phi\right) + \frac{\partial}{\partial x_j}\left(\rho U_j \Phi\right) = \frac{\partial}{\partial x_j}\left(\Gamma_\Phi \frac{\partial \Phi}{\partial x_j}\right) + S_\Phi \tag{6.33}$$

where $\Phi = 1$, U_i, h_s, and m_ℓ, yield, respectively, Equations 6.21, 6.25, 6.28, and 6.32. In Equation 6.33, each term is interpreted as follows:

$$\frac{\partial}{\partial t}\left(\rho \Phi\right) \equiv \text{Transient term}$$

$$\frac{\partial}{\partial x_j}\left(\rho U_j \Phi\right) \equiv \text{Convection term}$$

$$\frac{\partial}{\partial x_j}\left(\Gamma_\Phi \frac{\partial \Phi}{\partial x_j}\right) \equiv \text{Diffusion term}$$

$$S_\Phi \equiv \text{Source term}$$

Note that Equation 6.33 is valid whether the flow is laminar or turbulent, incompressible or compressible, or steady or unsteady.

6.4.2 Turbulence Modeling

There are two ways of numerically predicting turbulent flows, one by simulation; either direct numerical simulation (DNS) or large eddy simulation (LES), and another by statistical modeling. The first is too expensive and time-consuming for most practical designs, and the second is not universally accurate and reliable. We thus remain trapped between the proverbial rock and a hard place.

For statistical modeling of a turbulent flow, we decompose all its randomly varying properties into their statistically average values and their fluctuating parts. One such decomposition for $U = \overline{U} + u(t)$ is shown in Figure 6.7. In the top velocity plot the mean velocity $\overline{U}$ is time-independent, and the flow is considered to be stationary in the mean. In the bottom plot, the mean velocity obtained from ensemble averaging is varying with time.

6.4.2.1 Reynolds Equations: The Closure Problem

Reynolds averaging. Let us decompose $U_i(t)$, $P_s(t)$, and the general flow property $\Phi(t)$, respectively, as $U_i = \overline{U}_i + u_i$, $P_s = \overline{P}_s + p_s$, $\Phi = \overline{\Phi} + \varphi$. The time-averaging (also called Reynolds averaging) of these quantities yield their mean values as

$$\overline{U}_i = \frac{1}{t_2 - t_1}\int_{t_1}^{t_2} U_i \mathrm{d}t; \quad \overline{P}_s = \frac{1}{t_2 - t_1}\int_{t_1}^{t_2} P_s \,\mathrm{d}t; \quad \overline{\Phi} = \frac{1}{t_2 - t_1}\int_{t_1}^{t_2} \Phi \mathrm{d}t$$

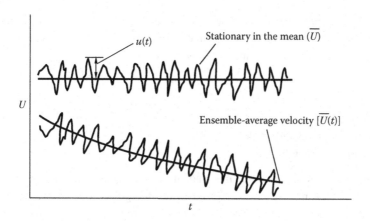

Figure 6.7 Velocity at a point in a turbulent flow.

The closure problem. We restrict the rest of our discussion of turbulence modeling to Navier-Stokes equations governing statistically stationary incompressible flows. The Reynolds averaging of Equation 6.25 with the source term S_{U_i} replaced by the pressure-gradient term, and dropping bars on the time-averaged quantities ($\overline{U}_i$, $\overline{P}_s$, $\overline{\Phi}$, etc.) in the following, yields

$$\frac{\partial}{\partial x_j}\left(\rho U_j U_i\right) + \frac{\partial}{\partial x_j}\left(\rho \overline{u_i u_j}\right) = -\frac{\partial P_s}{\partial x_i} + \frac{\partial}{\partial x_j}\left(\mu \frac{\partial U_i}{\partial x_j}\right)$$

$$\frac{\partial}{\partial x_j}\left(\rho U_j U_i\right) = -\frac{\partial P_s}{\partial x_i} + \frac{\partial}{\partial x_j}\left(\mu \frac{\partial U_i}{\partial x_j} - \rho \overline{u_i u_j}\right) \qquad (6.34)$$

where U_i and u_i are the mean and fluctuating parts, respectively, of the local instantaneous velocity and $-\rho \overline{u_j u_i}$ are the Reynolds stresses, which result from the time-averaging of the nonlinear convection terms in the Navier-Stokes equations. Clearly, additional equations are required to determine these stresses and thereby have a closed system. This is the task of turbulence modeling. Thus, the turbulence models essentially model the Reynolds stresses in terms of the mean flow quantities.

Boussinesq hypothesis. According to Boussinesq hypothesis, which invokes the gradient transport model as in a laminar flow, the Reynolds stresses are related to the mean velocity gradients by the following equation:

$$-\rho \overline{u_i u_j} = \mu_t\left(\frac{\partial U_i}{\partial x_j} + \frac{\partial U_j}{\partial x_i}\right) - \frac{2}{3}\rho k \delta_{ij} \qquad (6.35)$$

where

$\mu_t \equiv$ Turbulent viscosity
$k \equiv$ Turbulent kinetic energy
$\delta_{ij} \equiv$ Kronecker delta $= \begin{cases} 1 \text{ for } i = j \\ 0 \text{ for } i \neq j \end{cases}$

Note that the turbulent viscosity used in Equation 6.35, unlike a laminar (molecular) viscosity, is not a fluid property but that of a flow. However, by analogy with the molecular viscosity used in a laminar flow, we write

$$\mu_t \sim \rho \upsilon \ell \qquad (6.36)$$

where

$\upsilon \equiv$ Turbulent velocity scale
$\ell \equiv$ Turbulent length scale

According to Equation 6.36, the challenge of turbulence modeling to solve the second-order closure problem, discussed in the foregoing, boils down to finding the turbulent velocity and length scales. Although seldom used in current industrial design applications, for historical importance, two early turbulence models, namely, the Prandtl mixing-length model and the one-equation $k - L$ model are briefly presented in Sections 6.4.2.2 and 6.4.2.3.

6.4.2.2   Prandtl Mixing-Length Model

Prandtl proposed the following mixing-length hypothesis:

Turbulent length scale:

$$\ell = \ell_m$$

Turbulent velocity scale:

$$\upsilon = \ell_m \left| \frac{\partial U}{\partial y} \right|$$

Hence,

$$\mu_t = \rho \ell_m^2 \left| \frac{\partial U}{\partial y} \right| \qquad (6.37)$$

For a wall boundary layer, the mixing length ℓ_m in Equation 6.37 is specified in the inner and outer regions as follows:

$$\text{Inner region}: \ \ell_m = K_c y \left(1 - e^{-y^+/A^+} \right)$$

$$\text{Outer region}: \ \ell_m = 0.09 \delta$$

where

$\delta \equiv$ Velocity boundary layer thickness
$K_c \equiv$ von Karman constant
$A^+ = 26$
$y^+ = \dfrac{y \sqrt{\tau_w/\rho_w}}{\nu_w}$

Note that y^+ is a Reynolds number based on shear velocity $\left(\sqrt{\tau_w/\rho_w} \right)$ and the transverse distance from the wall.

6.4.2.3   One-Equation $k - L$ Model

In this model, while the turbulent velocity scale is obtained from the local value of the turbulent kinetic energy k by solving the corresponding differential transport equation, the turbulent length scale is directly specified.

Turbulent velocity scale: $v = k^{\frac{1}{2}}$

Turbulent length scale: $\ell = L$

$$\mu_t = C'_\mu \rho\, k^{\frac{1}{2}} L \tag{6.38}$$

where $C'_\mu = 0.548$. The transport equation for k is expressed as follows:

$$\frac{\partial}{\partial x_j}\left(\rho U_j k\right) = \frac{\partial}{\partial x_j}\left(\frac{\mu_t}{\sigma_k}\frac{\partial k}{\partial x_j}\right) + \rho\left(P_k - \varepsilon\right) \tag{6.39}$$

where σ_k(typically assigned a value 1.0) is the turbulent Prandtl number for k, P_k the production rate of k, and ε the dissipation rate of k. Both P_k and ε are evaluated as follows:

$$P_k = \mu_t\left(\frac{\partial U_i}{\partial x_j} + \frac{\partial U_j}{\partial x_i}\right)\frac{\partial U_i}{\partial x_j} \tag{6.40}$$

and

$$\varepsilon = C_\mathrm{D}\frac{k^{\frac{3}{2}}}{L} \tag{6.41}$$

where $C_\mathrm{D} = 0.164$. Note that we obtain $C'_\mu C_\mathrm{D} = C_\mu = 0.09$ where the coefficient C_μ is used in the two-equation $k - \varepsilon$ turbulence model (discussed in Section 6.4.2.4) to define the turbulent viscosity.

For a turbulent boundary layer in local equilibrium, it can be shown that

$$\left(C'_\mu/C_\mathrm{D}\right)^{\frac{1}{4}} L = \ell_\mathrm{m} \tag{6.42}$$

Specifying the mixing length ℓ_m as in the Prandtl mixing-length model, we use Equation 6.42 to specify L in the one-equation model.

6.4.2.4   Two-Equation $k-\varepsilon$ Model (BVM)

The conventional high-Reynolds-number two-equation $k - \varepsilon$ model embodies the Boussinesq eddy-viscosity hypothesis, which relates Reynolds stresses to mean-velocity gradients. This model, hereafter called the Boussinesq viscosity model (BVM) has been in widespread use for more than four decades and is generally now considered the starting point for most complex shear flow predictions in industrial design applications. Even when using a more advanced turbulence model in design, the baseline CFD predictions are often carried out using BVM. In this model, turbulent velocity scale, turbulent length scale, and eddy viscosity are computed using the following equations:

$$\text{Turbulent velocity scale : } v = k^{\frac{1}{2}}$$

$$\text{Turbulent length scale : } \ell = \frac{k^{\frac{3}{2}}}{\varepsilon}$$

$$\mu_t = C_\mu \rho \frac{k^2}{\varepsilon} \tag{6.43}$$

where

$\varepsilon \equiv$ Dissipation rate of turbulent kinetic energy (k)

$C_\mu \equiv$ Proportionality constant $= 0.09$

Equation 10.43 is based on the Kolmogorov-Prandtl relation, see Launder and Spalding (1974), and the fact that in a region of high turbulent Reynolds number, the dissipation (ε) is essentially controlled by large-scale turbulent motions through an energy cascade. Note that the transport equation for k in this model is the same as Equation 6.39 used in the one-equation $k - L$ model.

Although a transport equation for ε can be derived, the resulting form needs several modeling assumptions. In view of this, the practice has been to use an ε equation patterned along the lines of the k-equation, taking the following form:

$$\frac{\partial}{\partial x_j} \left(\rho U_j \varepsilon \right) = \frac{\partial}{\partial x_j} \left(\frac{\mu_t}{\sigma_\varepsilon} \frac{\partial \varepsilon}{\partial x_j} \right) + \rho \frac{\varepsilon}{k} \left(C_{\varepsilon_1} P_k - C_{\varepsilon_2} \varepsilon \right) \tag{6.44}$$

where σ_ε is the turbulent Prandtl number for ε and C_{ε_1} and C_{ε_2} are additional model constants.

Flow predictions are greatly influenced by the choice of constants that belong to a turbulence model. Rather than allowing them to be used arbitrarily to fit data, they are selected with hopes of having universality. The model constants in the BVM typically used by different investigators are summarized in Table 6.1. This table shows that there is a good agreement among different investigators on the values for these constants. The BVM model constants used in most industrial applications are those of Launder and Spalding (1974); a shortcoming of this choice, however, is the failure of the model to correctly predict the growth of an axisymmetric free jet.

Table 6.1 Values of model constants used in BVM

Investigators	C_μ	C_{ε_1}	C_{ε_2}	σ_k	σ_ε
Jones and Launder (1972)	0.09	1.45	1.90	1.0	1.30
Hanjalic and Launder (1972)	0.09	1.45	2.00	1.0	1.30
Wyngaard, Arya, and Cote (1974)	0.09	1.50	2.00	1.0	1.30
Launder and Spalding (1974)	0.09	1.44	1.92	1.0	1.30
Gibson and Launder (1976)	0.09	1.45	1.90	1.0	1.30
Pope and Whitelaw (1976)	0.09	1.45	1.90	1.0	1.30
Moon and Rudinger (1977)	0.09	1.44	1.70	1.0	1.30
Gosman et al. (1969)	0.09	1.42	1.92	1.0	1.22
Lilley and Rhode (1982)	0.09	1.44	1.92	1.0	1.23

6.4.2.5 Reynolds Stress Transport Model (RSTM)

Rodi (1980) presents a convenient method to derive the transport equations for Reynolds stresses. These equations are cast into the following form:

$$C_{\overline{u_i u_j}} - D_{\overline{u_i u_j}} = P_{ij} + \Phi_{ij} - \varepsilon_{ij} \tag{6.45}$$

where

Convection: $C_{\overline{u_i u_j}} = \dfrac{\partial}{\partial x_\ell} \left(U_\ell \overline{u_i u_j} \right)$

Diffusion: $D_{\overline{u_i u_j}} = -\dfrac{\partial}{\partial x_\ell} \left(\overline{u_i u_j u_\ell} + \dfrac{\overline{P_s u_i}}{\rho} \delta_{j\ell} + \dfrac{\overline{P_s u_j}}{\rho} \delta_{i\ell} \right)$

Production: $P_{ij} = -\left(\overline{u_i u_\ell} \dfrac{\partial U_j}{\partial x_\ell} + \overline{u_j u_\ell} \dfrac{\partial U_i}{\partial x_\ell} \right)$

Pressure–strain: $\varphi_{ij} = \dfrac{\overline{P_s}}{\rho} \overline{\left(\dfrac{\partial u_i}{\partial x_j} + \dfrac{\partial u_j}{\partial x_i} \right)}$

Dissipation: $\varepsilon_{ij} = 2\dfrac{\mu}{\rho} \overline{\left(\dfrac{\partial u_i}{\partial x_\ell} \dfrac{\partial u_j}{\partial x_\ell} \right)}$

For the second-order closure, the diffusion, dissipation, and pressure–strain terms in Equation 6.45 require modeling. In regions of high turbulent Reynolds number, the viscous diffusion term is negligible in comparison to the remaining diffusion terms. In addition, the small scales responsible for energy dissipation are essentially isotropic. Thus, Equation 6.45 assumes the following modeled form:

$$
\begin{aligned}
C_{\overline{u_i u_j}} = \; & C_s \frac{\partial}{\partial x_\ell} \left(\frac{k}{\varepsilon} \overline{u_\ell u_m} \frac{\partial \overline{u_i u_j}}{\partial x_m} \right) \\
& - \left(\overline{u_i u_\ell} \frac{\partial U_j}{\partial x_\ell} + \overline{u_j u_\ell} \frac{\partial U_i}{\partial x_\ell} \right) \\
& \underbrace{- C_1 \frac{\varepsilon}{k} \left(\overline{u_i u_j} - \frac{2}{3} k \delta_{ij} \right)}_{(\Phi_{ij})_1} \\
& \underbrace{- C_2 \left(P_{ij} - \frac{2}{3} P_k \delta_{ij} \right)}_{(\Phi_{ij})_2} \\
& - \frac{2}{3} \varepsilon \delta_{ij}
\end{aligned}
\tag{6.46}
$$

where $P_k = 0.5 P_{ii}$ and a gradient transport model as proposed by Daly and Harlow (1970) is used for the remaining diffusion terms. The pressure–strain term is split into two components such that $(\phi_{ij})_1$, which represents the contribution of the turbulence field alone, is modeled using the "return-to-isotropy" term of Rotta (1951) and $(\phi_{ij})_2$, which results from the interaction of the mean flow and the fluctuating field, represents the "rapid" part. This part is modeled according to the proposal of Naot, Shavit, and Wolfshtein (1973). Launder, Reece, and Rodi (1975) proposed a more elaborate model for $(\phi_{ij})_2$ whose leading term is identical to the one given here.

The turbulent kinetic energy k and its isotropic dissipation rate ε still appear as unknowns in Equation 6.46. Because $k = 0.5\overline{u_i u_i}$, its transport equation is already contained in Equation 6.46 and may be written as

$$\underbrace{\frac{\partial}{\partial x_j}\left(U_j k\right)}_{C_k} = \underbrace{C_s \frac{\partial}{\partial x_\ell}\left(\frac{k}{\varepsilon}\overline{u_\ell u_m}\frac{\partial k}{\partial x_m}\right)}_{D_k} \underbrace{- \overline{u_\ell u_m}\frac{\partial U_\ell}{\partial x_m}}_{P_k=0.5P_{ii}} - \varepsilon \tag{6.47}$$

As mentioned earlier, the transport equation for ε is simply a modeled equation assuming the form of the k equation. Accordingly, we write the ε equation as

$$\frac{\partial}{\partial x_j}\left(U_j \varepsilon\right) = C_\varepsilon \frac{\partial}{\partial x_\ell}\left(\frac{k}{\varepsilon}\overline{u_\ell u_m}\frac{\partial \varepsilon}{\partial x_m}\right) - C_{\varepsilon_1}\frac{\varepsilon}{k}P_k - C_{\varepsilon_2}\frac{\varepsilon^2}{k} \tag{6.48}$$

We now have a closed system of equations for the Reynolds stress transport model (RSTM) where the model constants in Equation 6.46 are typically assigned the following values:

$$C_s = 0.22, \ C_\varepsilon = 0.15, \ C_1 = 2.2, \text{ and } C_2 = 0.55$$

6.4.2.6   Algebraic Stress Model (ASM)

In the preceding, we have briefly outlined the RSTM in which the Reynolds stresses are governed by a set of six partial differential equations. Even though the mathematical system contains most of the desirable flow physics, being closely related to the Navier-Stokes equations, it still lacks necessary appeal for industrial flow computations. For a three-dimensional flow, one is required to solve eleven partial differential equations: four for the mean flow; six for the Reynolds stresses, and one for obtaining the length scale distribution. Because the turbulent field is always three-dimensional, even for a two-dimensional mean flow there is no substantial saving in terms of the total number of partial differential equations to be solved.

The convection and diffusion terms in Equation 6.45 render it to be a differential equation. Three main hypotheses have been proposed to replace these terms so that Equation 6.45 assumes an algebraic form. Launder (1971) proposed the following hypothesis:

$$C_{\overline{u_i u_j}} - D_{\overline{u_i u_j}} = 0$$

which implies an assumption of local equilibrium for the Reynolds stresses. Later, Mellor and Yamada (1974) assumed

$$C_{\overline{u_i u_j}} - D_{\overline{u_i u_j}} = \frac{2}{3}\left(C_k - D_k\right)\delta_{ij}$$

where the transport of Reynolds shear stresses is neglected. For the case of none-quilibrium shear flows, Rodi (1976) postulates that the net transport of stress component $\overline{u_i u_j}$ is proportional to the net transport of the turbulent kinetic energy k using $\overline{u_i u_j}/k$ as the constant of proportionality. Accordingly, we obtain

$$C_{\overline{u_i u_j}} - D_{\overline{u_i u_j}} = \frac{\overline{u_i u_j}}{k}\left(C_k - D_k\right) = \frac{\overline{u_i u_j}}{k}\left(P_k - \varepsilon\right)$$

Using this hypothesis in Equation 6.46, we obtain the following system of algebraic equations for Reynolds stresses:

$$\frac{\overline{u_i u_j} - \frac{2}{3}k\delta_{ij}}{k} = \left(\frac{1 - C_2}{C_1 - 1 + \lambda}\right)\left(\frac{P_{ij} - \frac{2}{3}k\delta_{ij}}{\varepsilon}\right) \tag{6.49}$$

where $\lambda = P_k/\varepsilon$, the ratio of local production of turbulent kinetic energy to its dissipation. While a general consensus on the model constants used in BVM prevails (see Table 6.1), no such agreement for C_1 and C_2 currently exists. By way of ASM calibration, one may consider variations around the nominal values of $C_1 = 2.2$ and $C_2 = 0.55$ proposed by Launder (1975). Sultanian (1984) and Sultanian, Neitzel, and Metzger (1986, 1987a, 1987b) found significantly superior performance of ASM over BVM in sudden-expansion pipe flows with and without swirl.

6.4.3 Boundary Conditions

All solutions of a given set of governing equations for the primitive variables and the turbulence model variable are subject to the specified boundary conditions at inflow, outflow, and wall boundaries. Often in a design environment, we may not know the detailed boundary conditions needed for a high-fidelity CFD solution. In such cases, it behooves us to perform sensitivity analyses to understand the variations in the computed CFD results as a result of uncertainties in critical boundary conditions.

In the following discussion, we will consider the BVM, that is, the high-Reynolds-number $k - \varepsilon$ turbulence model, which is widely used as the baseline turbulence model in most industrial CFD applications.

6.4.3.1 Inlet and Outlet Boundary Conditions

As a matter of CFD best practice used in an industrial design, inlets and outlets are assigned to the CFD calculation domain where the flow field is expected to be parabolic, that is, free from any reverse flow. At times, the calculation domain is modified with artificial extensions to achieve desirable inflow and outflow boundaries.

At inlets, we specify uniform or nonuniform profiles of all dependent variables of the mean flow either from available measurements or from other related analyses. For the $k - \varepsilon$ turbulence model, assuming isotropic turbulence at the inlet, k_{in} is computed by the equation

$$k_{in} = 1.5(Tu)^2 U_{in}^2 \tag{6.50}$$

where Tu is the average inlet turbulence intensity and U_{in} the mean inlet velocity. At each inlet, we specify ε_{in} in one of two ways:

Method 1: $$\varepsilon_{in} = C_\mu^{3/4} k_{in}^{3/2} / \ell_m \qquad (6.51)$$

Method 2: $$\varepsilon_{in} = \rho C_\mu k_{in}^2 / \mu_t \qquad (6.52)$$

In Equation 6.51, ℓ_m is the mixing length, which is determined from the inlet dimensions. In Equation 6.52, μ_t is the assumed inlet turbulent viscosity, say, a multiple of the fluid dynamic viscosity μ. In both equations, k_{in} is determined from Equation 6.50.

If an outlet is placed far enough downstream, the boundary conditions representing a fully-developed flow may be specified. Otherwise, all dependent variables are to be specified at the outlet. For computing compressible flows, specifying the total pressure at inlet and static pressure at outlet, if the outlet is not choked, works better than specifying mass flow rate through predetermined velocity and density distributions.

6.4.3.2   Wall Boundary Conditions: The Wall-Function Treatment

Although the near-wall region is characterized by steep gradients in mean flow variables and turbulence quantities, these quantities actually vanish right at the wall. In order to directly incorporate this simple wall boundary condition, the conservation equations must be integrated up to the wall. This requirement poses two main difficulties; first, both the BVM and ASM, in the form presented here, are not valid in the region of low turbulent Reynolds number that prevail near a wall; and second, a very fine grid is required near the wall so that the assumption of a linear profile for each quantity between grid points is valid for a proper numerical integration. The use of a wall function overcomes both these difficulties, because it directly links the near-wall equilibrium region (characteristic of all turbulent wall boundary layers where local production of turbulent kinetic energy balances its dissipation) with the wall.

When a turbulent boundary layer separates from the wall, either under an adverse pressure gradient or as a result of a step change in wall geometry, a stalled region of flow recirculation occurs at the wall. In this region, the turbulence energy production near the wall is negligible, and turbulence energy diffusion toward the wall nearly balances its local dissipation. In spite of some regions of local nonequilibrium, perhaps for reasons of simplicity or in the absence of better information, the wall-function approach as recommended by Launder and Spalding (1974) is widely used in the simulation of most industrial flows.

Logarithmic law of the wall. In a boundary layer, we define shear velocity U^* as $U^* = \sqrt{\tau_w / \rho}$ where τ_w is the wall shear stress. In terms of this shear velocity, we further define U^+ and y^+ as follows:

$$U^+ = \frac{U}{U^*}$$

and

$$y^+ = \frac{y U^*}{\nu}$$

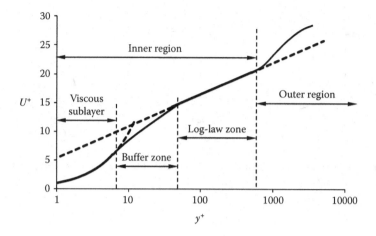

Figure 6.8 Velocity variation in the turbulent boundary layer on a flat plate.

Note that y^+ is a Reynolds number based on the shear velocity U^* and the distance y from the wall. Plotted in terms U^+ and y^+, Figure 6.8 shows the overall structure of a flat plate turbulent boundary layer, which is devoid of any influence from stream-wise pressure-gradient and streamline curvature. This boundary layer consists of two regions: inner region and outer region, which interfaces with the nearly potential outer flow. For the wall-function treatment, the inner region, which is further divided into three zones, is of interest. As shown in the figure, the innermost zone in direct contact with the wall is the viscous sublayer, which in the old literature on fluid mechanics was also called the laminar sublayer. The log-law zone at its outer edge interfaces with the outer region and at its inner edge connects with the viscous sublayer through the buffer zone. The log-law zone is considered somewhat universal in nature with the corresponding logarithmic law of the wall given by

$$U^+ = 5.5 \log_{10} y^+ + 5.45 = 2.388 \ln y^+ + 5.45$$

$$U^+ = \frac{1}{K_c} \ln \left(E_c y^+ \right) \tag{6.53}$$

where $K_c = 0.4187$ and $E_c = 9.793$.

Modified log-law for the velocity parallel to the wall. While solving for the velocity component parallel to the wall using a wall-function approach, one essentially applies the shear stress (or momentum flux) boundary condition for the near-wall control volume. Because the wall-function treatment in a turbulent flow CFD is targeted for all boundary layers, including those under nonzero stream-wise pressure gradients and with streamline curvature, the logarithmic-law of the wall given by Equation 6.53 is modified using the turbulence structure parameter, which in a turbulent boundary layer in local equilibrium is given by (based on measurements)

$$\frac{-\overline{uv}}{k} = C_\mu^{\frac{1}{2}} = 0.3 \tag{6.54}$$

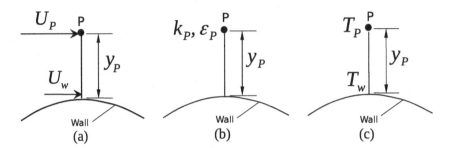

Figure 6.9 Wall boundary conditions: (a) velocity parallel to wall, (b) specifications of k_P and ε_P, and (c) temperature.

The shear velocity can now be expressed as

$$U^* = \sqrt{\tau_w/\rho_w} = \sqrt{-\overline{uv}} = C_\mu^{\frac{1}{4}} k^{\frac{1}{2}} \qquad (6.55)$$

For the near-wall node shown in Figure 6.9a, we can write

$$U^+ = \frac{(U_P - U_w)}{U^*} = \frac{1}{K_c} \ln\left(E_c y^+\right)$$

$$\frac{(U_P - U_w)U^*\rho}{\tau_w} = \frac{1}{K_c} \ln\left(E_c y^+\right)$$

In the aforementioned equation, substituting for U^* from Equation 6.55 and solving for τ_w yield

$$\tau_w = \frac{(U_P - U_w)K_c\rho C_\mu^{\frac{1}{4}} k_P^{\frac{1}{2}}}{\ln\left(E_c y^+\right)} \qquad (6.56)$$

The method to obtain k_P used in Equation 6.56 is discussed in the following.

Specifications of k_P and ε_P. From an equilibrium boundary layer consideration, ε_P is fixed at the near-wall node P (shown in Figure 6.9b), using the following relation

$$\varepsilon_P = \frac{C_\mu^{\frac{3}{4}} k_P^{\frac{3}{2}}}{K_c y_P} = \frac{U^{*3}}{K_c y_P} \qquad (6.57)$$

To obtain k_P, the k equation is solved assuming $k = 0$ at the wall, and the dissipation source term is approximated as

$$\int_0^{y_p} \varepsilon \, \mathrm{d}y = \frac{C_\mu k_P^{\frac{3}{2}} \ln\left(E_c y_P^+\right)}{K_c} \qquad (6.58)$$

Temperature. In the wall-function approach for the energy equation, the wall heat flux is calculated by the equation

$$\dot{q}_w = \frac{(T_w - T_P)\rho \, c_p \, C_\mu^{\frac{1}{4}} k_P^{\frac{1}{2}}}{Pr_t\left(U_P^+ + P_f\right)} \qquad (6.59)$$

where

$Pr_t \equiv$ Turbulent Prandtl number ≈ 0.9

$P_f \equiv$ Jaytilleke's P function given by

$$P_f = 9.24 \left\{ \left(\frac{Pr}{Pr_t}\right)^{\frac{3}{4}} - 1 \right\} \left\{ 1 + 0.28 \exp \left(-0.007 \frac{Pr}{Pr_t} \right) \right\}$$

6.4.3.3   Alternative Near-Wall Treatments

The wall-function treatment discussed in Section 6.4.3.2 has its limitations. The main drawback of the approach is that it is only valid for wall boundary layers that satisfy the local-equilibrium conditions, that is, the production-rate of turbulent kinetic energy is equal to its dissipation-rate. Complex geometries associated with industrial flow and heat transfer situations do not generally meet the restrictions of a wall-function treatment, whose continued use is primarily driven by a reduced model size as the first near-wall is placed in the y^+ range from 40 to 400 (see the log-law zone in Figure 6.8). Note that, while using wall-functions, any attempt to resolve the near-wall region with lower y^+ values for the node next to the wall will be counterproductive.

To overcome the limitations of wall-functions, alternative near-wall treatments have been developed and used, for example, two- and three-layer models and low-Reynolds number models. In the latter, the governing equations for both mean flow variables and turbulence model variables are integrated right up to the wall with the first layer of near-wall nodes placed at a y^+ of around unity.

Luo, Razinsky, and Moon (2012) compared the performance of several commonly-used turbulence models using both wall-functions and low-Reynolds modeling at the wall. It is important to note that, in addition to packing many more grid points near a wall to properly resolve the boundary layer, the turbulence models themselves are modified to model the low-Reynolds number region and the near-wall nonisotropic turbulence field. The standard BVM, discussed in Section 6.4.2.4, is not suitable as such to model the low-Reynolds number wall region even on a wall-integration grid. For a turbulent heat transfer prediction using a low-Reynolds number turbulence model with wall integration, the challenge of modeling a variable turbulent Prandtl number in the boundary layer remains.

6.4.4   Selection of a Turbulence Model

The accuracy of a turbulent flow CFD for a given set of boundary conditions depends mainly upon three factors: grid quality, accuracy of the numerical scheme, and the turbulence model used. While most CFD analysts in an industry rightly focus on generating a high-quality grid in their computation domain, and spend a good deal of time in obtaining a fully converged solution with a high-order numerical scheme, their choice of turbulence model is limited to the options available in the commercial CFD code used and the design cycle time available to conclude the analysis. If the CFD predictions fail to validate against the available test data, calculations are repeated with another turbulence model available in the CFD code. A lot of time and computing resources can be saved if the CFD analysts short-list the available turbulence models based on the flow and heat transfer physics of their design. In this section, we present a few examples illustrating the proper choice of a turbulence model depending on the flow and heat transfer physics we are trying to predict.

6.4.4.1   Turbulent Flow in a Noncircular Duct

As we discussed in Chapter 2, the turbulent flow in a noncircular duct features secondary flows of the second kind, which are nonexistent in a laminar flow in the duct. While the secondary flows are weak in magnitude, they have a significant effect on the wall shear stress and heat transfer. Because these secondary flows are driven by the gradients in the normal turbulent stresses at the wall, an isotropic turbulence model will fail to predict them, regardless of the quality of the CFD grid or the accuracy of the numerical scheme used. For example, the widely used two-equation, high Reynolds-number $k - \varepsilon$ turbulence model will not succeed in predicting these secondary flows in a noncircular duct. We must therefore choose a turbulence model, for example, the Reynolds stress model or its variants with the capability to model anisotropic turbulence.

6.4.4.2   Turbulent Flow in a Sudden Pipe Expansion

In Chapter 2, we discussed the key features of a turbulent flow in a sudden pipe expansion. Measurements in both water and air flows in this geometry show that reattachment occurs at around 8.6 step heights from the edge of the sudden expansion. One step height equals the difference between the downstream pipe radius and upstream pipe radius. CFD prediction of this flow with the standard $k - \varepsilon$ turbulence model yields a reattachment length of 6.5 step heights. As discussed by Sultanian (1984), the under-prediction of the reattachment length by the $k - \varepsilon$ turbulence model can be traced to its inability to accurately predict the attenuation of turbulence in the stabilizing curvature of the reattachment streamline. The higher production of turbulence predicted by the $k - \varepsilon$ turbulence model in the complex shear layer associated with the primary recirculation region results in a shorter reattachment length. Using an ASM, which is a variant of the RSTM, Sultanian, Neitzel, and Metzger (1986) successfully predict all the key features of a sudden expansion pipe flow, including the measured reattachment length of 8.6 step heights.

6.4.4.3   Swirling Turbulent Flow in a Sudden Pipe Expansion

Both the flow and heat transfer characteristics of a sudden expansion pipe flow change dramatically if the incoming flow is swirling. The reattachment length reduces as the swirl strength increases and the turbulence field becomes highly nonisotropic. As discussed by Sultanian (1984), any turbulence model that uses the assumption of isotropic turbulence will fail to accurately predict this flow field. As discussed in Chapter 2, for an incoming swirl exceeding a critical value, a region of vortex breakdown and flow recirculation occurs on the axis of the downstream pipe. This flow feature is widely used as an aerodynamic flame holder in the design of many modern gas turbine combustors. In order to accurately predict such a complex shear flow, we must at the minimum use an RSTM or its variant capable of predicting anisotropic turbulent stresses in the flow field.

6.4.4.4   Turbulent Flow and Heat Transfer in a Rotor Cavity

When it comes to modeling heat transfer in a complex shear flow, the near-wall modeling of turbulence becomes most important and challenging. The flow and heat transfer near a wall is dominated by low Reynolds numbers and variable turbulent Prandtl numbers for which we lack sufficient empirical data and correlations. The wall-function treatment presented in the foregoing section is valid only for a boundary layer in local equilibrium where the rate of turbulence production equals its rate of dissipation. In addition, the turbulent Prandtl number is assumed constant. In any other boundary layer, the application of wall functions to link the wall conditions to the near-wall nodes is not accurate. To circumvent the short-comings of the wall-function treatment, alternate methods of near-wall turbulence modeling such as two- and three-layer modeling have been proposed. Another approach is to use a low Reynolds-number turbulence model to integrate the governing mean flow and turbulence model equations right up to the wall. Sultanian and Nealy (1987), for example, use such an approach to predict heat transfer in a rotor cavity where the rotor walls are dominated by Ekman boundary layers (Lugt 1995).

6.4.5   Physics-Based Postprocessing of CFD Results

CFD uses many small control volumes and generates a detailed description of velocities, pressure, temperature, and other flow and heat transfer properties in each control volume. The last step, and perhaps the most important and nontrivial step, in a 3-D CFD analysis is to postprocess the computed results for design applications. This postprocessing is carried out with two initial objectives in mind. First, we use the results to get a clear qualitative understanding of the key features of the computed flow field by generating, for example, a plot of streamlines. Second, we obtain various integral quantities such as section-average values of static pressure, total pressure, static temperature, total temperature, shear force on the bounding walls, and the drag force on an internal design feature. For those who want to use the ultimate power of CFD analysis in a design, generating an entropy map from the computed results will clearly identify local areas of high entropy production. Improving these areas in design will certainly improve the component aerodynamic performance efficiency, which is difficult to achieve using other means.

6.4.5.1   Large Control Volume Analysis of CFD Results

Figure 6.10 shows large parallelepiped control volume drawn around a solid object around which a detailed 3-D CFD analysis has been performed. The simple geometry of the solid object shown in the figure is just for illustration. In an actual design, the object shape may be more complex. Let us assume that the main design objective of the CFD analysis is to evaluate the drag force acting on the solid object in the x-direction. A direct method to determine the drag force will be to integrate the x component of the forces from both pressure and shear stress distributions on the surface of the solid

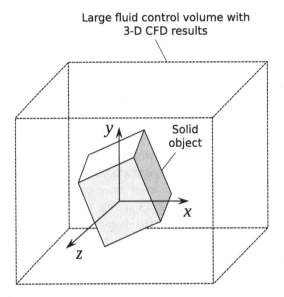

Large fluid control volume with
3-D CFD results

y

Solid
object

x

z

Figure 6.10 A large control volume representation of 3-D CFD results.

object. We can also use an indirect method of first postprocessing the CFD results to determine the x-momentum flux and surface force on each of the six faces of the large control volume, and then perform the x-momentum control volume analysis to find the drag force on the solid object. Except for some small numerical error introduced in the postprocessing of the CFD results, both methods should essentially yield the same result for the drag force. The second method, however, can also yield section-average values of the 3-D distribution of various quantities available from the CFD results. These section-average values are extremely useful in design applications to determine the overall performance of a flow component.

As an illustrative example of physics-based postprocessing of 3-D CFD results, let us consider a gas turbine exhaust diffuser for which we have completed the CFD analysis. The key aerodynamic performance parameter for a diffuser is its pressure recovery coefficient C_P, which is defined by the equation

$$C_P = \frac{\overline{P}_{s_{out}} - \overline{P}_{s_{in}}}{\overline{P}_{t_{in}} - \overline{P}_{s_{in}}} \tag{6.60}$$

where

$\overline{P}_{s_{in}} \equiv$ Section-average value of static pressure at diffuser inlet
$\overline{P}_{s_{out}} \equiv$ Section-average value of static pressure at diffuser outlet
$\overline{P}_{t_{in}} \equiv$ Section-average value of total pressure at diffuser inlet

Although we know the detailed distributions of all computed mean-flow quantities such as velocities, static pressure, and static temperature from the CFD results at each section, it is not obvious whether we should perform an area-weighted averaging or mass-weighted averaging to compute their section-average values. For a physics-based postprocessing of the CFD results to compute a section-average value, we must satisfy the following two criteria:

(1) Averaging must preserve the total flow of mass, momentum, angular momentum, energy, and entropy through the section, including the total surface force and torque at the section used in the CFD solution.

(2) The computed section-average quantity obtained either by area-weighted averaging or by mass-weighted averaging of the CFD results must be meaningful in an integral control volume analysis in which the section under consideration is a control surface.

Based on the second criterion, we see that the area-weighted averaging of the static pressure at a section will be a meaningful quantity, representing the total pressure force acting on the surface in a related control volume analysis. The section-average value of the total pressure obtained using either the area-weighted averaging or mass-weighted averaging of the CFD results will not be meaningful. To compute the section-average total pressure at a section, the following step-by-step physics-based method of postprocessing of CFD results is recommended. Without any loss of generality, let us assume that the section normal is along the x direction (the main flow direction). Let us also assume that the thermophysical properties are constant over the section.

Step 1: Calculate the cross-section area and the mass flow rate through the section:

$$A = \iint_A dA \quad \text{and} \quad \dot{m} = \iint_A \rho V_x dA$$

Step 2: Calculate the section-average static pressure by area-weighted averaging of the static pressure distribution from the CFD results:

$$\overline{P}_s = \frac{1}{A} \iint_A P_s dA \tag{6.61}$$

Step 3: Calculate the section-average static temperature by mass-weighted averaging of the static temperature distribution from the CFD results:

$$\overline{T}_s = \frac{1}{\dot{m}} \iint_A T_s \rho V_x dA \tag{6.62}$$

Step 4: Calculate the section-average specific kinetic energy by mass-weighted averaging of the kinetic energy flow through the section:

$$\frac{\overline{V}^2}{2} = \frac{1}{\dot{m}} \iint_A \left(\frac{V^2}{2}\right) \rho V_x dA \tag{6.63}$$

Step 5: Calculate the section-average total temperature as

$$\overline{T}_t = \overline{T}_s + \frac{\overline{V}^2}{2c_p} \tag{6.64}$$

where $\overline{T}_s$ and $\overline{V}^2$ are from Equations 6.62 and 6.63, respectively.

Step 6: Calculate the section-average total pressure as

$$\overline{P}_t = \overline{P}_s \left(\frac{\overline{T}_t}{\overline{T}_s} \right)^{\frac{\kappa}{\kappa-1}} \tag{6.65}$$

where $\overline{P}_s$, $\overline{T}_s$, and $\overline{T}_t$ are from Equations 6.61, 6.62, and 6.64, respectively.

Using this postprocessing method at diffuser inlet and outlet, we can easily compute the pressure recovery coefficient C_P from Equation 6.60. This method may similarly be used in other design applications involving 3-D CFD analysis.

6.4.5.2   Entropy Map Generation for Identifying Lossy Regions

In Section 6.4.5.1, we discussed postprocessing of 3-D CFD results to obtain section-average values to determine the overall equipment performance and efficiency. The knowledge of local areas of significant irreversibility present in the design, as obtained from the detailed CFD results, could help us redesign these areas, leading to an improved overall design.

The conventional approach to assessing the performance of a flow system is to find the loss of total pressure in the system. This approach serves well when dealing with incompressible flows, because the total pressure in such flows represent mechanical energy per unit volume, and its loss appears as increase in fluid internal energy, loosely called thermal energy, which we generally do not track in these flows. For a compressible flow system, the loss in total pressure is used to denote performance loss under adiabatic conditions with constant total temperature.

Entropy, being a scalar quantity and grounded in the second law of thermodynamics, serves as a more useful quantity to track irreversibility in a flow system. In terms of changes in total pressure and total temperature between two points in a flow system, we can compute entropy change by the equation

$$s_2 - s_1 = c_p \ln \left(\frac{T_{t_2}}{T_{t_1}} \right) - R \left(\frac{P_{t_2}}{P_{t_1}} \right) \tag{6.66}$$

For an adiabatic flow with $T_{t_2} = T_{t_1}$, Equation 6.66 yields

$$\frac{P_{t_2}}{P_{t_1}} - 1 = \frac{\Delta P_t}{P_{t_1}} = e^{-\left(\frac{\Delta s}{R} \right)} - 1 \tag{6.67}$$

which relates the increase in entropy to the loss in total pressure between any two points. Using static pressure and static temperature, the quantities directly available in the CFD results, we can easily compute the entropy at a point using the equation

$$s^* = \frac{s}{R} = \frac{c_p}{R} \ln \left(\frac{T_s}{T_{s_{ref}}} \right) - \left(\frac{P_s}{P_{s_{ref}}} \right) \tag{6.68}$$

where we have arbitrarily assumed zero entropy at the reference pressure $P_{s_{ref}}$ and reference temperature $T_{s_{ref}}$.

We use the following steps to generate an entropy map in a calculation domain to identify lossy regions:

Step 1: Obtain 3-D CFD results in the calculation domain earmarked for design improvement.

Step 2: Postprocess the 3-D CFD results to compute s^* with Equation 6.68 through-out the calculation domain.

Step 3: Calculate the section-average entropy $\bar{s}_{in}^*$ at the inlet by mass-weighted averaging of the inlet entropy distribution

$$\bar{s}_{in}^* = \left[\frac{1}{\dot{m}}\iint_A s^*\rho V_x dA\right]_{inlet} \tag{6.69}$$

Step 4: Plot the contours of $\bar{s}^* - \bar{s}_{in}^*$ in the calculation domain.

Step 5: The regions of high entropy production relative to the average inlet value are the regions to be improved in the next design iteration.

Thus, the entropy map provides an invaluable insight into the design space for local improvements in the regions of excessive entropy production. Further details in this promising area of CFD application to design optimization are given in Naterer and Camberos (2008) and Sciubba (1997).

6.5   Thermomechanical Analysis

So far in this chapter our focus has been on the accurate modeling of gas turbine internal flows and their energy exchange with the structural elements by convection and rotational work transfer. Durability considerations require that we must evaluate both thermal and mechanical stresses in critical gas turbine components, which have limited life under LCF and HCF. Accordingly, we now turn our attention to the methodology to compute these stresses in the complex structure of the whole engine during its diverse transient operations – a must for engine life prediction and management.

To develop an insightful understanding of mechanical and thermal stresses, let us consider the simple case of a constant-thickness rotor disk with its inner radius R_1 and outer radius R_2. The disk is rotating at an angular velocity Ω, and its radial temperature distribution is given by $T(r)$. For such a rotor disk, Hearn (1997) derived the following solutions for the radial and hoop (tangential) stresses in the disk:

$$\sigma_r = A - \frac{B}{r^2} - \left(\frac{3+v}{8}\right)\rho r^2\Omega^2 - \frac{E\alpha}{r^2}\int Trdr \tag{6.70}$$

$$\sigma_\theta = A + \frac{B}{r^2} - \left(\frac{1+3v}{8}\right)\rho r^2\Omega^2 - E\alpha T + \frac{E\alpha}{r^2}\int Trdr \tag{6.71}$$

where the integration constants A and B are to be determined from the boundary conditions $\sigma_r = 0$ at $r = R_1$ and $r = R_2$.

For a linear disk temperature profile $T = T_b + ar$, we obtain from Equation 6.70 the total radial stress $\sigma_r = \sigma_{r_{me}} + \sigma_{r_{th}}$ in the rotor disk where the radial mechanical stress $(\sigma_{r_{me}})$ as a result of the centrifugal body force from rotation is given by

$$\sigma_{r_{me}} = \left(\frac{3+v}{8}\right)\rho\Omega^2\left[R_1^2 + R_2^2 - \frac{R_1^2 R_2^2}{r^2} - r^2\right] \tag{6.72}$$

and the radial thermal stress $(\sigma_{r_{th}})$ is given by

$$\sigma_{r_{th}} = \frac{E\alpha a}{3}\left[\frac{R_1^2 R_2^2}{(R_1+R_2)}\left\{\frac{1}{R_1^2} - \frac{1}{r^2}\right\} + R_1 - r\right] \tag{6.73}$$

Similarly, we obtain from Equation 6.71 the total hoop stress $\sigma_\theta = \sigma_{\theta_{me}} + \sigma_{\theta_{th}}$ where the hoop mechanical stress $(\sigma_{\theta_{me}})$ is given by

$$\sigma_{\theta_{me}} = \frac{\rho\Omega^2}{8}\left[(3+v)\left\{R_1^2 + R_2^2 + \frac{R_1^2 R_2^2}{r^2}\right\} - (1+3v)r^2\right] \tag{6.74}$$

and the hoop thermal stress $(\sigma_{\theta_{th}})$ is given by

$$\sigma_{\theta_{th}} = \frac{E\alpha a}{3}\left[\frac{R_1^2 R_2^2}{(R_1+R_2)}\left\{\frac{1}{R_1^2} - \frac{1}{r^2}\right\} + R_1 - 2r\right] \tag{6.75}$$

From Equation 6.72, it can be easily verified that the maximum radial mechanical stress occurs at $r = \sqrt{R_1 R_2}$, given by

$$\sigma_{r_{me}(max)} = (3+v)\frac{\rho\Omega^2}{8}(R_2 - R_1)^2 \tag{6.76}$$

From Equation 6.74, we obtain the maximum hoop mechanical stress at the disk hub $r = R_1$ as

$$\sigma_{\theta_{me}(max)} = \frac{\rho\Omega^2}{4}\left[(3+v)R_2^2 + (1-v)R_1^2\right] \tag{6.77}$$

and the minimum hoop mechanical stress occurs at the disk rim $r = R_2$ as

$$\sigma_{\theta_{me}(min)} = \frac{\rho\Omega^2}{4}\left[(3+v)R_1^2 + (1-v)R_2^2\right] \tag{6.78}$$

Note from Equations 6.73 and 6.75 that both the radial and hoop thermal stresses depend only on the temperature gradient, which in this case equals the slope a of the linear temperature profile. The level of the uniform temperature in a constraint-free rotor disk produces no thermal stresses.

Each of the stresses normalized by its maximum value over the rotor disk is plotted in Figure 6.11 to show its variation with radius, which is normalized by the outer radius R_2. As can be seen from this figure, the radial mechanical stress, which is tensile in nature, varies with the disk radius like an inverted parabola with its peak skewed toward the inner radius. By contrast, the hoop mechanical stress decreases radially with its maximum value at the hub, featuring an initially parabolic variation followed by a linear

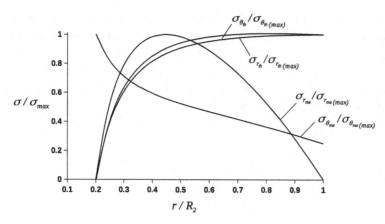

Figure 6.11 Radial variations of mechanical and thermal stresses in a rotor disk of constant thickness.

variation. Both the radial and hoop thermal stresses, arising from linearly increasing disk temperature, are nearly equal and tensile in nature, increasing monotonically from disk hub to rim. Although these stresses increase somewhat rapidly from the hub over the lower half of the disk, they become nearly uniform over the disk upper half.

If we had a radially decreasing disk temperature, both the thermal stresses would have been compressive in nature. This will have the beneficial effect of reducing the thermomechanical (mechanical plus thermal) radial and hoop stresses in the rotor disk.

The closed-form analytical solution obtained in the foregoing for the rotor disk with uniform thickness is not possible for a disk with varying thickness as typically found in gas turbines. For the thermomechanical stress and strain analysis of a real gas turbine structure, designers resort to numerical methods as implemented in the leading commercial codes such as ANSYS, NASTRAN, ABAQUS, and others. These codes primarily use finite element methods (FEM) for both thermal and thermomechanical analyses, commonly known as finite element analysis (FEA). In the following sections, we present an overview of FEM to mitigate the black-box syndrome in using one of these commercial codes. Readers who are interested in comprehensive details on FEM for developing their own software to perform these analyses may find the references cited in the bibliography useful.

Unlike the modeling of complex turbulent flow and heat transfer using turbulence and other physical models, it is fortuitous that the thermomechanical FEA is by and large linear. A nonlinear FEA arises from temperature dependent thermophysical properties and from the analysis that needs to be carried out in the plastic region. Another important feature of the thermomechanical FEA is that the thermal analysis, which yields temperature and heat flux distributions in the structure, can be performed independently. The results of this thermal analysis then become an input to the thermomechanical analysis, thus giving the distributions of combined mechanical and thermal stresses and strains in the structure. These two analyses may be sequentially iterated to account for various nonlinearities and coupling between the thermal and mechanical fields.

For simplicity of presentation within the scope of this book, let us consider a three-dimensional gas turbine structure made of a material of constant thermophysical properties. Both the mechanical and thermal loading of the structure is limited to its three-dimensional elastic deformations. For this structure, we write the equilibrium equations, strain-displacement equations, strain-stress-temperature relations, and heat conduction equation as follows:

Equilibrium Equations

$$\frac{\partial \tau_{xx}}{\partial x} + \frac{\partial \tau_{xy}}{\partial y} + \frac{\partial \tau_{xz}}{\partial z} + f_{Bx} = 0$$

$$\frac{\partial \tau_{yx}}{\partial x} + \frac{\partial \tau_{yy}}{\partial y} + \frac{\partial \tau_{yz}}{\partial z} + f_{By} = 0$$

$$\frac{\partial \tau_{zx}}{\partial x} + \frac{\partial \tau_{zy}}{\partial y} + \frac{\partial \tau_{zz}}{\partial z} + f_{Bz} = 0$$

Using Cartesian tensor notation ($i, j = 1,2,3$), the foregoing equilibrium equations can be expressed as

$$\tau_{ij,j} + f_{B_i} = 0 \tag{6.79}$$

where $\tau_{ij} = \tau_{ji}$ is the local stress tensor with six distinct components, and f_{B_i} is the body force per unit volume.

On a boundary surface, the equilibrium condition is given by

$$\tau_{ij}\hat{n}_j = f_{S_i} \tag{6.80}$$

where f_{S_i} is the stress vector on the surface and $\hat{n}_j$ is the outward-pointing unit normal vector ($\hat{n}_j = n_x\hat{i} + n_y\hat{j} + n_z\hat{k}$).

Strain-Displacement Equations

$$\varepsilon_{xx} = \frac{\partial \varphi_x}{\partial x} \quad \varepsilon_{xy} = \frac{\partial \varphi_y}{\partial x} + \frac{\partial \varphi_x}{\partial y}$$

$$\varepsilon_{yy} = \frac{\partial \varphi_y}{\partial y} \quad \varepsilon_{yz} = \frac{\partial \varphi_z}{\partial y} + \frac{\partial \varphi_y}{\partial z}$$

$$\varepsilon_{zz} = \frac{\partial \varphi_z}{\partial z} \quad \varepsilon_{zx} = \frac{\partial \varphi_x}{\partial z} + \frac{\partial \varphi_z}{\partial x}$$

Again, in tensor notation we can write

$$\varepsilon_{ij} = 2e_{ij} - \delta_{ij}e_{ij} \tag{6.81}$$

where

$$e_{ij} = \frac{1}{2}\left(\varphi_{i,j} + \varphi_{j,i}\right) \tag{6.82}$$

and δ_{ij} is the Kronecker delta, which is equal to 1 when $i = j$ and equal to 0 when $i \neq j$. In Equation 6.82, φ_i is the local displacement vector consisting of structural displacements in three orthogonal coordinate directions.

Stress–Strain–Temperature Relations

$$\varepsilon_{xx} = \frac{1}{E}\left[\tau_{xx} - v\left(\tau_{yy} + \tau_{zz}\right)\right] + \alpha\left(T - T_0\right)$$

$$\varepsilon_{yy} = \frac{1}{E}\left[\tau_{yy} - v\left(\tau_{zz} + \tau_{xx}\right)\right] + \alpha\left(T - T_0\right)$$

$$\varepsilon_{zz} = \frac{1}{E}\left[\tau_{zz} - v\left(\tau_{xx} + \tau_{yy}\right)\right] + \alpha\left(T - T_0\right)$$

$$\varepsilon_{xy} = \frac{\tau_{xy}}{G} \qquad \varepsilon_{yz} = \frac{\tau_{yz}}{G} \qquad \varepsilon_{zx} = \frac{\tau_{zx}}{G}$$

where the shear modulus G is related to the Young's modulus E and Poisson's ratio v by the equation

$$G = \frac{E}{2(1 + v)}$$

and α is the coefficient of thermal expansion.

In terms of e_{ij}, which is related to ε_{ij} by Equation 6.81, we can write the stress-strain-temperature relations as follows:

$$e_{ij} = \left(\frac{1 + v}{E}\right)\tau_{ij} - \frac{v}{E}\delta_{ij}\tau_{kk} + \alpha\delta_{ij}\Delta T \tag{6.83}$$

Note that the first two terms on the right-hand side of Equation 6.83 contribute to mechanical strains as a result of both normal and shear stresses in the material. The last term, which depends on the temperature gradients in the material, produces only normal strains and the corresponding normal thermal stresses.

According to Hughes and Gaylord (1964), we can express Equation 6.79 in terms of displacements in Cartesian tensor notation as

$$G\nabla^2\varphi_i + (\lambda + G)\frac{\partial e_{jj}}{\partial x_i} + f_{B_i} - \beta\nabla T = 0 \tag{6.84}$$

where

$$\lambda = \frac{vE}{(1 + v)(1 - 2v)} \quad \text{and} \quad \beta = \alpha(3\lambda + 2G)$$

Equation 6.84 reveals that the term $\beta\nabla T$ is akin to f_{B_i}, which is a body force per unit volume. Thus, in addition to the body forces and surface tractions applied to a structural component, the gradient of a potential $(-\beta T)$ acts as the body force and βT as the normal surface pressure, causing thermal displacements in the structure in addition to the mechanical ones.

Finally, the structural temperature field, which appears in Equations 6.83 and 6.84, is obtained from the following governing equation, which obeys the Fourier's law of heat conduction:

$$\rho c \frac{\partial T}{\partial t} = k \nabla^2 T \qquad (6.85)$$

where ρ is the material density, c the specific heat, and k the thermal conductivity.

Equations 6.79, 6.82, 6.83, and 6.85 provide sixteen equations for the six components of stress, the six components of strain, the three components of displacement, and temperature, which is a scalar quantity. Thus, they form a closed system of equations in which ten equations are partial differential equations, namely, three equilibrium equations, six strain-displacement equations, and one heat conduction equation. When using Equations 6.84 and 6.85, however, we have to deal with only four partial differential equations – three for displacements and one for temperature. Other quantities can easily be calculated from the distributions of these four variables (degrees of freedom) in the structure.

An analytical solution of the governing closed system of equations presented in the forgoing for thermomechanical analysis in a complex three-dimensional geometry with arbitrary boundary conditions is not possible. One must, therefore, resort to a numerical solution using a finite difference method (FDM) or a finite element method (FEM). Following the current design practice, we limit our discussion here to the finite element method for thermal and thermomechanical analysis of gas turbine components.

The overall scheme of a finite element method is shown in Figure 6.12. The central idea behind the method is to first discretize the domain geometry in terms of finite elements. These elements are connected at discrete nodes and edges (see Figure 6.13). The next step is to discretize the governing partial differential equations for each element using its shape function along with a variational method; for example, the Rayleigh–Ritz method, or alternatively, the Galerkin method, which is a method of weighted residuals. This step yields an algebraic equation at each node for each degree of freedom. These element equations are assembled together, yielding a large system of algebraic equations. This system is then solved numerically using either a direct method or an iterative method. Thus, we obtain the nodal values of all degrees of freedom (dependent variables) for further interpretation and design applications.

Figure 6.13a shows a solid body subjected to body force $f_B(x, y, z)$, surface force $f_S(x, y, z)$, and wall temperature distribution $T_w(x, y, z)$ for which we wish to perform a thermomechanical analysis using the finite element method. As shown in Figure 6.13b, the body geometry can be discretized using four-noded tetrahedral elements, one of which is shown in Figure 6.13c.

6.5.1   Thermal Finite Element Analysis

The thermal FEA in a structural component can be carried out independent of the thermomechanical analysis. We first present here the thermal FEA methodology, which is typically embedded in a commercial FEA code. Having discretized the solid body using finite elements, as shown in Figure 6.13b, our next step is to discretize Equation 6.85 for a general element, which may be inside the body or could have some of its sides represent parts of the body surface (wall). We limit our discussion here to the convective boundary conditions on the surface in contact with the cooling flow. In the indicial equation form, we can express this thermal boundary condition as

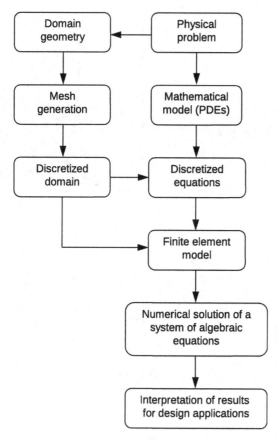

Figure 6.12 Finite element analysis (FEA) scheme.

$$k\frac{\partial T}{\partial x_i}\hat{n}_i + h(T_{\mathrm{w}} - T_{\mathrm{aw}}) = 0 \qquad (6.86)$$

where h is the heat transfer coefficient, T_{w} the wall temperature and T_{aw} the adiabatic wall temperature, which is used as the reference fluid temperature for convective heat transfer.

The key to equation discretization is to assume the element shape functions that determine how the dependent variable, which in this case is temperature T, varies over the element as a function of its nodal values. For example, for an element with n nodes, we can write

$$T(x, y, z) = N_i(x, y, z)\,T_i \qquad i = 1, 2, \ldots n \qquad (6.87)$$

where $N_i = 1$ at node i and zero at all other boundary nodes forming the element. Thus, the shape functions N_i are used for interpolation of temperature within the finite element from its nodal values

In matrix notation, we write Equation 6.87 as

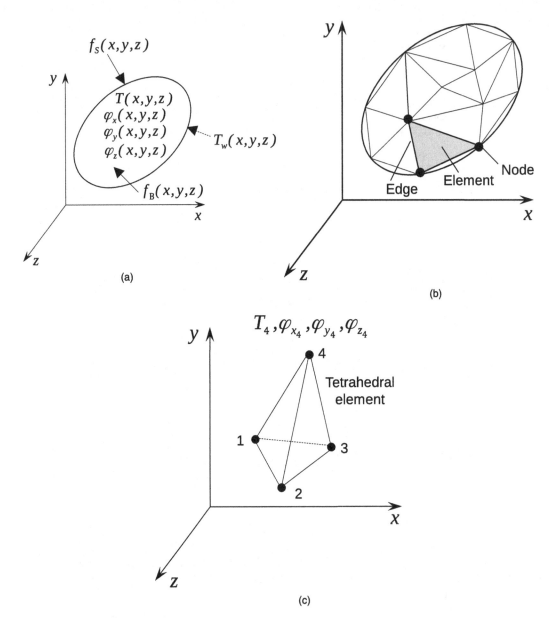

Figure 6.13 (a) A solid body for thermomechanical FEA, (b) finite element model, and (c) tetrahedral element with four nodes.

$$T(x,y,z) = [N]\{T\} = \begin{bmatrix} N_1 & N_2 & N_3 & \cdots & N_n \end{bmatrix} \begin{Bmatrix} T_1 \\ T_2 \\ T_3 \\ \cdots \\ T_n \end{Bmatrix} \qquad (6.88)$$

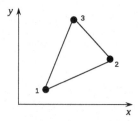

Figure 6.14 Triangular finite element.

where $[N]$ is the matrix of shape functions and $\{T\}$ is the vector of nodal temperatures for the element. To evaluate temperature gradients over the finite element, we differentiate Equation 6.88 to yield

$$\begin{bmatrix} \partial T/\partial x \\ \partial T/\partial y \\ \partial T/\partial z \end{bmatrix} = \begin{bmatrix} \partial N_1/\partial x & \partial N_2/\partial x & \partial N_3/\partial x & \cdots & \partial N_n/\partial x \\ \partial N_1/\partial y & \partial N_2/\partial y & \partial N_3/\partial y & \cdots & \partial N_n/\partial y \\ \partial N_1/\partial z & \partial N_2/\partial z & \partial N_3/\partial z & \cdots & \partial N_n/\partial z \end{bmatrix} \begin{Bmatrix} T_1 \\ T_2 \\ T_3 \\ \cdots \\ T_n \end{Bmatrix} = [B]\{T\}$$

(6.89)

where $[B]$ is the matrix for interpolating temperature gradients over the finite element.

To illustrate the evaluation of shape functions for a finite element, let us consider a triangular finite element shown in Figure 6.14. This element has three nodes, all in the xy plane. Let us assume that each shape function of the element is a bilinear function of x and y given by

$$N_i = a_i + b_i x + c_i y$$

(6.90)

Substitution of Equation 6.90 in Equation 6.87 yields the following system of three equations for the triangular finite element:

$$N_1: \quad a_1 + b_1 x_1 + c_1 y_1 = 1$$

$$N_2: \quad a_2 + b_2 x_2 + c_2 y_2 = 1$$

$$N_3: \quad a_3 + b_3 x_3 + c_3 y_3 = 1$$

At first it may appear that the aforementioned system of three equations and nine unknowns is indeterminate. However, with $N_i = 0$ at all nodes other than the node i, we generate six more equations, two for each N_i. For N_1, for example, we obtain the following closed system of three equations, which are written in the matrix form as

$$\begin{bmatrix} 1 & x_1 & y_1 \\ 1 & x_2 & y_2 \\ 1 & x_3 & y_3 \end{bmatrix} \begin{bmatrix} a_1 \\ b_1 \\ c_1 \end{bmatrix} = \begin{bmatrix} 1 \\ 0 \\ 0 \end{bmatrix}$$

(6.91)

Note that the determinant of the coefficient matrix in Equation 6.91 equals twice the area A of the triangular element, and it can be easily evaluated as

$$2A = x_1 (y_2 - y_3) + x_2 (y_3 - y_1) + x_3 (y_1 - y_2) \tag{6.92}$$

The solution of the system of three equations represented by Equation 6.91 finally yields

$$a_1 = \frac{x_2 y_3 - x_3 y_2}{2A} \quad b_1 = \frac{y_2 - y_3}{2A} \quad c_1 = \frac{x_3 - x_2}{2A}$$

By cyclic permutation, we easily obtain for N_2 and N_3

$$a_2 = \frac{x_3 y_1 - x_1 y_3}{2A} \quad b_2 = \frac{y_3 - y_1}{2A} \quad c_2 = \frac{x_1 - x_3}{2A}$$

$$a_3 = \frac{x_1 y_2 - x_2 y_1}{2A} \quad b_3 = \frac{y_1 - y_2}{2A} \quad c_3 = \frac{x_2 - x_1}{2A}$$

The matrix $[B]$ for the triangular element can be easily evaluated to yield

$$[B] = \begin{bmatrix} \partial N_1/\partial x & \partial N_2/\partial x & \partial N_3/\partial x \\ \partial N_1/\partial y & \partial N_2/\partial y & \partial N_3/\partial y \end{bmatrix} = \begin{bmatrix} b_1 & b_2 & b_3 \\ c_1 & c_2 & c_3 \end{bmatrix} \tag{6.93}$$

Using the Galerkin method, which is the method of weighted residuals with weighting functions the same as the shape functions, we can write Equation 6.85 as

$$\int_\Psi \left(\rho c \frac{\partial T}{\partial t} + k \nabla^2 T \right) N_i \, d\Psi = 0$$

$$\int_\Psi \rho c \frac{\partial T}{\partial t} N_i \, d\Psi + \int_\Psi k \nabla^2 T N_i \, d\Psi = 0$$

$$\int_\Psi \rho c \frac{\partial T}{\partial t} N_i \, d\Psi - \int_\Psi k \left(\frac{\partial^2 T}{\partial x^2} + \frac{\partial^2 T}{\partial y^2} + \frac{\partial^2 T}{\partial z^2} \right) N_i \, d\Psi = 0 \tag{6.94}$$

where Ψ stands for finite element volume. Invoking the integration by parts on the second integral of Equation 6.94 we obtain

$$-\int_\Psi k \left[\frac{\partial N_i}{\partial x} \frac{\partial T}{\partial x} + \frac{\partial N_i}{\partial y} \frac{\partial T}{\partial y} + \frac{\partial N_i}{\partial z} \frac{\partial T}{\partial z} \right] d\Psi + \int_S kN_i \left[\frac{\partial T}{\partial x} n_x + \frac{\partial T}{\partial y} n_y + \frac{\partial T}{\partial z} n_z \right] dS = 0 \tag{6.95}$$

In Equation 6.95, the surface integral over the finite element will in general involve a part of its surface interfacing internally with other finite elements and the remaining part externally with the fluid in convective heat transfer. We can, therefore, rewrite this integral as

$$\int_S kN_i \left[\frac{\partial T}{\partial x} n_x + \frac{\partial T}{\partial y} n_y + \frac{\partial T}{\partial z} n_z \right] dS = \int_{S_{\text{int}}} kN_i \left[\frac{\partial T}{\partial x} n_x + \frac{\partial T}{\partial y} n_y + \frac{\partial T}{\partial z} n_z \right] dS$$

$$+ \int_{S_{\text{ext}}} kN_i \left[\frac{\partial T}{\partial x} n_x + \frac{\partial T}{\partial y} n_y + \frac{\partial T}{\partial z} n_z \right] dS \tag{6.96}$$

Applying the convective boundary condition, Equation 6.86, to the second integral on the right-hand side of Equation 6.96 yields

$$\int\limits_{S_{\text{ext}}} kN_i \left[\frac{\partial T}{\partial x} n_x + \frac{\partial T}{\partial y} n_y + \frac{\partial T}{\partial z} n_z \right] \mathrm{d}S = - \int\limits_{S_{\text{ext}}} N_i h \left(T - T_{\text{aw}} \right) \mathrm{d}S \qquad (6.97)$$

Substituting the profile assumption of T, given by Equation 6.87, in Equation 6.94 with the inclusion of Equations 6.95, 6.96, and 6.97, we obtain

$$\int\limits_{V} \rho c N_i N_j \frac{\partial T_j}{\partial t} \mathrm{d}V + \int\limits_{V} k \left[\left(\frac{\partial N_i}{\partial x} \frac{\partial N_j}{\partial x} + \frac{\partial N_i}{\partial y} \frac{\partial N_j}{\partial y} + \frac{\partial N_i}{\partial z} \frac{\partial N_j}{\partial z} \right) T_j \right] \mathrm{d}V$$

$$- \int\limits_{S_{\text{int}}} kN_i \left[\left(\frac{\partial N_j}{\partial x} n_x + \frac{\partial N_j}{\partial y} n_y + \frac{\partial N_j}{\partial z} n_z \right) T_j \right] \mathrm{d}S \qquad (6.98)$$

$$+ \int\limits_{S_{\text{ext}}} N_i h \left(N_j T_j - T_{\text{aw}} \right) \mathrm{d}S = 0$$

where i and j represent nodes of the finite element. We can re-write Equation 6.98 in a compact matrix form as

$$[C_e]\{\dot{T}\} + [K_e]\{T\} = \{R_e\} + \{I_e\} \qquad (6.99)$$

where

$$[C_e] = \int\limits_{V} \rho c [N]^{\mathrm{T}} [N] \mathrm{d}V$$

$$[K_e] = \int\limits_{V} k [B]^{\mathrm{T}} [B] \mathrm{d}V + \int\limits_{S_{\text{ext}}} h [N]^{\mathrm{T}} [N] \mathrm{d}S$$

$$\{R_e\} = \int\limits_{S_{\text{ext}}} h T_{\text{aw}} [N]^{\mathrm{T}} \mathrm{d}S$$

$$\{I_e\} = - \int\limits_{S_{\text{int}}} \{q\}^{\mathrm{T}} \{n\} [N]^{\mathrm{T}} \mathrm{d}S$$

$$\{q\} = -k[B]\{T\}$$

Note that Equation 6.99 pertains to finite elements individually, and we have added subscript e to indicate that this is an element equation. For the thermal finite element model, we take a global view of all the elements combined. Because the heat flux across the interface between two finite elements will be equal with a zero sum, the quantity $\{I_e\}$ will have no contribution to the global FEM equation. We can, therefore, express the global system of algebraic equations for thermal FEA as

$$\left[C_g \right]\{\dot{T}\} + \left[K_g \right]\{T\} = \left\{ R_g \right\} \qquad (6.100)$$

where, for example, $\left[K_g \right]$ is the global stiffness matrix (the name originates from structural FEA) and is the sum of element stiffness matrices $\left[K_g \right] = \sum [K_e]$, and

$\left\{ R_g \right\}$ is the assemblage of resultant vector of nodal forcing parameters. It should be noted that the elements of $\left[K_e \right]$ matrices common to other elements through the nodal connections should be summed up algebraically during the assembly process.

When we numerically solve Equation 6.100 with the initial condition $T(x, y, z, 0) = T_o(x, y, z)$ specified at all nodes and the convective boundary conditions at the boundary nodes specified in terms of heat transfer coefficient $h(x, y, z)$ and adiabatic wall temperature $T_{aw}(x, y, z)$, we obtain the unsteady thermal solution, giving time-varying temperature at node in the model.

6.5.2   Thermomechanical Finite Element Analysis

In the thermal FEA presented in Section 6.5.1, we have at each node only one degree of freedom, namely, temperature. At each node in a thermomechanical FEA, however, we have four degrees of freedom. These are three displacements, one along each coordinate direction, and temperature. A practical strategy used in the thermomechanical FEA is to first perform the thermal FEA, maybe on a coarser finite element model, and then map the nodal temperatures on to a finer finite element model typically used for a mechanical FEA to delineate areas of high stress concentration. In this section, we will first present the methodology of a mechanical FEA and then modify the resulting finite element equation using the additional body force, called thermal body force, derived from the nodal temperatures.

In what follows, we present a step-wise methodology for the derivation of finite element equations for computing the displacement vector at each node of a finite element. In this derivation, we will use the Rayleigh-Ritz method, which is based on variational principle, instead of the Galerkin method we have used in Section 6.5.1 for the thermal FEA.

Step 1: Element Shape Function

Using the shape function $N(x, y, z)$, we can express the element displacement vector $\{\varphi(x, y, z)\}$ with components $\varphi_x(x, y, z)$, $\varphi_y(x, y, z)$, and $\varphi_z(x, y, z)$ in terms of displacements at each node as

$$\{\varphi(x, y, z)\} = [N(x, y, z)]\{\phi\} \tag{6.101}$$

where $\{\phi\} = \left\{ \phi_{1x} \quad \phi_{1y} \quad \phi_{1z} \quad \phi_{2x} \quad \phi_{2y} \quad \phi_{2z} \quad \cdots \quad \phi_{nx} \quad \phi_{ny} \quad \phi_{nz} \right\}^T$ for an element having n nodes.

In the full matrix form, Equation 6.101 can be expressed as

$$\begin{Bmatrix} \varphi_x \\ \varphi_y \\ \varphi_z \end{Bmatrix} = \begin{bmatrix} N_1 & 0 & 0 & N_2 & 0 & 0 & \cdots & N_n & 0 & 0 \\ 0 & N_1 & 0 & 0 & N_2 & 0 & \cdots & 0 & N_n & 0 \\ 0 & 0 & N_1 & 0 & 0 & N_2 & \cdots & 0 & 0 & N_n \end{bmatrix} \begin{Bmatrix} \phi_{1x} \\ \phi_{1y} \\ \phi_{1z} \\ \phi_{2x} \\ \phi_{2y} \\ \phi_{1z} \\ \cdots \\ \phi_{nx} \\ \phi_{ny} \\ \phi_{nz} \end{Bmatrix} \tag{6.102}$$

Step 2: Element Strain-Displacement Equations

The discretized equations relating element strains and nodal displacements can be written as

$$\varepsilon_{xx} = \frac{\partial}{\partial x}\left(N_i\phi_{ix}\right) \quad \varepsilon_{yy} = \frac{\partial}{\partial y}\left(N_i\phi_{iy}\right) \quad \varepsilon_{zz} = \frac{\partial}{\partial z}\left(N_i\phi_{iz}\right)$$

$$\varepsilon_{xy} = \frac{\partial}{\partial x}\left(N_i\phi_{iy}\right) + \frac{\partial}{\partial y}\left(N_i\phi_{ix}\right) \quad \varepsilon_{yz} = \frac{\partial}{\partial y}\left(N_i\phi_{iz}\right) + \frac{\partial}{\partial z}\left(N_i\phi_{iy}\right) \quad \varepsilon_{zx} = \frac{\partial}{\partial z}\left(N_i\phi_{ix}\right) + \frac{\partial}{\partial x}\left(N_i\phi_{iz}\right)$$

These equations can be written in a compact matrix form as

$$\{\varepsilon\} = [B]\{\phi\} \tag{6.103}$$

where the components of matrix $[B]$ are the spatial derivatives of the components of the shape function $[N]$. Equation 6.103 relates the element strains to nodal displacements.

Step 3: Element Stress-Displacement Equations

For a finite element, stresses are related to strains by the equation

$$[\tau] = [D]\{\varepsilon\} \tag{6.104}$$

where the elasticity matrix $[D]$ is given by

$$[D] = \frac{E}{(1+v)(1-2v)} \begin{bmatrix} (1-v) & v & v & 0 & 0 & 0 \\ v & (1-v) & v & 0 & 0 & 0 \\ v & v & (1-v) & 0 & 0 & 0 \\ 0 & 0 & 0 & \frac{(1-2v)}{2} & 0 & 0 \\ 0 & 0 & 0 & 0 & \frac{(1-2v)}{2} & 0 \\ 0 & 0 & 0 & 0 & 0 & \frac{(1-2v)}{2} \end{bmatrix}$$

$$\tag{6.105}$$

Substituting Equation (6.103) in Equation (6.104) yields

$$[\tau] = [D][B]\{\phi\} \tag{6.106}$$

which relates the element stresses to nodal displacements.

Step 4: Element Equilibrium Equations Using Rayleigh-Ritz Method

The strain energy stored in a solid deformed by external forces can be computed by the equation

$$\tilde{U} = \frac{1}{2}\int_{\Psi} \{\varepsilon\}^{\mathrm{T}}\{\tau\}\mathrm{d}\Psi \tag{6.107}$$

which with the substitution of $\{\varepsilon\}$ and $\{\tau\}$ in terms of nodal displacements from Equations 6.103 and 6.104, respectively, yields

$$\tilde{U} = \frac{1}{2}\int_{\mathcal{V}} ([B]\{\phi\})^{\mathrm{T}}[D][B]\{\varphi\}\mathrm{d}\mathcal{V}$$

$$\tilde{U} = \frac{1}{2}\int_{\mathcal{V}} \{\phi\}^{\mathrm{T}}[B]^{\mathrm{T}}[D][B]\{\phi\}\mathrm{d}\mathcal{V} \tag{6.108}$$

For deformations under the body force $\{f_{\mathrm{B}}\}$ and surface force $\{f_{\mathrm{S}}\}$, which acts on the part of the surface boundary represented by the finite element, we can express the work done as

$$\tilde{W} = \int_{\mathcal{V}} \{\varphi\}^{\mathrm{T}}\{f_{\mathrm{B}}\}\mathrm{d}\mathcal{V} + \int_{S} \{\varphi\}^{\mathrm{T}}\{f_{\mathrm{S}}\}\mathrm{d}S$$

which upon substituting the nodal displacements becomes

$$\tilde{W} = \int_{\mathcal{V}} ([N]\{\phi\})^{\mathrm{T}}\{f_{\mathrm{B}}\}\mathrm{d}\mathcal{V} + \int_{S} ([N]\{\phi\})^{\mathrm{T}}\{f_{\mathrm{S}}\}\mathrm{d}S$$

$$\tilde{W} = \int_{\mathcal{V}} \{\phi\}^{\mathrm{T}}[N]^{\mathrm{T}}\{f_{\mathrm{B}}\}\mathrm{d}\mathcal{V} + \int_{S} \{\phi\}^{\mathrm{T}}[N]^{\mathrm{T}}\{f_{\mathrm{S}}\}\mathrm{d}S \tag{6.109}$$

Using Equations 6.108 and 6.109, we can express the element potential energy as

$$\tilde{P} = \tilde{U} - \tilde{W}$$

$$= \frac{1}{2}\int_{\mathcal{V}} \{\phi\}^{\mathrm{T}}[B]^{\mathrm{T}}[D][B]\{\phi\}\,\mathrm{d}\mathcal{V} - \int_{\mathcal{V}} \{\phi\}^{\mathrm{T}}[N]^{\mathrm{T}}\{f_{\mathrm{B}}\}\mathrm{d}\mathcal{V} - \int_{S} \{\phi\}^{\mathrm{T}}[N]^{\mathrm{T}}\{f_{\mathrm{S}}\}\mathrm{d}S \tag{6.110}$$

For the equilibrium condition, the Rayleigh-Ritz method requires the minimization of the potential energy with respect to the nodal displacements

$$\frac{\partial \tilde{P}(\phi)}{\partial \{\phi\}} = 0$$

which for Equation 6.110 yields

$$\int_{\mathcal{V}} [B]^{\mathrm{T}}[D][B]\{\phi\}\,\mathrm{d}\mathcal{V} - \int_{\mathcal{V}} [N]^{\mathrm{T}}\{f_{\mathrm{B}}\}\mathrm{d}\mathcal{V} - \int_{S} [N]^{\mathrm{T}}\{f_{\mathrm{S}}\}\mathrm{d}S = 0$$

$$\int_{\mathcal{V}} [B]^{\mathrm{T}}[D][B]\{\phi\}\,\mathrm{d}\mathcal{V} = \int_{\mathcal{V}} [N]^{\mathrm{T}}\{f_{\mathrm{B}}\}\mathrm{d}\mathcal{V} + \int_{S} [N]^{\mathrm{T}}\{f_{\mathrm{S}}\}\mathrm{d}S \tag{6.111}$$

Equation 6.111 can be rewritten as the following element equation in a compact form:

$$[K_{\mathrm{e}}]\{\phi\} = \{F_{\mathrm{e}}\} \tag{6.112}$$

where $[K_{\mathrm{e}}]$ is the element stiffness matrix and $[F_{\mathrm{e}}]$ the nodal force matrix, expressed as follows:

$$[K_e] = \int_V [B]^T [D] [B] \{\phi\} \, dV$$

$$[F_e] = \int_V [N]^T \{f_B\} \, dV + \int_S [N]^T \{f_S\} \, dS$$

Equation 6.112 includes the influence from the mechanical forces only. For thermo-mechanical FEA, we augment this equation to include "thermal forces" $\{Q_e\}$ as a body force arising from a nonuniform temperature distribution over the element and write

$$[K_e]\{\phi\} = \{F_e\} + \{Q_e\} \tag{6.113}$$

where

$$\{Q_e\} = \int_V \alpha [B]^T [D] \{T_e\} \, dV \tag{6.114}$$

Finally, for the global equilibrium of the entire finite element model, we assemble the element matrices and element displacement vector of Equation 6.113 and write the global equation as

$$\left[K_g\right]\left\{\phi_g\right\} = \left\{F_g\right\} + \left\{Q_g\right\} \tag{6.115}$$

where, for example, $\left[K_g\right] = \sum [K_e]$. The numerical solution of this equation provides the distribution of three displacement components at each node of the finite element model. Using this distribution, we then compute (postprocess) throughout the model other quantities, namely the distributions of stress and strain.

6.6 Validation with Engine Test Data

As opposed to verification, which ensures that the design equations are correctly coded in a design code or the methodology process captured in a checklist has been adhered to, validation ensures acceptable comparison between the engine test data and correspond-ing predictions. While verification is prerequisite to validation, the latter is not a one-time deal. It must be performed continuously not only for different operating points a gas turbine engine but across the entire product line.

The validation process typically followed in a gas turbine OEM is shown in Figure 6.15. Using the current WEM methodology, pretest predictions are carried out first. The prediction results become helpful in the design of engine instrumentation layout for the planned engine tests at various operating points. Once the tests are completed and data gathered, it becomes necessary to carry out the posttest prediction for the exact operating points and boundary conditions used during testing. These prediction results are meticulously compared with the test data, eliminating those that are unreliable. If the comparison is satisfactory, the WEM methodology used in the posttest predictions becomes the new design practice for future design applications.

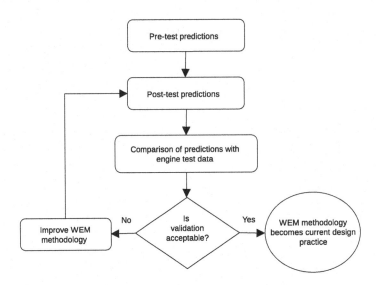

Figure 6.15 WEM methodology validation process.

Otherwise, the methodology is reinforced with improvements in physics-based modeling and adjustments in various empirical constants.

A fully validated WEM methodology is the main workhorse of a gas turbine design system. With well-defined interface boundary conditions, the methodology is equally applicable to the robust design of major gas turbine components, for example, internally cooled turbine vanes and blades. Key design implications of WEM are (1) improved component life prediction, (2) improved performance of various seals, and (3) reduction in cooling and sealing flows for higher cycle efficiency.

6.7   Concluding Remarks

As discussed in this chapter, WEM embodies the multiphysics (fluid flow, heat transfer, and thermomechanical) modeling of the entire gas turbine engine without explicitly including the aerodynamic design of compressor and turbine primary flow paths. While making use of various concepts and methods presented in previous chapters, we have introduced here a new analytical formulation, including its closed-form solution, of a multisurface forced vortex convection link to accurately compute changes in coolant fluid temperature over the link by way of convective heat transfer from the linked rotor and stator surfaces and work transfer (windage) from rotor surfaces. The network of convection links acts a buffer layer between the flow network model and structural heat transfer model. Note that the convection links get their mass flow rates from the associated flow network whose nodal temperatures, in turn, are iteratively updated by these convection links.

Theoretically, CFD offers the best capability for a 3-D thermofluids analysis is gas turbine design. However, for internal flow systems, it is not yet widely accepted as a reliable, fast, and cost-effective design method. In this chapter, however, we have provided helpful suggestions to leverage the key strengths of continuously improving CFD technology in physics-based design of gas turbine internal flow systems. First, 3-D CFD can be used effectively as a quick flow visualization tool. The enhanced understanding of the flow field from the CFD results can be used to reinforce flow network modeling of a design. Unlike CFD, FNM allows easy adjustment of empirical correlations for an acceptable product validation. For the short design cycle time available, a design engineer will most likely prefer a fine-tuned flow network model, which is considered a reduced-order model, to using a full 3-D CFD model for a robust design, requiring multiple probabilistic analyses. The second powerful application of 3-D CFD discussed here pertains to generating an entropy map from the CFD results. Such a map is extremely valuable in designing out local areas of unacceptable irreversibility, thereby improving the overall design performance. The details on CFD discussed in this chapter will allow the readers to use one or more commercial CFD codes more judiciously, rather than just a black box. For further study in this area, readers are referred to the references and bibliography given here.

We have provided in this chapter a brief overview of thermomechanical FEA, primarily for the benefit of readers who are most familiar with the thermofluids analysis methods. When it comes to mechanical design of various components of a gas turbine engine, each component structural integrity and durability become of paramount importance and cannot be compromised in favor of cycle performance for engine higher efficiency. The thermomechanical FEA using a commercial code is the basis of almost all mechanical design. As discussed in this chapter, the thermal FEA precedes a thermomechanical FEA. The thermal FEA is generally performed by a heat transfer engineer, often using the same commercial code. According to the process outlined in this chapter, in order to execute the thermal FEA the heat transfer engineer will need to develop both the network of convection links and flow network model, requiring a huge amount of manually-input data on geometry, modeling constants, and boundary conditions. It is no wonder, heat transfer and SAS engineers are often unfairly blamed for delaying the engine design cycle time. In contrast, a thermomechanical FEA is performed quickly!

In summary, WEM embodies a fully integrated procedure that takes into account the interaction between the secondary air system and the changes in the seal clearances and leakage gaps as a result of the thermal and mechanical loads, which vary during engine transients. A continuously validated WEM enables gas turbine design engineers to accurately predict the life (LCF and HCF) of various critical components, ensuring their structural integrity and durability so that the entire engine operation becomes highly reliable through a cost-effective maintenance schedule.

Project

6.1 Armand (1995) presents a finite-difference-based structural optimization methodology for a gas turbine rotor disk. In this project, you are asked to reinforce this methodology by disk temperatures obtained using the methodology of transient numerical heat transfer of a rotor disk presented in Appendix C. Develop a general-purpose, stand-alone computer code in FORTRAN, or the language of your choice, that combines the two methodologies for a reduced-order, thermomechanical analysis of a rotor disk operating under a given transient mission to determine the variation of the clearance s between the blade tip and casing , as shown in Figure 6.16.

Figure 6.16 shows schematically a rotor disk, which is convectively cooled on both its forward and aft faces with equal coolant air mass flow rate of $\dot{m}_c = 1.0$ kg/s at the inlet temperature of $T_{t_{c_{in}}} = 400$ K. Fifty blades are mounted equidistant on the disk. Each blade may be assumed to be a circular truncated cone with dimensions: $D_{b_1} = 15$ mm; $D_{b_2} = 5$ mm; $H_b = 20$ mm. The casing inner and outer radii under cold conditions are $R_1 = 322$ mm and $R_2 = 327$ mm, respectively.

Additional data and assumptions for the project are given as follows:

Disk geometry:

Radius (mm)	Thickness (mm)
40	70
80	70
100	20
270	20
280	40
300	40

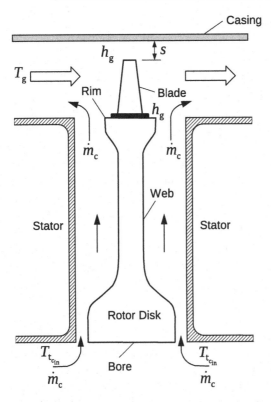

Figure 6.16 Cooled rotor disk with blades in a casing for clearance prediction during a transient operation (Project 6.1).

Thermophysical properties of disk, blade, and casing material:

Density: $\rho_m = 8220 \text{ kg/m}^3$
Young's modulus of elasticity: $E = 139 \times 10^9 \text{ Pa}$
Poisson's ratio: $v = 0.4$
Thermal conductivity: $k_m = 24.9 \text{ W/(m K)}$
Specific heat: $c_m = 645 \text{ J/(kg K)}$
Coefficient of thermal expansion: $\alpha = 15.1 \times 10^{-6} \text{ 1/K}$

Thermal boundary conditions:

- Disk bore surface is insulated, and so are the stators forming the disk cavities.
- Transient convective boundary conditions on both disk faces are to be based on the heat transfer coefficients computed using the Dittus-Boelter equation $Nu = 0.023 Re^{0.8} Pr^{0.333}$ where the characteristic velocity corresponds to the local tangential velocity of the coolant air.
- Assume a constant swirl factor of 0.5 for the coolant air flow in the cavity between the rotor disk and the adjacent stator. The minimum gap between them at the bore is 3 mm.
- On the entire disk rim surface and casing inner surface, use convective boundary conditions, which linearly ramp up from $T_g = 500$ K and $h_g = 500 \text{ W/(m}^2\text{ K)}$ to $T_g = 1500$ K and $h_g = 2500 \text{ W/(m}^2\text{ K)}$ in 20 seconds from a cold start with the entire system at 298 K.
- Neglect blade heat transfer but include its centrifugal load acting on the rotor disk at its rim surface. The disk bore surface is free from any mechanical load.
- Use $T_{amb} = 298$ K and $h_{amb} = 50 \text{ W/(m}^2\text{ K)}$ as convective boundary conditions on the casing outer surface.

Transient mission:

Carry out the transient mission analysis for 100 seconds where the rotor disk linearly ramps up from rest to 10,000 rpm in 25 seconds and remains at this value during the rest of the mission.

References

Armand, S. C. 1995. *Structural Optimization Methodology for Rotating Disks of Aircraft Engines*. NASA TM 4693.

Bradshaw, P. 1997. Understanding and prediction of turbulent flow – 1996. *Int. J. Heat Fluid Flow*. 18:45–54.

Daly, B. J., and F. H. Harlow. 1970. Transport equations of turbulence. *Phys. Fluids*. 13: 2634–2649.

Durbin, P. A., and T.I-P. Shih. 2005. An overview of turbulence modeling. In B. Sunden and M. Faghri, eds., *Modeling and Simulation of Turbulent Heat Transfer*. Ashurst, UK: WIT Press.

Gosman, A. D., W. M. Pun, A. K. Runchal, D. B. Spalding, and M. Wolfshtein. 1969. *Heat and Mass Transfer in Recirculating Flows*. London: Academic Press.

Gibson, M. M., and B. E. Launder. 1976. On the calculation of horizontal turbulent shear flows under gravitational influence. *ASME J. Heat Transfer*. 98: 81–87.

Hanjalic, K., and B. E. Launder. 1972. A Reynolds stress model of turbulence and its application to thin shear flows. *J. Fluid Mech.* 52: 609–638.

Hearn, E. J. 1997. *Mechanics of Materials 2, Third Edition: The Mechanics of Elastic and Plastic Deformation of Solids and Structural Materials.* Oxford: Butterworth-Heinemann.

Hughes, W. F., and E. W. Gaylord. 1964. *Basic Equations of Engineering Science. Schaum's Outline Series.* New York: McGraw-Hill.

Jones, W. P., and B. E. Launder. 1972. The prediction of laminarization with a two-equation model of turbulence. *Int. J. Heat Mass Transfer.* 15: 301–314.

Launder, B. E. 1971. *An Improved Algebraic Stress Model of Turbulence.* Mechanical Engineering Department, Imperial College, London, Report TM/TN/A8.

Launder, B. E. 1975. On the effects of gravitational field on turbulent transport of heat and momentum. *J. Fluid Mech.* 67: 569–581.

Launder, B. E., and D. B. Spalding. 1974. The numerical computation of turbulent flows. *Comp. Methods Appl. Mech. Eng.* 3: 269–289.

Launder, B. E., G. J. Reece, and W. Rodi. 1975. Progress in the development of a Reynolds-stress turbulence closure. *J. Fluid Mech.* 68: 537–566.

Leschziner, M. 2016. *Statistical Turbulence Modeling for Fluid Dynamics – Demysti ed: An Introductory Text for Graduate Engineering Students.* London: Imperial College Press.

Lilley, D. B., and D. L. Rhode. 1982. *STARPIC: A Computer Code for Swirling Turbulent Axisymmetric Recirculating Flows in Practical Isothermal Combustor Geometries.* NASA CR-3442.

Luo, J., E. H. Razinsky, and H.-K. Moon. 2012. Three-Dimensional RANS prediction of gas-side heat transfer coefficients on turbine blade and endwall. *J. Turbomachinery.* 135 (2): 1–11.

Meller, G., and T. Yamada. 1974. A hierarchy of turbulence models for planetary boundary layers. *J. Atmos. Sci.* 31: 1971–1982.

Moon, L. F., and G. Rudinger. 1977. Velocity distribution in an abruptly expanding circular duct. *ASME J. Fluids Eng.* 99: 226–230.

Naot, D., A. Shavit, and M. Wolfshtein. 1973. Two-point correlation model and redistribution of Reynolds stresses. *Phys. Fluids.* 16: 738–743.

Naterer, G. F., and J. A. Camberos. 2008. *Entropy-Based Design and Analysis of Fluids Engineering Systems.* Boca Raton, FL: CRC Press.

Patankar, S. V. 1980. *Numerical Heat Transfer and Fluid Flow.* Boca Raton, FL: CRC Press.

Pletcher, R. H., J. C. Tannehill, and D. A. Anderson. 2012. *Computational Fluid Mechanics and Heat Transfer, Third Edition.* Boca Raton, FL: CRC Press.

Pope, S. B., and J. H. Whitelaw. 1976. The calculation of near-wake flows. *J. Fluid Mech.* 73: 9–32.

Rodi, W. 1976. A new algebraic relation for calculating the Reynolds stresses. *ZAMM* 56: 219–221.

Rodi, W. 1980. *Turbulence Models and Their Applications in Hydraulics – a State of the Art Review.* Deft, Netherlands: International Association for Hydraulic Research.

Rotta, J. C. 1951. Statistische theorie nchthomogener turbulence. *Z.F. Physiks.* 129: 547–572. 131: 51–77.

Sciubba, E. 1997. Calculating entropy with CFD. *Mech. Eng.* 119(10): 86–88.

Sultanian, B. K. 1984. Numerical modeling of turbulent swirling flow downstream of an abrupt pipe expansion. PhD diss., Arizona State University.

Sultanian, B. K., and D. A. Nealy. 1987. Numerical modeling of heat transfer in the flow through a rotor cavity. In D.E. Metzger, ed., *Heat Transfer in Gas Turbines*, HTD-Vol. 87, 11–24. New York: The ASME

Sultanian, B. K., G. P. Neitzel, and D. E. Metzger. 1986. A study of sudden expansion pipe flow using an algebraic stress transport model. Paper presented at the AIAA/ASME 4th Fluid Mechanics, Plasma Dynamics and Lasers Conference, Atlanta.

Sultanian, B. K., G. P. Neitzel, and D. E. Metzger. 1987a. Comment on the flow field in a suddenly enlarged combustion chamber. *AIAA J.* 25(6): 893–895.

Sultanian, B. K., G. P. Neitzel, and D. E. Metzger. 1987b. Turbulent flow prediction in a sudden axisymmetric expansion. In C. J. Chen, L. D. Chen, and F. M. Holly, eds., *Turbulence measurements and Flow Modeling*. New York: Hemisphere.

Thompson, J. F., B. K. Soni, and N. P. Weatherill, eds. 1998. *Handbook of Grid Generation*. Boca Raton, FL: CRC Press.

Wilcox, D. C. 2006. *Turbulence Modeling for CFD*, 3rd edn. La Cañada Flintridge, CA: DCW Industries.

Wyngaard, J. C., S. P. Arya, and O. R. Cote. 1974. Some aspects of the structure of convective planetary boundary layers. *J. Atmos. Sci.* 31: 747–754.

Bibliography

Bradshaw, P. 1971. *An Introduction to Turbulence*. New York: Pergamon Press.

Cebeci, T., and P. Bradshaw. 1984. *Physical and Computational Aspects of Convective Heat Transfer*. New York: Springer

Durbin, P. A. 1991. Near-wall turbulence modeling without damping functions. *Theor. Comput. Fluid Dyn.* 3: 1–13.

Durbin, P. A. 1993. A Reynolds-stress model for near-wall turbulence. *J. Fluid Mech.* 249: 465–498.

Durbin, P. A., and G. Medic. 2007. *Fluid Dynamics with a Computational Perspective*. Cambridge: Cambridge University Press.

Engelman, M. S. 1993. CFD – an industrial perspective. In M. D. Gunzburger and R. A. Nicolaides, eds., *Incompressible Computational Fluid Dynamics*. Cambridge: Cambridge University Press.

Ferziger, J. H., and M. Peric. 2013. *Computational Methods for Fluid Dynamics*, 3rd edn. New York: Springer.

Gilham, S., I. R. Cowan, and E. S. Kaufman. 1999. Improving gas turbine power plant safety: The application of computational fluid dynamics to gas leaks. Proceedings of the Inst. *Mech. Eng. Part A: J. Power Energy.* 213 (6): 475–489.

Hinze, J. O. 1975. *Turbulence*, 2nd edn. New York: McGraw-Hill.

Hsu, T.-R. 1986. *The Finite Element Method in Thermomechanics*. Boston: Allen & Unwin.

Hunt, J. C. R., and J. C. Vassilicos. 1991. Kolmogorov's contributions to the physical and geometrical understanding of small-scale turbulence and recent developments. In J. C. R. Hunt, O. M. Phillips, and D. Williams, eds., *Turbulence and Stochastic Processes: Kolmogorov's Ideas 50 Years On*. Cambridge: Cambridge University Press.

Hussain, F., and M. V. Melander. 1992. Understanding turbulence via vortex dynamics. In T. B. Gatski, S. Sarkar, and C. G. Speziale, eds., *Studies in Turbulence*. New York: Springer-Verlag.

Landahl, M. T., and E. Mollo-Christensen. 1992. *Turbulence and Random Process in Fluid Mechanics*, 2nd edn. Cambridge: Cambridge University Press.

Muller, Y. 2008. Secondary air system model for integrated thermomechanical analysis of a jet engine. ASME Paper No. GT2008-50078.

Piomelli, U. 1993. Application of large eddy simulations in engineering: an overview. In B. Galperin and S. A. Orszag, eds., *Large Eddy Simulation of Complex Engineering and Geophysical Flows*. Cambridge: Cambridge University Press.

Pope, S. B. 2000. *Turbulent Flows*. Cambridge: Cambridge University Press.

Ponnuraj, B., B. K. Sultanian, A. Novori, and P. Pecchi. 2003. 3D CFD analysis of an industrial gas turbine compartment ventilation system. Proceedings of ASME IMECE, Washington, DC.

Reddy, V. V., K. Selvam, and R. D. Prosperis. 2016. Gas turbine shutdown thermal analysis and results compared with experimental data. ASME Paper No. GT2016–56601.

Santon, R. C., J. W. Kidger, and C. J. Lea. 2002. Safety developments in gas turbine power applications. Proceedings of ASME Turbo Expo 2002. ASME Paper GT2002–30469. Amsterdam, Netherlands.

Shih, T.I.-P., and B. K. Sultanian. 2001. Computations of internal and film cooling. In B. Sunden and M. Faghri, eds., *Heat Transfer in Gas Turbines*. Ashurst, UK: WIT Press.

Shih, T.I.-P., and P. Durbin. 2014. Modeling and simulation of turbine cooling. In T. Shih and V. Yang, eds., *Turbine Aerodynamics, Heat Transfer, Materials, and Mechanics*. AIAA Progress Series, AIAA, Vol. 243.

Sultanian, B. K., and H. C. Mongia. 1986. Fuel nozzle air flow modeling. AIAA/ASME/SAE/ASEE 22nd Joint Propulsion Conference, Huntsville, Alabama. Paper No. AIAA-86–1667.

Sultanian, B. K., S. Nagao, and T. Sakamoto. 1999. Experimental and three-dimensional CFD investigation in a gas turbine exhaust system. ASME, *J. Eng. Gas Turbines Power*. 121: 364–374.

Tennekes, H., and J. L. Lumley. 1972. *A First Course in Turbulence*. Cambridge, MA: MIT Press.

Timoshenko, S., and J. Goodier. 1970. *Theory of Elasticity*. New York: McGraw-Hill.

Townsend, A. A. 1976. *The Structure of Turbulent Shear Flow*. Cambridge: Cambridge University Press.

Wang, J. X. 2005. *Engineering Robust Design with Six Sigma*. New York: Prentice Hall.

Zahavi, E., and V. Torbilo. 1996. *Fatigue Design: Life Expectancy of Machine Parts*. Boca Raton, FL: CRC Press.

Nomenclature

A	Section area; surface area; area of a 2-D finite element
A^{+}	Prandtl mixing length model constant
ASM	Algebraic stress model
BVM	Boussinesq viscosity model
c	Specific heat of structural material
c_p	Specific heat at constant pressure
$C_{\mathrm{D}}, C_{\mu'}$	Model constants in the one-equation $(k-L)$ turbulence model
$C_\mu, C_{\varepsilon_1}, C_{\varepsilon_2}$	Model constants in the two-equation $(k-\varepsilon)$ turbulence model
$C_s, C_\varepsilon, C_1, C_2$	Model constants in turbulence models RSTM and ASM
C_k	Convection of turbulent kinetic energy k

C_P	Pressure recovery coefficient
$C\overline{u_i u_j}$	Convection of $\overline{u_i u_j}$
CV	Control volume
CFD	Computational fluid dynamics
CHT	Conjugate heat transfer
D_k	Diffusion of turbulent kinetic energy k
$D\overline{u_i u_j}$	Diffusion of $\overline{u_i u_j}$
E	Young's modulus of elasticity
E_c	Constant in the logarithmic law of the wall
f_B	Body force per unit volume
f_S	Surface force per unit area
FDM	Finite difference method
FEA	Finite element analysis
FEM	Finite element method
FNM	Flow network model
G	Shear modulus
$\dot{G}$	Rate of uniform heat generation over duct length L
h	Specific enthalpy; heat transfer coefficient
HCF	High cycle fatigue
HTC	Heat transfer coefficient
k	Turbulent kinetic energy; solid thermal conductivity
$\tilde{k}$	Fluid thermal conductivity
K_c	von Karman constant
ℓ	Turbulent length scale
ℓ_m	Mixing length
L	Mixing length in the one-equation $k - L$ turbulence model; length
LCF	Low cycle fatigue
$\dot{m}$	Mass flow rate
m_ℓ	Mass fraction of chemical species ℓ
N	Number of sublinks (control volumes); finite element shape function
NTU	Number of transfer units
ODE	Ordinary differential equation
OEM	Original equipment manufacturer
p	Fluctuating part of P
P	Pressure
$\tilde{P}$	Potential energy of a finite element
P_f	Jaytilleke's P-function
P_k	Production rate of k
P_{ij}	Production of $\overline{u_i u_j}$
Pr	Prandtl number
Pr_t	Turbulent Prandtl number
$\dot{q}$	Heat flux
$\dot{Q}$	Heat transfer rate
r	Radius

R	Gas constant; radius
	Generation rate per unit volume (from chemical reaction)
RSTM	Reynolds stress transport model
s	Specific entropy
S	Source term in the common-form transport equation; surface area
$\tilde{S}$	Source terms other than pressure gradient terms in the momentum equation
S_f	Swirl factor
t	Time
T	Temperature
Tu	Turbulence intensity
T_{aw}	Adiabatic wall temperature
u	Fluctuating part of U
u_i	Fluctuating velocities in tensor notation
U	Velocity component along x direction
$\tilde{U}$	Internal energy of a finite element
U^+	Dimensionless velocity in a wall boundary layer
U^*	Shear velocity
U_i	Velocities in tensor notation
v	Fluctuating part of velocity along y direction
V	Total velocity; velocity component along y direction
$\forall$	Volume
$\vec{V}$	Velocity vector
W	Velocity component along z direction
$\tilde{W}$	Work done by external forces on a finite element
$\dot{W}$	Rate of uniform work transfer over duct length L
WEM	Whole engine modeling
x	Cartesian coordinate x
x_i	Cartesian coordinates in tensor notation
y	Cartesian coordinate y
y^+	Dimensionless transverse distance in a wall boundary layer
z	Cartesian coordinate z

Subscripts and Superscripts

b	Disk bore
c	Convection; coolant
e	Finite element
ext	External (boundary)
g	Global
h	For enthalpy; solution of the homogeneous part of ODE
i	Convection link with inflow into junction J
in	Inlet
int	Internal
j	Convection link with outflow from junction J

J	Junction
ℓ	Chemical species ℓ
m	Mean
me	Mechanical
m_ℓ	For mass fraction of chemical species ℓ
max	Maximum
min	Minimum
out	Outlet
P	Particular solution of ODE; New-wall node P
r	Radial direction
ref	Reference state
s	Static
t	Total (stagnation)
th	Thermal
U_i	For U_i
w	Wall
x	Component in x-coordinate direction
y	Component in y-coordinate direction
z	Component in z-coordinate direction
$(^-)$	Time-average or ensemble-average
$(^*)$	Dimensionless
θ	Tangential direction

Greek Symbols

α	Coefficient of thermal expansion
Γ	Diffusion coefficient
δ	Velocity boundary layer thickness
δ_{ij}	Kronecker delta = 1 for $i = j$ and = 0 for $i \neq j$
ε	Dissipation rate of k; strain
$\tilde{\varepsilon}$	Dimensionless heat generation rate, $\tilde{\varepsilon} = \dot{G}/\left\{ \dot{m}c_p \left(T_w - T_{t_{in}} \right) \right\}$
$\hat{\varepsilon}$	Dimensionless heat generation rate through windage
ε_{ij}	Dissipation rate of $\overline{u_i u_j}$
η	Number of transfer unit, $\eta = \left(hA_w \right)/\left(\dot{m}c_p \right)$
θ	Dimensionless temperature
κ	Ratio of specific heats
λ	Ratio P_k/ε
μ	Dynamic viscosity
μ_t	Turbulent (or eddy) viscosity
ν	Kinematic viscosity ($\nu = \mu/\rho$), Poisson's ratio
ν_t	Kinematic eddy viscosity ($\nu_t = \mu_t/\rho$)
ξ	Dimensionless distance along x direction ($\xi = x/L$)
σ	Normal stress

σ_k	Turbulent Prandtl number for k
σ_ε	Turbulent Prandtl number for ε
ρ	Density
τ	Shear stress
υ	Turbulent velocity scale
φ	Local structural displacement
ϕ	Fluctuating part of Φ; structural displacement in a finite element
ϕ_{ij}	Pressure-strain term in RSTM
Φ	General transport variable of the common-form governing transport equation
Ω	Angular velocity

Appendix A Review of Necessary Mathematics

A.1 Suf x Notation and Tensor Algebra

In Cartesian coordinates, the unit vectors along the x, y, and z coordinate directions are denoted by $\hat{i}$, $\hat{j}$, and $\hat{k}$, respectively. In Cartesian tensor, we name the x, y, and z coordinates as x_1, x_2, and x_3, respectively, with the corresponding unit vectors denoted by $\hat{e}_1$, $\hat{e}_2$, and $\hat{e}_3$. A typical vector $\vec{a}$ with components a_1, a_2, and a_3 in Cartesian coordinates can be expressed as

$$\vec{a} = a_1\hat{i} + a_2\hat{j} + a_3\hat{k} = a_1\hat{e}_1 + a_2\hat{e}_2 + a_3\hat{e}_3$$

A.1.1 Summation Convention

When an index is repeated precisely twice in a term, it implies summation over all possible values of the index, which has values 1, 2, and 3 in Cartesian coordinates. Accordingly, we can express our vector $\vec{a}$ in the compact tensor notation as

$$\vec{a} = a_i\hat{e}_i$$

Note that the summation convention possesses the commutative and distributive properties as shown here.

Commutative:

$$a_ib_i = b_ia_i$$

Distributive:

$$a_i(b_i + c_i) = a_ib_i + a_ic_i$$

A.1.2 Free and Dummy Indices

In the term on the left-hand side of the following equation, the index j is repeated twice, implying summation. This index is called the dummy index. By contrast, the index m can have any value, and is called the free index:

$$a_{mj}x_j = c_m$$

This equation can be alternatively written as

$$a_{kq}x_q = c_k$$

317

Both of these equations are a compact form (in tensor notation) of the following three equations:

$$a_{11}x_1 + a_{12}x_2 + a_{13}x_3 = c_1$$

$$a_{21}x_1 + a_{22}x_2 + a_{23}x_3 = c_2$$

$$a_{31}x_1 + a_{32}x_2 + a_{33}x_3 = c_3$$

A.1.3   Two Special Symbols

Kronecker delta:

$$\delta_{ij} = \hat{e}_i \bullet \hat{e}_j = \begin{cases} 1 & \text{if } i = j \\ 0 & \text{if } i \neq j \end{cases}$$

$$\delta_{ij} = \begin{bmatrix} \delta_{11} & \delta_{12} & \delta_{13} \\ \delta_{21} & \delta_{22} & \delta_{23} \\ \delta_{31} & \delta_{32} & \delta_{33} \end{bmatrix} = \begin{bmatrix} 1 & 0 & 0 \\ 0 & 1 & 0 \\ 0 & 0 & 1 \end{bmatrix}$$

The alternating symbol:

$$\varepsilon_{ijk} = \hat{e}_i \bullet \left(\hat{e}_j \times \hat{e}_k \right) = \begin{cases} 1 & \text{if } i, j, k \text{ are a cyclic permutation of } 1, 2, \text{and } 3 \\ -1 & \text{if } i, j, k \text{ are a anticyclic permutation of } 1, 2, \text{and } 3 \\ 0 & \text{if any index is repeated} \end{cases}$$

According to the definition of ε_{ijk}, we can write the following identities:

$$\varepsilon_{ijk} = \varepsilon_{kij} = \varepsilon_{jki} \quad \text{and} \quad \varepsilon_{jik} = -\varepsilon_{ijk}$$

A.2   Gradient, Divergence, Curl, and Laplacian

A.2.1   Gradient

Define:

$$\nabla = \frac{\partial}{\partial x}\hat{i} + \frac{\partial}{\partial y}\hat{j} + \frac{\partial}{\partial z}\hat{k} = \hat{e}_i \frac{\partial}{\partial x_i}$$

For example, we can write the gradient of a scalar Φ as

$$\nabla\Phi = \frac{\partial\Phi}{\partial x}\hat{i} + \frac{\partial\Phi}{\partial y}\hat{j} + \frac{\partial\Phi}{\partial z}\hat{k} = \frac{\partial\Phi}{\partial x_1}\hat{e}_1 + \frac{\partial\Phi}{\partial x_2}\hat{e}_2 + \frac{\partial\Phi}{\partial x_3}\hat{e}_3 = \hat{e}_i \frac{\partial\Phi}{\partial x_i}$$

In cylindrical coordinates, the above equation becomes

$$\nabla\Phi = \frac{\partial\Phi}{\partial r}\hat{e}_r + \frac{1}{r}\frac{\partial\Phi}{\partial\theta}\hat{e}_\theta + \frac{\partial\Phi}{\partial z}\hat{e}_z$$

where $\hat{e}_r$, $\hat{e}_\theta$, and $\hat{e}_z$ are the unit vectors along the r, θ, and z coordinate directions, respectively.

A.2.2 Divergence

We can write the divergence of a vector $\vec{V} = u\hat{i} + v\hat{j} + w\hat{k} = \hat{e}_i u_i$ as

$$\nabla \bullet \vec{V} = \frac{\partial u}{\partial x} + \frac{\partial v}{\partial y} + \frac{\partial w}{\partial z} = \frac{\partial u_1}{\partial x_1} + \frac{\partial u_2}{\partial x_2} + \frac{\partial u_3}{\partial x_3} = \frac{\partial u_i}{\partial x_i}$$

In cylindrical coordinates, this equation becomes

$$\nabla \bullet \vec{V} = \frac{1}{r}\frac{\partial (r u_r)}{\partial r} + \frac{1}{r}\frac{\partial u_\theta}{\partial \theta} + \frac{\partial u_z}{\partial z}$$

A.2.3 Curl

We can write the curl of a vector $\vec{V} = u\hat{i} + v\hat{j} + w\hat{k} = \hat{e}_i u_i$

$$\nabla \times \vec{V} = \begin{bmatrix} \hat{i} & \hat{j} & \hat{k} \\ \dfrac{\partial}{\partial x} & \dfrac{\partial}{\partial y} & \dfrac{\partial}{\partial z} \\ u & v & w \end{bmatrix} = \begin{bmatrix} \hat{e}_1 & \hat{e}_2 & \hat{e}_3 \\ \dfrac{\partial}{\partial x_1} & \dfrac{\partial}{\partial x_2} & \dfrac{\partial}{\partial x_3} \\ u_1 & u_2 & u_3 \end{bmatrix}$$

$$\nabla \times \vec{V} = \left(\frac{\partial w}{\partial y} - \frac{\partial v}{\partial z}\right)\hat{i} + \left(\frac{\partial u}{\partial z} - \frac{\partial w}{\partial x}\right)\hat{j} + \left(\frac{\partial v}{\partial x} - \frac{\partial u}{\partial y}\right)\hat{k}$$

$$\nabla \times \vec{V} = \left(\frac{\partial u_3}{\partial x_2} - \frac{\partial u_2}{\partial x_3}\right)\hat{e}_1 + \left(\frac{\partial u_1}{\partial x_3} - \frac{\partial u_3}{\partial x_1}\right)\hat{e}_2 + \left(\frac{\partial u_2}{\partial x_1} - \frac{\partial u_1}{\partial x_2}\right)\hat{e}_3 = \hat{e}_i \varepsilon_{ijk} \frac{\partial u_k}{\partial x_j}$$

In cylindrical coordinates, this equation becomes

$$\nabla \times \vec{V} = \left(\frac{1}{r}\frac{\partial u_z}{\partial \theta} - \frac{\partial u_\theta}{\partial z}\right)\hat{e}_r + \left(\frac{\partial u_r}{\partial z} - \frac{\partial u_z}{\partial r}\right)\hat{e}_\theta + \left(\frac{1}{r}\frac{\partial (r u_\theta)}{\partial r} - \frac{1}{r}\frac{\partial u_r}{\partial u_\theta}\right)\hat{e}_z$$

A.2.4 Laplacian

We can write the Laplacian of a scalar Φ as

$$\nabla^2 \Phi = \nabla \bullet \nabla \Phi = \frac{\partial^2 \Phi}{\partial x^2} + \frac{\partial^2 \Phi}{\partial y^2} + \frac{\partial^2 \Phi}{\partial z^2} = \frac{\partial^2 \Phi}{\partial x_1^2} + \frac{\partial^2 \Phi}{\partial x_2^2} + \frac{\partial^2 \Phi}{\partial x_3^2} = \frac{\partial}{\partial x_i}\frac{\partial \Phi}{\partial x_i}$$

In cylindrical coordinates, this equation becomes

$$\nabla^2 \Phi = \frac{\partial^2 \Phi}{\partial r^2} + \frac{1}{r}\frac{\partial \Phi}{\partial r} + \frac{1}{r^2}\frac{\partial^2 \Phi}{\partial \theta^2} + \frac{\partial^2 \Phi}{\partial z^2} = \frac{1}{r}\frac{\partial}{\partial r}\left(r\frac{\partial \Phi}{\partial r}\right) + \frac{1}{r^2}\frac{\partial^2 \Phi}{\partial \theta^2} + \frac{\partial^2 \Phi}{\partial z^2}$$

A.3   Dyad in Total Derivative

The total or substantial derivative of the velocity vector $\vec{V}$ can be written in vector form as

$$\frac{\mathrm{D}\vec{V}}{\mathrm{D}t} = \frac{\partial \vec{V}}{\partial t} + \left(\vec{V} \cdot \nabla\right)\vec{V}$$

which makes it independent of the coordinate system. In this equation, $\vec{V} \cdot \nabla$ is called a dyad, which is easily handled in the tensor notation. Using the vector identity, we can express the dyad in the equation as

$$\left(\vec{V} \cdot \nabla\right)\vec{V} = \nabla\left(\frac{1}{2}V^2\right) - \vec{V} \times \left(\nabla \times \vec{V}\right)$$

A.4   Total Derivative

The total or substantial or material derivative following a particle (Lagrangian viewpoint) in a fluid flow can be written in various forms (vector, differential, and tensor) as given in the following.

Cartesian coordinates:

$$\frac{\mathrm{D}\vec{V}}{\mathrm{D}t} = \frac{\partial \vec{V}}{\partial t} + \left(\vec{V} \cdot \nabla\right)\vec{V}$$

$$\frac{\mathrm{D}\vec{V}}{\mathrm{D}t} = \left(\frac{\partial u}{\partial t} + u\frac{\partial u}{\partial x} + v\frac{\partial u}{\partial y} + w\frac{\partial u}{\partial z}\right)\hat{i} + \left(\frac{\partial v}{\partial t} + u\frac{\partial v}{\partial x} + v\frac{\partial v}{\partial y} + w\frac{\partial v}{\partial z}\right)\hat{j}$$
$$+ \left(\frac{\partial w}{\partial t} + u\frac{\partial w}{\partial x} + v\frac{\partial w}{\partial y} + w\frac{\partial w}{\partial z}\right)\hat{k}$$

Cartesian tensor notation:

$$\frac{\mathrm{D}\vec{V}}{\mathrm{D}t} = \frac{\partial u_i}{\partial t} + u_j\frac{\partial u_i}{\partial x_j}$$

Cylindrical coordinates:

$$\frac{\mathrm{D}\vec{V}}{\mathrm{D}t} = \left(\frac{\partial u_r}{\partial t} + u_r\frac{\partial u_r}{\partial r} + \frac{u_\theta}{r}\frac{\partial u_r}{\partial \theta} + u_z\frac{\partial u_r}{\partial z} - \frac{u_\theta^2}{r}\right)\hat{e}_r$$
$$+ \left(\frac{\partial u_\theta}{\partial t} + u_r\frac{\partial u_\theta}{\partial r} + \frac{u_\theta}{r}\frac{\partial u_\theta}{\partial \theta} + u_z\frac{\partial u_\theta}{\partial z} + \frac{u_r u_\theta}{r}\right)\hat{e}_\theta$$
$$+ \left(\frac{\partial u_z}{\partial t} + u_r\frac{\partial u_z}{\partial r} + \frac{u_\theta}{r}\frac{\partial u_z}{\partial \theta} + u_z\frac{\partial u_z}{\partial z}\right)\hat{e}_z$$

A.5   Vector Identities

In the following vector identities, Φ is a scalar and $\vec{A}$, $\vec{B}$, and $\vec{C}$ are vectors:

$$\nabla \times \nabla \Phi = 0$$

$$\nabla \bullet \left(\nabla \times \vec{A} \right) = 0$$

$$\nabla \bullet \left(\Phi \vec{A} \right) = \Phi \left(\nabla \bullet \vec{A} \right) + \vec{A} \bullet \nabla \Phi$$

$$\nabla \times \left(\Phi \vec{A} \right) = \nabla \Phi \times \vec{A} + \Phi \left(\nabla \times \vec{A} \right)$$

$$\left(\vec{A} \bullet \nabla \right) \vec{A} = \frac{1}{2} \nabla \left(\vec{A} \bullet \vec{A} \right) - \vec{A} \times \left(\nabla \times \vec{A} \right)$$

$$\nabla \times \left(\nabla \times \vec{A} \right) = \nabla \left(\nabla \bullet \vec{A} \right) - \nabla^2 \vec{A}$$

$$\nabla \bullet \left(\vec{A} \times \vec{B} \right) = \vec{B} \bullet \left(\nabla \times \vec{A} \right) - \vec{A} \bullet \left(\nabla \times \vec{B} \right)$$

$$\nabla \times \left(\vec{A} \times \vec{B} \right) = \vec{A} \left(\nabla \bullet \vec{B} \right) - \vec{B} \left(\nabla \bullet \vec{A} \right) - \left(\vec{A} \bullet \nabla \right) \vec{B} + \left(B \bullet \nabla \right) \vec{A}$$

For further details on the mathematical topics presented in this appendix, and for additional topics, interested readers may refer to the references listed in the bibliography given here.

Bibliography

Aris, R. 1962. *Vectors, Tensors, and the Basic Equations of Fluid Mechanics*. Englewood Cliffs, NJ: Prentice Hall.

Hughes, W. A., and Gaylord, E. W. 1964. *Basic Equations of Engineering Science*. New York: McGraw-Hill.

Appendix B Equations of Air Thermophysical Properties

Zografos, Martin, and Sunderland (1987) provide polynomials to compute dynamic viscosity, thermal conductivity of air, and specific heat at constant pressure of air as a function of static temperature T_s on the Kelvin scale as follows:

Dynamic viscosity (μ) in units of N s/m^2 = kg/(m s) = Pa s:

$$\mu = 4.112985 \times 10^{-6} + 5.052295 \times 10^{-8} T_s - 1.43462 \times 10^{-11} T_s^2 + 2.591403 \times 10^{-15} T_s^3$$

Thermal conductivity (k) in units of W/(m K):

$$k = -7.488 \times 10^{-3} + 1.708186 \times 10^{-4} T_s - 23.757762 \times 10^{-6} T_s^2 + 220.11791 \times 10^{-12} T_s^3$$
$$- 945.995536 \times 10^{-16} T_s^4 + 1579.657437 \times 10^{-20} T_s^5$$

Specific heat at constant pressure (c_p) in units of J/(kg K):

$$c_p = 1061.332 - 0.432819 T_s + 1.02344 \times 10^{-3} T_s^2 - 6.47474 \times 10^{-7} T_s^3$$
$$+ 1.3864 \times 10^{-10} T_s^4$$

Additionally, we can compute air density and Prandtl number as

Density (ρ) in units of kg/m^3:

$$\rho = \frac{P_s}{R T_s}$$

where P_s is the static pressure in Pascal, and the gas constant has the value $R = 287$ J/(kg K).

Prandtl number (Pr):

$$Pr = \frac{\mu c_p}{k}$$

Reference

Zografos, A. I., W. A. Martin, and J. E. Sunderland. 1987. Equations of properties as a function of temperature for seven fluids. Comput. Methods Appl. Mech. Eng. 61: 177–187.

Appendix C Transient Heat Transfer in a Rotor Disk

C.1   Introduction

Compressor and turbine rotor disks of a gas turbine rotating at very high angular velocity are subjected to both thermal and centrifugal loads and are considered critical as their failure is potentially catastrophic for the entire engine. For initial sizing of a rotor disk, its one-dimensional transient heat transfer analysis becomes necessary to carry out its thermomechanical analysis for assessing its mechanical integrity and operational life during the entire envelope of engine operational characteristics. In this appendix, we present a numerical method to compute the transient temperature response of a rotor disk whose thickness varies arbitrarily from bore to rim. The formulation allows arbitrary rim and bore thermal boundary conditions, including arbitrary convective boundary conditions on disk faces. We, however, assume no temperature variation across the disk thickness. The proposed numerical scheme used a finite-volume approach, being unconditionally stable and fully implicit in time.

C.2   Energy Balance on a Disk Element

Figure C.1a shows a rotor disk whose thickness varies from bore to rim. With no variation of disk temperature along its thickness, convective heat transfer at its forward and aft faces, at any time τ, an energy balance on a small disk element control volume shown in Figure C.1b, which spans Δr in the radial direction and $\Delta\theta$ in the circumferential direction, yields

$$\rho_{\mathrm{m}} c_{\mathrm{m}} \left(t + \frac{\mathrm{d}t}{\mathrm{d}r} \frac{\Delta r}{2} \right) \left(r + \frac{\Delta r}{2} \right) \Delta r \Delta\theta \frac{\partial T_{\mathrm{m}}(r,\tau)}{\partial \tau}$$

$$= \dot{q}_{\mathrm{cond}} t r \Delta\theta - \left\{ \dot{q}_{\mathrm{cond}} t r \Delta\theta + \frac{\partial}{\partial r} \left(\dot{q}_{\mathrm{cond}} t r \right) \Delta r \Delta\theta \right\}$$

$$- \dot{q}_{\mathrm{conv}_1} \left(r + \frac{\Delta r}{2} \right) \Delta r \Delta\theta - \dot{q}_{\mathrm{conv}_2} \left(r + \frac{\Delta r}{2} \right) \Delta r \Delta\theta$$

which, after neglecting second-order terms and for $\Delta r \to 0$, simplifies to

$$\rho_{\mathrm{m}} c_{\mathrm{m}} t \frac{\partial T_{\mathrm{m}}(r,\tau)}{\partial \tau} = -\frac{1}{r} \frac{\partial}{\partial r} \left(t r \dot{q}_{\mathrm{cond}} \right) - \dot{q}_{\mathrm{conv}_1} - \dot{q}_{\mathrm{conv}_2} \tag{C.1}$$

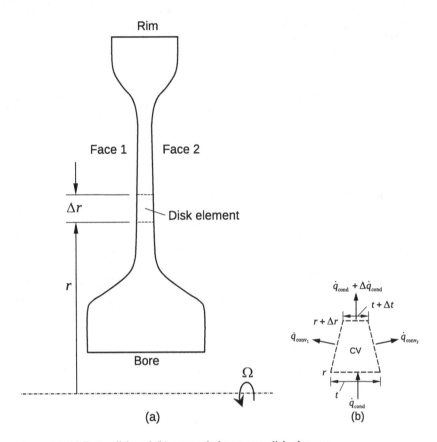

Figure C.1 (a) Rotor disk and (b) energy balance on a disk element.

where

$c_m \equiv$ Temperature-dependent specific heat of disk material
$\dot{q}_{cond} \equiv$ Radially varying conduction heat flux
$\dot{q}_{conv_1} \equiv$ Local convection heat flux at disk forward face
$\dot{q}_{conv_2} \equiv$ Local convection heat flux at disk aft face
$t \equiv$ Radially varying disk thickness
$T_m \equiv$ Disk temperature
$\rho_m \equiv$ Density of disk material
$\tau \equiv$ Time

From Fourier's law of heat condition, we write

$$\dot{q}_{cond} = -k_m \frac{\partial T_m}{\partial r} \tag{C.2}$$

where k_m is the temperature-dependent disk material thermal conductivity.
We write the convection flux on each disk face as

$$\dot{q}_{conv_1} = h_1 \left(T_m - T_{aw_1} \right) \tag{C.3}$$

$$\dot{q}_{\mathrm{conv}_2} = h_2 \left(T_{\mathrm{m}} - T_{\mathrm{aw}_2} \right) \tag{C.4}$$

where h_1 and h_2 are heat transfer coefficients on disk forward and aft faces, respectively, with corresponding adiabatic wall temperatures T_{aw_1} and T_{aw_2}.

Substituting Equations C.2–C.4 in Equation C.1 finally yields

$$\rho_{\mathrm{m}} c_{\mathrm{m}} t \frac{\partial T_{\mathrm{m}}}{\partial \tau} = \frac{1}{r} \frac{\partial}{\partial r} \left(t r k_{\mathrm{m}} \frac{\partial T_{\mathrm{m}}}{\partial r} \right) - h_1 \left(T_{\mathrm{m}} - T_{\mathrm{aw}_1} \right) - h_2 \left(T_{\mathrm{m}} - T_{\mathrm{aw}_2} \right) \tag{C.5}$$

which is a governing nonlinear second-order partial differential equation for time-dependent radial variation of axisymmetric disk temperature. The temperature-dependent thermophysical properties of the disk material render Equation C.5 nonlinear. For an arbitrary variation of disk thickness, this equation can only be solved numerically, as discussed in Section C.4.

Because the convective boundary conditions on disk faces are already included in Equation C.5, we need only one initial condition and two thermal boundary conditions for its solution.

Initial condition. At the start of the solution, we specify the radial temperature distribution in the rotor disk as

$$\tau = 0 \ : \ T_{\mathrm{m}}(r, 0) = T_{\mathrm{m}_0}(r) \tag{C.6}$$

Boundary conditions at disk bore. At the rotor disk bore, we may specify one of the following three types of thermal boundary condition:

Type 1: Specified surface temperature

$$T_{\mathrm{m}} \left(r_{\mathrm{bore}}, \tau \right) = T_{\mathrm{w}_{\mathrm{bore}}} (\tau) \tag{C.7}$$

where $T_{\mathrm{w}_{\mathrm{bore}}}$ is the disk bore surface temperature.

Type 2: Specified heat flux

$$-\left(k_{\mathrm{m}} \frac{\partial T_{\mathrm{m}}(r, \tau)}{\partial r} \right)_{r=r_{\mathrm{bore}}} = \dot{q}_{\mathrm{bore}}(\tau) \tag{C.8}$$

Note that, for disk bore cooling, $\dot{q}_{\mathrm{bore}}(\tau)$ is negative, and for heating, it is positive.

Type 3: Convective boundary condition

$$\left(k_{\mathrm{m}} \frac{\partial T_{\mathrm{m}}(r, \tau)}{\partial r} \right)_{r=r_{\mathrm{bore}}} = h_{\mathrm{bore}} \left(T_{\mathrm{w}_{\mathrm{bore}}} - T_{\mathrm{aw}_{\mathrm{bore}}} \right) \tag{C.9}$$

Boundary conditions at disk rim. At the rotor disk bore, we may specify one of the following three types of thermal boundary condition:

Type 1: Specified surface temperature

$$T_{\mathrm{m}} \left(r_{\mathrm{rim}}, \tau \right) = T_{\mathrm{w}_{\mathrm{rim}}} (\tau) \tag{C.10}$$

where $T_{\mathrm{w}_{\mathrm{rim}}}$ is the disk rim surface temperature.

Type 2: Specified heat flux

$$\left(k_{\mathrm{m}}\frac{\partial T_{\mathrm{m}}(r,\tau)}{\partial r}\right)_{r=r_{\mathrm{rim}}} = \dot{q}_{\mathrm{rim}}(\tau) \tag{C.11}$$

Note that, for disk rim cooling, $\dot{q}_{\mathrm{rim}}(\tau)$ is negative, and for heating, it is positive.

Type 3: Convective boundary condition

$$-\left(k_{\mathrm{m}}\frac{\partial T_{\mathrm{m}}(r,\tau)}{\partial r}\right)_{r=r_{\mathrm{rim}}} = h_{\mathrm{rim}}\left(T_{\mathrm{w}_{\mathrm{rim}}} - T_{\mathrm{aw}_{\mathrm{rim}}}\right) \tag{C.12}$$

C.3   Analytical Solution

In this section we develop an analytical solution of Equation C.5 for the particular case of transient heat transfer in a rotor disk with variable thickness and variable convective heat transfer coefficient on both disk faces under the following assumptions:

1. Disk thickness variation: $t = \dfrac{\pi_1}{r}$, where π_1 is a specified constant.
2. $h_1 = h_2 = h = \dfrac{\pi_1 \pi_2 k_{\mathrm{m}}}{2r}$, where π_2 is a specified constant.
3. Disk bore surface is insulated.
4. The disk is immerged in a fluid at constant temperature T_∞.
5. The heat transfer coefficient at the disk rim is h_{rim}.
6. Thermophysical properties of disk material are constant.

Under these assumptions, Equation C.5 reduces to

$$\frac{1}{\alpha}\frac{\partial \hat{T}_{\mathrm{m}}}{\partial \tau} = \frac{\partial^2 \hat{T}_{\mathrm{m}}}{\partial r^2} - \pi_2 \hat{T}_{\mathrm{m}} \tag{C.13}$$

where $\hat{T}_{\mathrm{m}} = T_{\mathrm{m}} - T_\infty$ and $\alpha = \dfrac{k_{\mathrm{m}}}{\rho_{\mathrm{m}} c_{\mathrm{m}}}$.

Now our task is to solve Equation C.13 subject to the following initial and boundary conditions:

$\tau = 0$:

$$\hat{T}_{\mathrm{m}}(r,0) = \hat{T}_{\mathrm{m}_0} \tag{C.14}$$

Disk bore $\left(r = r_{\mathrm{bore}}\right)$:

$$\left(\frac{\partial \hat{T}_{\mathrm{m}}}{\partial r}\right)_{r=r_{\mathrm{bore}}} = 0 \tag{C.15}$$

Disk rim $\left(r = r_{\mathrm{rim}}\right)$:

$$\left(\frac{\partial \hat{T}_{\mathrm{m}}}{\partial r}\right)_{r=r_{\mathrm{rim}}} = \frac{-h_{\mathrm{rim}}\hat{T}_{\mathrm{m}_{\mathrm{rim}}}}{k_{\mathrm{m}}} \tag{C.16}$$

Using the method of separation of variables, we can write the solution of Equation C.13 as

$$\hat{T}_{\mathrm{m}}(r,\tau) = S_r(r)S_\tau(\tau) \tag{C.17}$$

whose substitution in Equation C.13 yields

$$\frac{S_r}{\alpha}\frac{\mathrm{d}S_\tau}{\mathrm{d}\tau} = S_\tau \frac{\mathrm{d}^2 S_r}{\mathrm{d}r^2} - \pi_2 S_r S_\tau$$

$$\frac{1}{\alpha S_\tau}\frac{\mathrm{d}S_\tau}{\mathrm{d}\tau} = \frac{1}{S_r}\frac{\mathrm{d}^2 S_r}{\mathrm{d}r^2} - \pi_2 = -\lambda$$

where $-\lambda$ is the separation constant, which leads to two ordinary differential equations:

$$\frac{\mathrm{d}S_\tau}{\mathrm{d}\tau} + \alpha\lambda S_\tau = 0 \tag{C.18}$$

$$\frac{\mathrm{d}^2 S_r}{\mathrm{d}r^2} + \beta^2 S_r = 0 \tag{C.19}$$

where $\beta^2 = \lambda - \pi_2$ such that $\lambda > \pi_2$.

The general solution of Equation C.18 can be written as

$$S_\tau = C_1 e^{-\alpha\lambda\tau} \tag{C.20}$$

and that of Equation C.19 as

$$S_r = C_2 \cos{(\beta r)} + C_3 \sin{(\beta r)} \tag{C.21}$$

Now we can write the general solution of Equation C.13:

$$\hat{T}_\mathrm{m}(r,\tau) = C_1 e^{-\alpha\lambda\tau}\left(C_2 \cos{(\beta r)} + C_3 \sin{(\beta r)}\right)$$

$$\hat{T}_\mathrm{m}(r,\tau) = e^{-\alpha\lambda\tau}\left(\tilde{C}_1 \cos{(\beta r)} + \tilde{C}_2 \sin{(\beta r)}\right) \tag{C.22}$$

where $\tilde{C}_1 = C_1 C_2$ and $\tilde{C}_2 = C_1 C_3$, which along with the characteristic parameter β need to be evaluated from the given initial and boundary conditions, Equations C.14, C.15, and C.16, yielding the final solution

$$\hat{T}_\mathrm{m}(r,\tau) = \sum_{n=1}^{\infty} e^{-\alpha\lambda_n\tau}\left(\frac{2\sin{\left(\beta_n r_\mathrm{rim}\right)}\cos{\left(\beta_n r\right)}}{\beta_n r_\mathrm{rim} + \sin{\left(\beta_n r_\mathrm{rim}\right)}\cos{\left(\beta_n r_\mathrm{rim}\right)}}\right) \tag{C.23}$$

where

$$\cot{\left(\beta_n r_\mathrm{rim}\right)} = \frac{\beta_n r_\mathrm{rim}}{Bi} \tag{C.24}$$

$$\lambda_n = \pi_2 + \beta_n^2 \tag{C.25}$$

$$Bi = \frac{h_\mathrm{rim} r_\mathrm{rim}}{k_\mathrm{m}} \tag{C.26}$$

Note that Equation C.24 is transcendental in β_n. For $Bi = 1$, the first ten roots of this equation are tabulated here:

n	$\beta_n r_{\text{rim}}$
1	0.8603336
2	3.4256185
3	6.4372982
4	9.5293344
5	12.6452872
6	15.7712849
7	18.9024100
8	22.0364967
9	25.1724463
10	28.3096429

C.4   Numerical Solution

In Section C.2, we performed the energy balance on a disk element and derived Equation C.5, which is a nonlinear second-order partial differential equation to compute transient temperature in the disk resulting from one-dimensional radial heat conduction with simultaneous convection on disk surface. Because the analytical solution of Equation C.5 with realistic disk geometry and thermal boundary conditions is not possible, we resort to its numerical solution, the methodology for which is presented in this section.

For deriving the discretized algebraic equations corresponding to Equation C.5, we again use the control volume approach of energy balance on a disk element without making it differential. Toward this approach, the rotor disk is first divided into N finite control volumes, as shown in Figure C.2a. We place a node at the center of each control volume. To facilitate the application of thermal boundary conditions on the disk bore and rim surfaces, we create two fictitious boundary nodes, which are placed across the disk faces at a distance equal to that of the adjacent interior node, as shown in the figure, resulting in $N + 2$ total number of nodes. In the control volume approach presented by Patankar (1980), the first node and the last node are placed directly on the end surfaces.

The energy balance on the control volume associated with an interior node i, shown in Figure C.2b, yields

$$\delta m_i c_{m_i} \frac{T_{m_i} - T_{m_i}(\tau)}{\Delta \tau} = \frac{A_{i_1}\left(k_{m_i} + k_{m_{i-1}}\right)}{2(r_i - r_{i-1})}\left(T_{m_{i-1}} - T_{m_i}\right)$$

$$-\frac{A_{i_2}\left(k_{m_i} + k_{m_{i+1}}\right)}{2(r_{i+1} - r_i)}\left(T_{m_i} - T_{m_{i+1}}\right) \qquad \text{(C.27)}$$

$$- h_{i_1} S_{i_1}\left(T_{m_i} - T_{\text{aw}_{i_1}}\right) - h_{i_2} S_{i_2}\left(T_{m_i} - T_{\text{aw}_{i_2}}\right)$$

where, except for $T_{m_i}(\tau)$, which is known at the current time, all other quantities pertain to the time $\tau + \Delta\tau$. Other quantities in Equation C.27 are defined as follows:

$A_{i_1} \equiv$ Area of conduction from node $i - 1$ to node i

$A_{i_2} \equiv$ Area of conduction from node i to node $i + 1$

$c_{\mathrm{m}} \equiv$ Temperature-dependent specific heat of disk material

$h_{i_1} \equiv$ Heat transfer coefficient on Face 1 of node i control volume

$h_{i_2} \equiv$ Heat transfer coefficient on Face 2 of node i control volume

$k_{\mathrm{m}_i}, k_{\mathrm{m}_{i-1}}, k_{\mathrm{m}_{i+1}} \equiv$ Disk material thermal conductivities evaluates at temperatures

$\quad T_{\mathrm{m}_i}, T_{\mathrm{m}_{i-1}}, T_{\mathrm{m}_{i+1}}$, respectively

$r_i \equiv$ Radius of node i

$r_{i-1} \equiv$ Radius of node $i - 1$

$r_{i+1} \equiv$ Radius of node $i + 1$

$S_{i_1} \equiv$ Convection surface area on Face 1 of node i control volume

$S_{i_2} \equiv$ Convection surface area on Face 2 of node i control volume

$T_{\mathrm{m}_i}, T_{\mathrm{m}_{i-1}}, T_{\mathrm{m}_{i+1}} \equiv$ Disk temperature at node i, $i - 1$, and $i + 1$, respectively

$T_{\mathrm{aw}_{i_1}} \equiv$ Adiabatic wall temperature on Face 1 of node i control volume

$T_{\mathrm{aw}_{i_2}} \equiv$ Adiabatic wall temperature on Face 2 of node i control volume

$\delta m_i \equiv$ Mass of node i control volume

$\tau \equiv$ Time

$\Delta\tau \equiv$ Time step

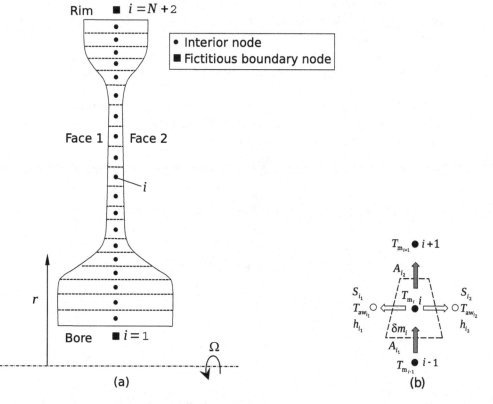

Figure C.2 (a) Rotor disk control volumes and nodes (b) energy balance on an interior control volume for numerical solution.

We can write Equation C.27 in the following form:

$$b_i T_{m_{i-1}} + d_i T_{m_i} + a_i T_{m_{i+1}} = c_i \qquad \text{(C.28)}$$

where $i = 2, 3, \ldots, N + 1$ and

$$b_i = -\frac{A_{i_1}\left(k_{m_i} + k_{m_{i-1}}\right)}{2\left(r_i - r_{i-1}\right)}$$

$$d_i = \frac{A_{i_1}\left(k_{m_i} + k_{m_{i-1}}\right)}{2\left(r_i - r_{i-1}\right)} + \frac{A_{i_2}\left(k_{m_i} + k_{m_{i+1}}\right)}{2\left(r_{i+1} - r_i\right)} + h_{i_1} S_{i_1} + h_{i_2} S_{i_2} + \frac{\delta m_i c_{m_i}}{\Delta \tau}$$

$$a_i = -\frac{A_{i_2}\left(k_{m_i} + k_{m_{i+1}}\right)}{2\left(r_{i+1} - r_i\right)}$$

$$c_i = \frac{\delta m_i c_{m_i} T_{m_i}(\tau)}{\Delta \tau} + h_{i_1} S_{i_1} T_{aw_{i_1}} + h_{i_2} S_{i_2} T_{aw_{i_2}}$$

Note that the system of equations that result from Equation C.28 is not closed for it has N number of equations and $N + 2$ unknown temperatures. To close this system, we need two more equations from the thermal boundary conditions specified at disk bore and rim surfaces, relating the temperatures at fictitious boundary nodes to those at their next interior nodes.

Let us now establish equations relating temperatures at bore-side and rim-side fictitious nodes to those at their neighboring interior nodes, which are shown in Figure C.3.

Disk bore. Three types of thermal boundary conditions can in general be specified at the disk bore wall, allowing us to develop an equation to relate the fictitious node 1 to interior node 2 in the form of Equation C.28.

Type 1: Specified surface temperature $T_{w_{bore}}$

$$T_{w_{bore}} = \frac{T_{m_1} + T_{m_2}}{2}$$

$$T_{m_1} + T_{m_2} = 2T_{w_{bore}} \qquad \text{(C.29)}$$

Comparing Equation C.29 with Equation C.28 for $i = 1$ yields

$$b_1 = 0, \ \ d_1 = 1, \ \ a_1 = 1, \ \ \text{and} \ \ c_1 = 2T_{w_{bore}} \qquad \text{(C.30)}$$

Type 2: Specified heat flux $\dot{q}_{bore}$

$$k_{m_{bore}}\left(\frac{T_{m_1} - T_{m_2}}{r_2 - r_1}\right) = \dot{q}_{bore}$$

$$T_{m_1} - T_{m_2} = \frac{\left(r_2 - r_1\right)\dot{q}_{bore}}{k_{m_{bore}}} \qquad \text{(C.31)}$$

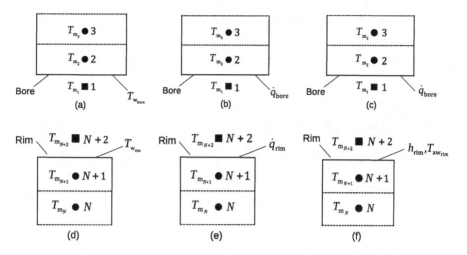

Figure C.3 (a) Type 1 BC on disk bore, (b) Type 2 BC on disk bore, (c) Type 3 BC on disk bore, (d) Type 1 BC on disk rim, (e) Type 2 BC on disk rim, and (f) Type 3 BC on disk rim.

Comparing Equation C.31 with Equation C.28 for $i = 1$ yields in this case

$$b_1 = 0, \ d_1 = 1, \ a_1 = -1, \ \text{and} \ c_1 = \frac{(r_2 - r_1)\dot{q}_{\text{bore}}}{k_{m_{\text{bore}}}} \tag{C.32}$$

Note that, for disk bore cooling, $\dot{q}_{\text{bore}}$ is negative, and for heating, it is positive.

Type 3: Convective boundary condition with specified h_{bore} and $T_{\text{aw}_{\text{bore}}}$

$$k_{m_{\text{bore}}}\left(\frac{T_{m_2} - T_{m_1}}{r_2 - r_1}\right) = h_{\text{bore}}\left(\frac{T_{m_1} + T_{m_2}}{2} - T_{\text{aw}_{\text{bore}}}\right)$$

$$\left(\frac{2k_{m_{\text{bore}}}}{r_2 - r_1}\right)\left(T_{m_2} - T_{m_1}\right) = h_{\text{bore}}\left(T_{m_1} + T_{m_2} - 2T_{\text{aw}_{\text{bore}}}\right)$$

$$\left(\frac{2k_{m_{\text{bore}}}}{r_2 - r_1} + h_{\text{bore}}\right)T_{m_1} - \left(\frac{2k_{m_{\text{bore}}}}{r_2 - r_1} - h_{\text{bore}}\right)T_{m_2} = 2h_{\text{bore}}T_{\text{aw}_{\text{bore}}} \tag{C.33}$$

Comparing Equation C.33 with Equation C.28 for $i = 1$ yields in this case

$$b_1 = 0, \ d_1 = \frac{2k_{m_{\text{bore}}}}{r_2 - r_1} + h_{\text{bore}}, \ a_1 = -\frac{2k_{m_{\text{bore}}}}{r_2 - r_1} + h_{\text{bore}}, \ \text{and} \ c_1 = 2h_{\text{bore}}T_{\text{aw}_{\text{bore}}} \tag{C.34}$$

Disk rim. At the disk rim also, in general, we can specify one of the three types of thermal boundary conditions. For each type, let us develop an equation to relate the fictitious node $N + 2$ to interior node $N + 1$ in the form of Equation C.28.

Type 1: Specified surface temperature $T_{w_{rim}}$

$$T_{w_{rim}} = \frac{T_{m_{N+1}} + T_{m_{N+2}}}{2}$$

$$T_{m_{N+1}} + T_{m_{N+2}} = 2T_{w_{rim}} \tag{C.35}$$

Comparing Equation C.35 with Equation C.28 for $i = N + 2$ yields

$$b_{N+2} = 1, \ d_{N+2} = 1, \ a_{N+2} = 0, \ \text{and} \ c_{N+2} = 2T_{w_{rim}} \tag{C.36}$$

Type 2: Specified heat flux $\dot{q}_{rim}$

$$k_{m_{rim}} \left(\frac{T_{m_{N+2}} - T_{m_{N+1}}}{r_{N+2} - r_{N+1}} \right) = \dot{q}_{rim}$$

$$-T_{m_{N+1}} + T_{m_{N+2}} = \frac{\left(r_{N+2} - r_{N+1} \right) \dot{q}_{rim}}{k_{m_{rim}}} \tag{C.37}$$

Comparing Equation C.37 with Equation C.28 for $i = N + 2$ yields in this case

$$b_{N+2} = -1, \ d_{N+2} = 1, \ a_{N+2} = 0, \ \text{and} \ c_{N+2} = \frac{\left(r_{N+2} - r_{N+1} \right) \dot{q}_{rim}}{k_{m_{rim}}} \tag{C.38}$$

Note that, for disk rim cooling, $\dot{q}_{rim}$ is negative, and for heating, it is positive.

Type 3: Convective boundary condition with specified h_{rim} and $T_{aw_{rim}}$

$$k_{m_{rim}} \left(\frac{T_{m_{N+1}} - T_{m_{N+2}}}{r_{N+2} - r_{N+1}} \right) = h_{rim} \left(\frac{T_{m_{N+1}} + T_{m_{N+2}}}{2} - T_{aw_{rim}} \right)$$

$$\left(\frac{2k_{m_{rim}}}{r_{N+2} - r_{N+1}} \right) \left(T_{m_{N+1}} - T_{m_{N+2}} \right) = h_{rim} \left(T_{m_{N+1}} + T_{m_{N+2}} - 2T_{aw_{rim}} \right)$$

$$-\left(\frac{2k_{m_{rim}}}{r_{N+2} - r_{N+1}} - h_{rim} \right) T_{m_{N+1}} + \left(\frac{2k_{m_{rim}}}{r_{N+2} - r_{N+1}} + h_{rim} \right) T_{m_{N+2}} = 2h_{rim} T_{aw_{rim}} \tag{C.39}$$

Comparing Equation C.39 with Equation C.28 for $i = N + 2$ yields in this case

$$b_{N+2} = -\frac{2k_{m_{rim}}}{r_{N+2} - r_{N+1}} + h_{rim}, \ d_{N+2} = \frac{2k_{m_{rim}}}{r_{N+2} - r_{N+1}} + h_{rim}, \ a_{N+2} = 0,$$

and

$$c_{N+2} = 2h_{rim} T_{aw_{rim}} \tag{C.40}$$

Now we have all the needed equations for each of the three thermal boundary conditions at disk bore and rim to include two more equations, one for $i = 1$ and the other for $i = N + 2$, in the system of equations, which arise from Equation C.28. This, at each time step, results in a closed (number of equations equals number of unknowns) tridiagonal system of equations whose solution by the Thomas algorithm is presented in Appendix E.

Reference

Patankar, S. V. 1980. *Numerical Heat Transfer and Fluid Flow.* Boca Raton, FL: Taylor & Francis.

Appendix D Regula Falsi Method

D.1 Regula Falsi Method

In gas turbine internal flow systems modeling and many other engineering applications, we often need to find the root x of the equation $f(x) = 0$ where $f(x)$ is an analytic function. When we are given an equation of the form $g_1(x) = g_2(x)$, where both $g_1(x)$ and $g_2(x)$ are analytic, the equation is called the transcendental equation and may the rewritten in the standard form $f(x) = g_2(x) - g_1(x) = 0$. The Regula Falsi method, presented in detail in Carnahan, Luther, and Wilkes (1969), is a powerful technique to iteratively solve $f(x) = 0$ for its root that lies within the limits $x_L \leq x \leq x_R$ such that $f(x_L) < 0$ and $f(x_R) > 0$. Note that in MS Excel, one can use the "Goal Seek" function to find the root of $f(x) = 0$.

Knowing x_L, x_R, $f(x_L)$, and $f(x_R)$, the Regula Falsi algorithm can be summarized as follows:

1. Evaluate x by the equation

$$x_C = \frac{x_L f_R - x_R f_L}{f_R - f_L} \tag{D.1}$$

where f_L and f_R represent values of $f(x)$ evaluated at x_L and x_R, respectively.

2. Evaluate $f_C = f(x_C)$. If $f_C \leq E$, where E is an acceptable error in $f(x) = 0$, we have obtained x_C as the desired root of the equation, and the iteration can be terminated. Otherwise, go to step 3.
3. If $f_L f_C < 0$, set $x_R = x_C$ and $f_R = f_C$; otherwise, set $x_L = x_C$ and $f_L = f_C$.
4. Repeat steps from 1 to 3 until the convergence condition in step 2 is satisfied.

The subroutine REGULA listed at the end of this appendix uses the foregoing algorithm of the Regula Falsi method.

Example D.1 For the total-pressure mass flow function $\hat{F}_{f_t} = 0.4$ and $\kappa = 1.4$, use the Regula Falsi method to find the Mach number from the equation

$$\hat{F}_{f_t} = M \sqrt{\frac{\kappa}{\left(1 + \frac{\kappa-1}{2}M^2\right)^{\frac{\kappa+1}{\kappa-1}}}}$$

Tabulate values of Mach number and function $f(M)$ for the first five iterations and graphically represent the iteration process for the first two iterations.

Solution: First, we cast the given equation in the following form:

$$f(M) = M\sqrt{\frac{\kappa}{\left(1 + \frac{\kappa-1}{2}M^2\right)^{\frac{\kappa+1}{\kappa-1}}}} - \hat{F}_{f_t} = 0$$

Using $\hat{F}_{f_t} = 0.4$ and $\kappa = 1.4$ in this equation, and starting with $M_L = 0.1$ and $M_R = 1$, we obtained the following results for the first five iterations of the Regula Falsi algorithm:

Iteration	M_L	f_L	M_R	f_R	M_C	f_C
1	0.10000	−0.28239	1.00000	0.28473	0.54814	0.14441
2	0.10000	−0.28239	0.54814	0.14441	0.39651	0.02754
3	0.10000	−0.28239	0.39651	0.02754	0.37016	0.00386
4	0.10000	−0.28239	0.37016	0.00386	0.36652	0.00051
5	0.10000	−0.28239	0.36652	0.00051	0.36603	0.00007

The results show that Mach number of 0.36603 obtained in just five iterations is within 0.02 percent of the exact value of 0.36596.

Figure D.1 shows how the Regula Falsi method works for the first two iterations.

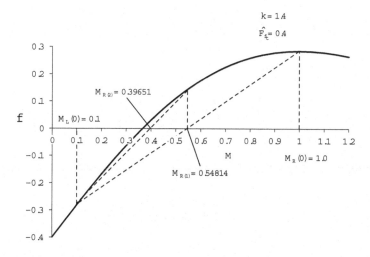

Figure D.1 Computation of Mach number for a given total-pressure mass flow function using Regula Falsi method (Example D.1).

D.2 REGULA Subroutine

The call statement for subroutine REGULA is of the form

```
CALL REGULA (N, XL, XR, XC)
```

Listing of Subroutine REGULA

```
      SUBROUTINE REGULA (N, XL, XR, XC)
      IMPLICIT DOUBLEPRECISION (A-H, O-Z)
C
C USING REGULA FALSI METHOD, THIS SUBROUTINE FINDS THE SOLUTION
C OF A TRANSCENDENTAL EQUATION.
C
      TOL = 1.0D-5
C
      CALL FUNCTION (XL, FL)
      CALL FUNCTION (XR, FR)
      IF (FL*FR .GT. 0.) THEN
      PRINT*, "REVISE VALUES OF XL AND XR TO YIELD FL*FR < 0."
      RETURN
      ELSE
      END IF
C
C ------ COMPUTE THE TANSCENDENTAL EQUATION ROOT ------
      DO I = 1, N
      XC = (XL*FR - XR*FL)/(FR - FL)
C
      CALL FUNCTION (XC, FC)
C
C ------ CHECK FOR CONVERGENCE ------
      IF (F .LT. TOL) RETURN
C
C ------ KEEP RIGHT OR LEFT SUBINTERVAL ------
      IF (FC*FL .LT. 0.0) THEN
      XR = XC
      FR = FC
      ELSE
      XL = XC
      FL = FC
      END IF
      END DO
C
      RETRUN
C
      END
```

Reference

Carnahan, B., H. A. Luther, and J. O. Wilkes. 1969. *Applied Numerical Methods*. New York: John Wiley & Sons.

Appendix E Thomas Algorithm for Solving a Tridiagonal System of Linear Algebraic Equations

E.1 Thomas Algorithm

We brie y present here the Thomas algorithm, which is widely used to solve a tridiagonal system of linear algebraic equations. Additional details with its applications are given in Patankar (1980).

Let us consider N nodes arranged along the west-east direction numbered $1, 2, 3, \ldots, N$. From a control volume analysis, the nodal values ϕ_i are governing by the following linear algebraic equation:

$$b_i \phi_{i-1} + d_i \phi_i + a_i \phi_{i+1} = c_i \tag{E.1}$$

where $i = 1, 2, 3, \ldots, N$.

For each value of i, except for $i = 1$ with $b_1 = 0$ and $i = N$ with $a_N = 0$, Equation E.1 relates ϕ_i to its before-node value ϕ_{i-1} and its after-node value ϕ_{i+1}. Furthermore, when write Equation E.1 all values of i, the resulting system of equations assumes the following matrix equation where the nonzero elements of the coef cient matrix fall along three diagonals, hence it is called tridiagonal matrix. In this matrix, the coef cient vector d_i occupies the main diagonal, vector b_i the lower or before diagonal, and vector a_i the upper or after diagonal — the choice of symbols for these vectors re ects this fact.

$$
\begin{bmatrix}
d_1 & a_1 & & & & \\
b_2 & d_2 & a_2 & & & \\
& b_3 & d_3 & a_3 & & \\
& & \ddots & \ddots & \ddots & \\
& & & b_{N-1} & d_{N-1} & a_{N-1} \\
& & & & b_N & d_N
\end{bmatrix}
\begin{bmatrix}
\phi_1 \\
\phi_2 \\
\phi_3 \\
\vdots \\
\vdots \\
\phi_{N-1} \\
\phi_N
\end{bmatrix}
=
\begin{bmatrix}
c_1 \\
c_2 \\
c_3 \\
\vdots \\
\vdots \\
c_{N-1} \\
c_N
\end{bmatrix}
\tag{E.2}
$$

The rst equation in Equation E.2 involves only ϕ_1 and ϕ_2, which implies that we can express ϕ_1 in terms of ϕ_2. Marching forward, while expressing value at the preceding node in terms of the value at the current node, we obtain a relation between the values at the current node and the node that follows. Recognizing this fact, let us assume the following algebraic equation form for this relation.

$$\phi_i = \alpha_i \phi_{i+1} + \beta_i \tag{E.3}$$

which also yields

$$\phi_{i-1} = \alpha_{i-1} \phi_i + \beta_{i-1} \tag{E.4}$$

Using Equation E.4 to eliminate ϕ_{i-1} in Equation E.1, we obtain

$$b_i\left(\alpha_{i-1}\phi_i + \beta_{i-1}\right) + d_i\phi_i + a_i\phi_{i+1} = c_i$$
$$\phi_i = \frac{-a_i}{d_i + b_i\alpha_{i-1}}\phi_{i+1} + \frac{c_i - b_i\beta_{i-1}}{d_i + b_i\alpha_{i-1}} \tag{E.5}$$

Comparing Equations E.4 and E.5 yields

$$\alpha_i = \frac{-a_i}{d_i + b_i\alpha_{i-1}} \tag{E.6}$$

$$\beta_i = \frac{c_i - b_i\beta_{i-1}}{d_i + b_i\alpha_{i-1}} \tag{E.7}$$

For $i = 1$ with $b_1 = 0$, Equations E.6 and E.7 yield

$$\alpha_1 = \frac{-a_1}{d_1} \quad \text{and} \quad \beta_1 = \frac{c_1}{d_i},$$

and for $i = N$ with $a_N = 0$, we obtain $\alpha_N = 0$.

Substitution of $\alpha_N = 0$ in Equation E.3 yields $\phi_N = \beta_N$, which, through back substitution in this equation, allows us to unpack all remaining values of $\phi_{N-1}, \phi_{N-2}, \dots, \phi_1$.

Thus, we can summarize the Thomas algorithm for solving the tridiagonal system of linear algebraic equations represented by Equation E.1 as follows:

1. Compute $\alpha_1 = \frac{-a_1}{d_1}$ and $\beta_1 = \frac{c_1}{d_i}$.
2. Use the recurrence Equations E.6 and E.7 to compute α_i and β_i for $i = 2, 3, \dots, N$.
3. Set $\phi_N = \beta_N$.
4. Use Equation E.3 to compute $\phi_{N-1}, \phi_{N-2}, \dots, \phi_1$ through back substitution.

E.2　THOMAS Subroutine

The call statement for subroutine THOMAS is of the form

```
CALL THOMAS (NJ, AA, BB, CC, DD, TM)
```

Listing of Subroutine THOMAS

```
SUBROUTINE THOMAS (N, A, B, C, D, PHI)
IMPLICIT DOUBLE PRECISION (A-H, O-Z)
```

```
      PARAMETER (ND = 100)
      DOUBLE PRECISION A(ND), B(ND), C(ND), D(ND), PHI(ND)
      DOUBLE PRECISION ALPHA(ND), BETA(ND)
C
C     USING THOMAS ALGORITHM, THIS SUBROUTINE SOLVES A
C     TRIDIGONAL SYSTEM OF LINEAR ALGEBRAIC EQUATIONS.
C
      SMALL = 1.0D-20
      ALPHA(1) = -A(1)/D(1)
      BETA(1) = C(1)/D(1)
C
C     ------ COMPUTE COEFFICIENTS OF RECURRENCE FORMULA ------
      DO I = 2, N
      DENOM = D(I) + B(I)*ALPHA(I-1)
      IF (DENOM .LE. SMALL) DENOM = SMALL
      ALPHA(I) = -A(I)/DENOM
      BETA(I) = (C(I) - B(I)*BETA(I-1))/DENOM
      END DO
C
      PHI(N) = BETA(N)
      NM1 = N - 1
C
C     ------ BACK SUBTITUTION TO OBTAIN REMAINING PHI'S ------
      DO I = NM1, 1, -1
      PHI(I) = ALPHA(I)*PHI(I+1) + BETA(I)
      END DO
      RETRUN
C
      END
```

Reference

Patankar, S. V. 1980. *Numerical Heat Transfer and Fluid Flow*. Boca Raton, FL: Taylor & Francis.

Appendix F Solution of an Overdetermined System of Linear Algebraic Equations

F.1 Introduction

In gas turbine internal flow systems modeling and many other engineering applications that require finding a regression equation as a curve-fit to a large number of experimental or numerical data, we often need to solve an overdetermined system of linear algebraic equations. In such a system, the number of equations is more than the number of unknowns. Of course, a method capable of solving such a system can also provide a direct solution when the system is closed with equal number of equations and unknowns. When the coefficient matrix depends on the solution itself, we have a system of nonlinear equations, having no known direct solution method. The numerical solution of such a system requires iterative direct solutions of a series of intermediate systems of linear equations, updating the coefficient matrix as needed from iteration to iteration. Hopefully, such an iterative solution method leads to the correct converged solution of the system of nonlinear equations on hand.

F.2 Linear Least-Squares Data Fitting

Sultanian (1980) used the robust Householder reflection method for linear least-squares curve fitting. The method was first proposed by Golub (1965). In this section, we provide some introductory details to facilitate the use of the computer code CURVE listed at the end.

Suppose we are given m data points $\left(\theta_i; \mu_i\right)$, which may be experimental or numerical, where $_i$ are the values of the independent variable like dimensionless temperature, $_i$ the values of the dependent variable like fluid viscosity, and the subscript $i = 1, 2, \ldots, M$. We do not know the exact analytical relation to compute $_i$ for a given $_i$. The following regression or curve-fit equation provides approximations to $_i$:

$$\hat{\mu}(\theta) = C_1\phi_1(\theta) + C_2\phi_2(\theta) + \cdots + C_N\phi_N(\theta) \tag{F.1}$$

where we must have $M \geq N$. Typically, in a regression problem, M is much greater than N. Note that ϕ_j may be nonlinear functions of θ, however, the equation is linear because the coefficients C_j appear linearly. Our task is to determine these coefficients in the least-squares sense.

Applying Equation F.1 to each data point, we obtain the following matrix equation:

$$
\begin{bmatrix}
a_{11} & a_{12} & \cdots & a_{1N} \\
a_{21} & a_{22} & \cdots & a_{2N} \\
\vdots & \vdots & \vdots & \vdots \\
a_{M1} & a_{M2} & \cdots & a_{MN}
\end{bmatrix}
\begin{bmatrix}
C_1 \\ C_2 \\ \vdots \\ C_N
\end{bmatrix}
=
\begin{bmatrix}
\mu_1 \\ \mu_2 \\ \vdots \\ \mu_M
\end{bmatrix}
\tag{F.2}
$$

which can be written in a compact form as $[A]\{X\} = \{B\}$ where $[A]$ is a $M \times N$ coefficient matrix with its elements given by $a_{ij} = \phi_j(\theta_i)$. The column vector $\{X\}$, which is a $N \times 1$ matrix, represents the column vector $\{C\}$ on the left-hand side of Equation F.2, while the column vector $\{B\}$, which is a $M \times 1$ matrix, represents the column vector $\{\mu\}$ on the right-hand side of the equation.

F.3   Computer Program HOME

The computer program listing at the end consists of the main program HOME and subroutines REDUCE and PIVOT. The main program is currently designed for curve-fitting of data using a polynomial function, internally generating the coefficient matrix $[A]$. Taking data points as input, the program returns the coefficients of the polynomial function as the curve-fit equation. One can easily modify the main program for different basis functions for curve-fitting or directly input the elements of the coefficient matrix for solving a closed system of linear algebraic equations.

Subroutine REDUCE computes various characteristic parameters of Householder reflections and reduces the coefficient matrix into an upper triangular matrix. To minimize the round-off errors in numerical computation, at each step, the subroutine PIVOT chooses the column the coefficient matrix with the largest sum of squares to be reduced next.

Listing of Program HOME

```
C       MAIN PROGRAM ----- HOME ------
C
        PROGRAM HOME
        IMPLICIT DOUBLE PRECISION (A-H, O-Z)
        PARAMETER (ND = 100)
        DOUBLE PRECISION A(ND, ND), X(ND), B(ND), THETA(ND)
        DOUBLE PECISION ICONT(ND), DUMP (ND)
        COMMON A, B, X, M, N, ICONT, RELERR
        CHARACTER TITLE*10
        CHARACTER INFLNM*80
C
C       ------ INPUT FOR INDEPENDENT VARIABLE ARE STORED IN THETA(I)
C       INPUT FOR DEPENDENT VARIABLE ARE STORED IN B(I) ------
```

```
      RELERR = 0.
      PRINT*," "
      PRINT*,"INPUT NUMBER OF DATA POINTS"
      READ*, M
      PRINT*," "
      PRINT*,"INPUT NUMBER OF UNKNOWN COEFFICIENTS FOR"
      PRINT*,"THE POLYNOMIAL FIT (NOTE: ONE MORE THAN THE"
      PRINT*,"DEGREE OF THE POLYNOMIAL)"
      READ*, N
      PRINT *," "
C
      INFL = 1
      PRINT*,"ENTER THE NAME OF YOUR INPUT DATA FILE"
      READ (5,'(A)') INFLNM
C
      OPEN (UNIT=INFL, FILE=INFLNM)
C
      READ (INFL,'(A)') TITLE
      READ (INFL,'(A)') TITLE
      READ (INFL,*) (THETA(I), B(I), I = 1, M)
C
   3  FORMAT (2A9)
   4  FORMAT ("X = ")
   5  FORMAT (8F10.4)
   6  FORMAT ("Y = ")
   7  FORMAT ("YC = ")
C
      OPEN (UNIT=8, FILE='OUTFILE')
C
      WRITE (8, 4)
      WRITE (8, 5) (THETA(I), I = 1, M)
      WRITE (8, 6)
      WRITE (8, 5) (B(I), I = 1, M)
C
      DO 10 I = 1, M
      A(I,1) = 1.0
      DO 10 J = 2, N
  10  A(I,J) = THETA(I)**(J-1)
C
      CALL REDUCE
C
C     ------ CALCULATE VALUES USING THE CURVE-FIT EQUATION ------
      DO 23 I = 1, M
      SUM = X(1)
      DO 25 J = 2, N
```

```
          SUM = SUM + X(J)*THETA(I)**(J - 1)
   25  CONTINUE
          DUMP(I) = SUM
   23  CONTINUE
C
          WRITE (8, 4)
          WRITE (8, 5) (THETA(I), I = 1, M)
          WRITE (8, 7)
          WRITE (8, 5) (DUMP(I), I = 1, M)
          WRITE (8,15) (X(I), I = 1, N)
          PRINT*," "
          PRINT*,"RESULTS ARE PRINTED IN OUTFILE.
          "PRINT*,"------------------------------"
   15  FORMAT (//, 4(E13.7, 5X))
          STOP
C
          END

          SUBROUTINE REDUCE
C
C      SOLUTION OF AN OVERDETERMINED SYSTEM OF EQUATIONS USING
C      HOUSEHOLDER REFLECTION METHOD
C      SYSTEM: AX = B, WHERE A(M,N), X(N), B(M), AND
C      M IS GREATER THAN OR EQUAL TO N
C
          IMPLICIT DOUBLE PRECISION (A-H, O-Z)
          PARAMETER (ND = 100)
          DOUBLE PRECISION A(ND, ND), X(ND), B(ND), ICONT(ND), V(ND)
          COMMON A, B, X, M, N, ICONT, RELERR
          ERROR = 0.0
C
C      ------ MERGING OF COLUMN VECTOR B(M) WITH MATRIX A(M,N) ------
          N1 = N + 1
          DO 2 I = 1, M
    2  A(I,N1) = B(I)
          DO 3 J = 1, N
    3  ICONT(J) = J
          DO 55 J = 1, N
          SUM = 0.
          IF (J .GE. N) GO TO 4
C
C      ------ PIVOTING THE COLUMN WITH THE LARGEST SUM OF SQUARES
C      OF MATRIX A (M, N) ------
          JPVT = J
C
```

```
      CALL PIVOT (JPVT)
C
C       ------ COMPUTATION OF CHARACTERISTIC PARAMETERS
        OFHOUSEHOLDER REFLECTIONS ------
   4  IF (A(J, J)) 7, 8, 8
   7  SIGN = -1.0
      GO TO 11
   8  SIGN = 1.
  11  DO 9 I = J, M
   9  SUM = SUM + A(I, J)*A(I, J)
      ALPHA = SIGN*DSQRT (SUM)
      DO 12 I = 1, M
      IF (I-J) 13, 14, 15
  13  V(I) = 0.
      GO TO 12
  14  V(I) = A(I, J) + ALPHA
      GO TO 12
  15  V(I) = A(I, J)
  12  CONTINUE
      BETA = ALPHA*(A(J, J) + ALPHA)
C
C       ------ REDUCTION PROCESS FOR AUGMENTED MATRIX A (M, N) ------
      K = J + 1
      A(J, J) = -ALPHA
  18  A(I, J) = 0.0
      DO 16 L = K, N1
      SUM = 0.
      DO 17 I = 1, M
  17  SUM = SUM + V(I)*A(I,L)
      GAMA = SUM/BETA
      DO 16 I = 1, M
  16  A(I, L) = A(I, L) - GAMA*V(I)
  55  CONTINUE
C
C       ------ LARGEST DIAGONAL ELEMENT
      SIGMAX = DABS (A(N, N))
      DO 60 I = 1, N
      IF (DABS (A(I, I)) .GT. SIGMAX) SIGMAX = DABS (A(I, I))
  60  X(I) = 0.
      TAU = RELERR*SIGMAX
      TAU = 1.E-30
C       ------ COMPUTATION OF ELEMENTS (COEFF) OF COLUMN
C       VECTOR X(N) ------
      IF (DABS (A(N, N)) .LT. TAU) GO TO 65
      X(N) = A(N, N1)/A(N, N)
```

```
C
  65   DO 45 I = 2, N
       L = N - I + 1
       L1 = L + 1
       SUM = 0.
       IF (DABS (A(L, L)) .LT. TAU) GO TO 45
       DO 40 J = L1, N
  40   SUM = SUM + X(J)*A(L, J)
       X(L) = (A(L, N1) - SUM)/A(L, L)
  45   CONTINUE
C
C      ------ REARRANGEMENT OF COEFFICIENTS IN
C      COLUMN VECTOR X(N) ------
       DO 30 J = 1, N
       DO 35 I = 1, N
       IF (ICONT(I) .EQ. J) GO TO 30
  35   CONTINUE
  30   B(J) = X(I)
       DO 25 I = 1, N
  25   X(I) = B(I)
       RETURN
       END


       SUBROUTINE PIVOT(K)
C
C      PIVOTING THE COLUMN WITH THE LARGEST SUM OF SQUARES OF
C      MATRIX A(M,N) . THE SUBROUTINE STARTS PIVOTING FROM
C      K TH COLUMN OF MATRIX A (M,N).
C
       IMPLICIT DOUBLE PRECISION (A-H, O-Z)
       PARAMETER (ND = 100)
       DOUBLE PRECISION CC(ND), DD(ND), A(ND,ND), ICONT(ND),
       COLUMN(ND)
       COMMON A, CC, DD, M, N, ICONT, RELERR
       CMAX = 0.
       INDEX = K
       DO 5 J = K, N
       SUM = 0.
       DO 6 I = 1, M
   6   SUM = SUM + A(I, J)*A(I, J)
   5   COLUMN(J) = SUM
C
       DO 7 J = K, N
       IF (COLUMN(J) .LE. CMAX) GO TO 7
       CMAX = COLUMN (J)
```

```
          INDEX = J
     7    CONTINUE
          IF (INDEX .EQ. K)GO TO 8
          J = INDEX
C
C    ------ EXCHANGE OF COLUMNS ------
          DO 9 I = 1, M
          DUMP = A(I, J)
          A(I, J) = A(I, K)
     9    A(I, K) = DUMP
C
C    ------ UPDATING THE UNKNOWN X(N) COLUMN VECTOR ------
          IDUMP = ICONT(J)
          ICONT(J) = ICONT(K)
          ICONT(K) = IDUMP
     8    RETURN
C
          END
```

Reference

Golub, G. M. 1965. Numerical methods for solving linear least-squares problem. *Numer. Mathematics*. 7: 206–216.

Sultanian, B. K. 1980. HOME – a program package on Householder reflection method for linear least-squares data fitting. *J. Instit. Eng. (I)*. 60(pt Et. 3):71–75.

Epilogue
Current Research Work and Challenges

Historical Perspective

Gas turbines are ubiquitous as an energy conversion device for both aircraft propulsion and marine- and land-based power generation in simple and combined-cycle operations. The growing emphasis on a renewable energy source, such as wind, solar energy, and sea, has not dampened the need for continued technological innovation toward higher-efficiency gas turbines, meeting durability constraints at lower cost. In aviation, gas turbines continue to dominate with their remarkable record of reliability, which is the most important design consideration for the safety of all passengers and crew members.

Traditionally, a lot of emphasis has been placed on primary flowpath design leading to the development of 3-D airfoils using 3-D CFD technology. A number of textbooks on the subject are in print today, and the new titles and new editions of the popular ones continue to appear each year to support the related graduate and undergraduate courses on turbomachinery in the mechanical and aerospace engineering departments of universities around the world.

It is a bit ironic that the important area of secondary flow systems for internal cooling and sealing of gas turbines has not received the same attention. Although the area is vital for both engine cycle performance and life assessment of critical parts at an OEM, much is carried out using homegrown methods and tools, which are integrated into evolving design practices. Because these design practices are largely based on product experience and continuous validation, they generally involve a number of fuzz factors imbedded into various empirical correlations, most of which are deemed proprietary and at times not physics-based. Very little of this knowledge ever gets into the open literature for future growth and training of the new generation of engineers. This is also the reason behind the persisting lack of university-level courses in this area.

Over the last half century, the University of Sussex and the University of Bath have played a leading role in advancing the understanding and technology in the fascinating area of gas turbine internal-air systems through the passion of Professors Fred Bailey and Mike Owen. Research monograms in two volumes – see Owen and Rogers (1989, 1995) – were the first to be published in this area. Unfortunately, both these volumes are no longer in print. Covering many of the topics of gas turbine internal-air systems, Professor Ed Greitzer – see Greitzer, Tan, and Graf (2004) – published a textbook for a graduate course at MIT. This well-written book with its strong academic focus remains somewhat less appealing to practicing gas turbine engineers. More recently, Professor Peter Child – see Childs (2011) – published a title on rotating flows inclusive of their occurrences in both gas turbines and atmosphere. The book, while discussing many concepts of gas turbine internal-air systems, provides a rich bibliography of various research publications, both old and contemporary. In the foreword to this title, Professor Mike Owen writes, "The student familiar with conventional non-rotating flows has, like Alice, to go through the looking glass into a strange world that is counter-intuitive to the uninitiated."

Industry versus University Perspective

As to the research activities in this somewhat elusive area of gas turbine design, the industry and university continue to maintain two different perspectives. All experimental and analytical research work at an OEM of gas turbines is directed towards reinforcing their existing design practices and tools, making them ever more predictive. As a result of the proprietary nature of most industry research, not much of the related results get published in the open literature. By contrast, universities prefer to engage in research activities that will lead to conference presentation and journal publications. As a result of these somewhat mutually exclusive objectives, although some of the university research gets published, it does not become a part of an engineering design practice in gas turbine internal flow systems. To make matters worse, because of publication pressure, some of the research carried out at universities simplifies the problem to such an extent that the results are no longer useful for design applications. For example, eliminating disk rotation and flow swirl in an investigation where these are the primary drivers will not be acceptable to gas turbine design engineers. As another example, investigating hot gas ingestion under axisymmetric flowpath conditions, with or without rotation, will be akin to making the problem simpler to the point that the results lose their practical appeal, while missing the primary mechanism driving this phenomenon.

ASME Turbo Expo

ASME Turbo Expo provides the most popular avenue to present and publish more than 90 percent of the global research work carried out in the area of gas turbine secondary air systems under the track "Internal Air Systems and Seals," organized by the Heat Transfer Committee jointly with the Turbomachinery Committee. Approximately fifty papers are presented each year under various sessions: (1) Air System Analysis, (2) Air System Components, (3) Hot Gas Ingestion or Rim Seals, (4) Rotor-Stator Cavity, (5) Rotor-Rotor Cavity or Rotating Cavity, (6) Labyrinth Seals, (7) Brush Seals, (8) Leaf, Shaft, and Strip Seals, (9) Pre-Swirler System, and (10) Lube Oil Systems. In the review process, some of the papers are accepted for publications in either the *ASME Journal of Engineering for Gas Turbines and Power* or the *ASME Journal of Turbomachinery*.

Current Research Topics

Based on the five-year trend of papers presented in Internal Air Systems and Seals at ASME Turbo Expo, the research topic of hot gas ingestion or rim seals remains the most active one. Investigations involving specific geometric features find limited applications in design, because no single universal correlation can be developed. Experimental data from these investigations may, however, be useful for validation of an analytical or numerical prediction method.

The flow and heat transfer modeling of compressor drum cavity, which is a rotating cavity with centrifugally-driven buoyant convection, with or without a bore cooling flow, has been the subject of several recent investigations. The flow and heat transfer modeling of these rotating cavities from the viewpoint of design applications remains challenging, both experimentally and numerically. There seems to be a renewed focus on the development of more innovative seals of all types. Other traditional topics of internal air systems, mentioned earlier, continue to be investigated to reinforce better understanding toward their physics-based, reduced-order modeling in design involving multiple components, while capturing their coupled effects.

Challenges

Today's challenges in understanding and predictive modeling of gas turbine internal flow systems are not the same as they were a couple of decades ago. In spite of significant progress made in this area so far, much remains to be done to support today's advanced gas turbines, which are to be developed under a compressed design cycle time, for various applications. A time-invariant fact of secondary air systems is that they are inherently complex and counterintuitive in their flow and heat transfer physics as a result of the presence of rotation associated with the flow and one or more bounding walls. For example, the uncertainty associated with the discharge coefficient of a rotating orifice is far greater than that of a stationary orifice. The rotation of a duct with otherwise uniform 1-D compressible flow develops stratification in its cross-sectional static pressure distribution as a result of Coriolis forces with simultaneous changes in heat transfer characteristic around the duct wall. The turbulence field in most internal flow system components is highly nonisotropic, making CFD predictions with RANS-type isotropic turbulence models highly unreliable.

Because the continued research work in gas turbine internal air systems and seals is of an applied nature as opposed to fundamental research pursued in other scientific disciplines, it eventually get directed by the evolving needs of gas turbine design. All OEM-funded research at universities tends to establish a synergy between the industry needs and university research activities to directly or indirectly support them. So, to understand the future direction of research in this area, we must first understand the evolving gas turbine industry needs and trends.

Under fierce global completion, gas turbine OEM's today are facing a unique challenge of developing a robust design, which is simultaneously cheap, fast, and accurate. Toward this objective, multiphysics analyses needed in design involve integrated heat transfer, fluid mechanics, and thermomechanical (solid domain) analysis to realize the whole engine modeling – a must for LCF/HCF life assessment and management of engine critical parts. Almost gone are the days when engineers remained focused on their discipline-specific expertise, such as heat transfer engineers only dealing with heat transfer aspects of design, SAS engineers dealing with secondary air system aspects, and mechanical design engineers finally integrating all in the final design of parts to be released for manufacturing. In today's gas turbine engineering, a design is to be performed in a highly collaborative environment with a common design tool or platform integrating various disciplines. Because flow and heat transfer are closely coupled in a secondary air system design, it behooves both heat transfer engineers and SAS engineers to perform this design actively as one discipline, eventually performing the whole engine modeling by carrying out thermomechanical analysis. For mechanical design engineers to grow upstream to carry out both heat transfer and SAS analysis is an unlikely scenario.

In addition to the challenging trend of gas turbine designers performing multi-disciplinary analysis for the whole engine modeling, the trend is to develop reduced-order modeling for various SAS components to realize a robust or probabilistic design, which requires thousands of deterministic runs for the numerical DOE. This requirement precludes the direct use of CFD as an integrated design tool. Instead, the CFD technology can be used to generate response surface equations, where empirical correlations are inadequate, to reinforce the reduced-order modeling methodology. For example, the development of a CFD-based response surface equation to compute discharge coefficient of a generalized orifice will be a helpful contribution to SAS modeling.

The rotation in gas turbine internal flow systems results in Ekman boundary layers, which behave differently from the traditional boundary layers. The effect of geometry in these flows is not easily discernible, as it is generally done using the concept of mean hydraulic diameter for the flow and heat transfer in a duct with arbitrary cross-section. Developing a way to account for geometry in various SAS components in a universal correlation remains a challenge for ongoing research in this area.

In the reduced-order modeling of various rim seal designs using single or multiple orifices, the primary challenge boils down to assigning a discharge coefficient for each orifice, for which no universal empirical correlation possibly exits today. Developing such a correlation experimentally does not seem feasible. The only other possibility is to leverage the powerful CFD technology for this challenging task.

A perennial challenge to accurately predict heat transfer coefficients from empirical correlations to simulate convective heat transfer boundary conditions in various SAS components continue to persist. A well-established practical strategy to overcome this problem in gas turbine design is to use one or more of the well-known and closely related empirical correlations and adjust their coefficients by matching predictions with thermal survey data from extensive engine tests. Because RANS-based near-wall turbulence modeling for heat transfer is rather weak as a result of the varying turbulent Prandtl number in the boundary layer, the related CFD predictions are often not reliable and consistent. The only other option to change this status quo in this area is to resort to simulation-type CFD modeling using LES, including its variants, and DNS; the latter could be viewed as the test bed for numerical experiments, harnessing ever-growing computing power and resources.

References

Childs, P. R. N. 2011. *Rotating Flow*. Burlington, MA: Elsevier.

Greitzer, E. M., C. S. Tan, and M. B. Graf. 2004. *Internal Flow: Concepts and Applications*. Cambridge: Cambridge University Press.

Owen, J. M., and R. H. Rogers. 1989. *Flow and Heat Transfer in Rotating Disc Systems, Vol. 1: Rotor-Stator Systems*. Taunton, UK: Research Studies Press.

Owen, J. M., and R. H. Rogers. 1995. *Flow and Heat Transfer in Rotating Disc Systems, Vol. 2: Rotating Cavities*. Taunton, UK: Research Studies Press.

Index

Printed in the United States
by Baker & Taylor Publisher Services